"十二五"国家重点图书出版规划项目

风力发电工程技术丛书

现代英汉风力发电工程

左潞 袁越 霍志红 张玮 编

中国水利水电出版社
www.waterpub.com.cn

内容提要

本书是《风力发电工程技术丛书》之一，收录的词汇以直接与风力发电工程有关的词汇为主，以间接有关的基础工业词汇为辅，以直接参与风力发电学习、研究、设计、生产和使用的科研人员为主要服务对象，为促进和扩大我国风力发电行业国际间的科技及经贸交流发挥作用。

本书可供从事风力发电的科研技术人员以及在校学生查阅。

图书在版编目（CIP）数据

现代英汉风力发电工程词汇 / 左潞等编. -- 北京：中国水利水电出版社，2014.8
（风力发电工程技术丛书）
ISBN 978-7-5170-2484-2

Ⅰ.①现… Ⅱ.①左… Ⅲ.①风力发电—词汇—英、汉 Ⅳ.①TM614-61

中国版本图书馆CIP数据核字(2014)第212369号

书　　名	风力发电工程技术丛书 现代英汉风力发电工程词汇
作　　者	左潞　袁越　霍志红　张玮　编
出版发行	中国水利水电出版社 （北京市海淀区玉渊潭南路1号D座　100038） 网址：www.waterpub.com.cn E-mail: sales@waterpub.com.cn 电话：（010）68367658（发行部）
经　　售	北京科水图书销售中心（零售） 电话：（010）88383994、63202643、68545874 全国各地新华书店和相关出版物销售网点
排　　版	中国水利水电出版社微机排版中心
印　　刷	北京鑫丰华彩印有限公司
规　　格	184mm×260mm　16开本　24.75印张　638千字
版　　次	2014年8月第1版　2014年8月第1次印刷
印　　数	0001—3000册
定　　价	72.00元

凡购买我社图书，如有缺页、倒页、脱页的，本社发行部负责调换

版权所有·侵权必究

《风力发电工程技术丛书》
编 委 会

顾　　问	陆佑楣　张基尧　李菊根　晏志勇　周厚贵 施鹏飞
主　　任	徐　辉　毕亚雄
副 主 任	汤鑫华　陈星莺　李　靖　陆忠民　吴关叶 李富红
委　　员	（按姓氏笔画排序）

马宏忠　王丰绪　王永虎　尹廷伟　申宽育
冯树荣　刘　丰　刘　玮　刘志明　刘作辉
齐志诚　孙　强　孙志禹　李　炜　李　莉
李同春　李承志　李健英　李睿元　杨建设
吴敬凯　张云杰　张燎军　陈　刚　陈党慧
林毅峰　易跃春　周建平　郑　源　赵生校
赵显忠　胡立伟　胡昌支　俞华锋　施　蓓
洪树蒙　祝立群　袁　越　黄春芳　崔新维
彭丹霖　董德兰　游赞培　蔡　新　糜又晚

丛书主编　郑　源　张燎军

主要参编单位 （排名不分先后）

河海大学
中国长江三峡集团公司
中国水利水电出版社
水资源高效利用与工程安全国家工程研究中心
华北电力大学
水电水利规划设计总院
水利部水利水电规划设计总院
中国能源建设集团有限公司
上海勘测设计研究院
中国水电顾问集团华东勘测设计研究院有限公司
中国水电顾问集团西北勘测设计研究院有限公司
中国水电顾问集团中南勘测设计研究院有限公司
中国水电顾问集团北京勘测设计研究院有限公司
中国水电顾问集团昆明勘测设计研究院有限公司
长江勘测规划设计研究院
中水珠江规划勘测设计有限公司
内蒙古电力勘测设计院
新疆金风科技股份有限公司
华锐风电科技股份有限公司
中国水利水电第七工程局有限公司

丛书总策划 李 莉

编委会办公室

主　　　任　胡昌支
副 主 任　王春学　李 莉
成　　　员　殷海军　丁 琪　高丽霄　王 梅　白 杨
　　　　　　汤何美子

《现代英汉风力发电工程词汇》
编辑出版人员名单

责 任 编 辑　丁 琪　李 莉
封 面 设 计　李 菲
版 式 设 计　黄云燕
责 任 校 对　张 莉　梁晓静
责 任 印 制　孙长福

前言

风能是可再生能源中发展最快的清洁能源。风能的利用是一个系统工程，其中涉及气象学、流体力学、固体力学、电子技术、机械工程、电气工程、海洋工程、材料科学、环境科学等多种学科和专业。风力发电是风能利用的主要方式，是目前全球可再生能源中各种技术发展最快、技术最为成熟的能源。随着世界各国的风力发电工业蓬勃发展，新理论、新技术、新产品、新设备、新材料不断涌现，相应的英文新词汇也在日益增加。我国的风力发电发展更为迅猛，2010年我国的风电安装总量排名第一，并成为了国际风电产业中心，风电产品也开始进入国际市场，风电从业人员国际间的经贸科技交流日益频繁，设备和资料大量引进，专业新技术新设备广泛应用，整个行业迫切需要一本收词较新、较全的专业英语词汇，而社会上有关这方面的专业词汇书较少，难以满足广大读者需要，为此，河海大学组织专业人士编撰了这本《现代英汉风力发电工程词汇》，相信它的出版能对广大专业人士有所帮助，为促进和扩大我国风电行业的国际科技及经贸交流发挥作用。相信该词汇将成为我国广大从事风力发电行业人员的一本必备和实用的工具书。

本书收录的词汇以直接与风力发电工程有关的词汇为主，以间接有关的基础工业词汇为辅，以直接参与风力发电学习、研究、设计、生产和使用单位的科技人员为主要服务对象，又可供广大工业界科技人员查阅。本词汇共收词20000多条，收词参考国际风工程学会（IAWE）、美国机械工程师学会（ASME）、美国机电学会（AES）、美国国家标准学会（ANSI）及我国国家质量技术监督局颁发的风力发电术语，并收集了国外近年的专业书籍和网上的最新专业词汇等。由于风力发电技术发展迅速，新词汇不断涌现，本书难免有遗漏与不当之处，请读者多提宝贵意见和建议。

编者

2013年6月5日

目录

前言

A 2	H 160	O 230	V 346
B 44	I 170	P 240	W 356
C 60	J 182	Q 260	X 370
D 90	K 184	R 264	Y 372
E 112	L 188	S 282	Z 376
F 128	M 202	T 320	
G 150	N 222	U 340	

附录

377

附录一
希腊字母表
377

附录二
化学元素表
377

附录三
世界部分国家和地区名称及其符号表
379

附录四
蒲福（Beaufort）风力等级表
380

附录五
风工程常用无量纲数
380

附录六
单位换算表
381

现代英汉风力发电工程

A Modern English-Chinese Dictionary of Wind Power Engineering

AAAS (American Association for the Advancement of Science) 美国科学发展协会
AACC (American Automatic Control Council) 美国自动控制委员会
AAE (American Association of Engineers) 美国工程师协会
AAS (American Academy of Sciences) 美国科学院
abate 减少；抑制
abatement 减少；降低
abatement of pollution 消除污染
abatement of smoke 消除烟尘
abatement of wind 风力减弱
abat-vent 转向装置；折流板；固定百叶窗；通风帽
ability 能力；性能
ablative insulating quality 隔热性
ablution 清除；清洗
abnormal 不正常的；不规则的；异常的
abnormal error 不规则误差
abnormal flow 不正常流动；不规则流动
abnormal load 不规则载荷
abnormal operation test 非正常工作试验
abnormal sound 异声；不正常声音；异响
abnormal state 异常状态

abnormal temperature 异常温度
abnormal test condition 特殊试验条件
abort light 紧急故障信号
abort sensing 故障测定
abort situation 故障位置
about pitch axis balance 对俯仰轴的平衡
above 在……之上
above critical 超临界的；临界以上的
above mean sea level 平均海拔高度以上
above sea level 海拔
abradability 磨损度
abrasion 机械磨蚀；磨损；磨耗
abrasion loss 磨耗量
abrasion marks 磨痕
abrasion resistance 抗磨损性
abrasion run 磨耗运行
abrasion test 耐磨实验
abrasion value 磨耗量
abrasion wear 磨损
abrasion-proof 耐磨的
abrasion-resistance steel 耐磨钢
abrasion-resistant reinforcement 耐磨损加强
abrasive 粗糙的；摩擦的
abrasive action 磨损作用
abrasive cloth 砂布

abrasive disc 磨盘；磨片
abrasive resistance 耐磨性；耐磨强度
abridged 删减的；削减的
abridged general view 示意图
abrupt 陡坡的；突然的
abrupt curve 陡变曲线
abruption 断裂
abscissa 横坐标；横轴线
abscissa axis 横坐标
abscissa of convergence 收敛横坐标
absolute 绝对的
absolute accuracy 绝对精确度
absolute altitude 绝对高度；标高；海拔
absolute atmosphere (ATA) 绝对大气压
absolute black body 绝对黑体
absolute calibration 绝对校准
absolute capacity 绝对容量
absolute coefficient 绝对系数
absolute construction 独立结构
absolute coordinate 绝对坐标
absolute data 绝对数据
absolute density 绝对密度
absolute dimension 绝对尺寸；绝对坐标值
absolute dry condition 绝对干燥状态
absolute elongation 绝对伸长
absolute encoder 绝对式编码器
absolute error (AE) 绝对误差
absolute frequency 绝对频率
absolute height 绝对高度
absolute humidity 绝对湿度
absolute humidity of gas 气体的绝对湿度
absolute intensity 绝对强度
absolute manometer 绝对压力计
absolute maximum temperature 绝对最高温度

absolute minimum temperature 绝对最低温度
absolute moisture 绝对湿度
absolute moisture content 绝对含湿量；绝对水分含量
absolute monthly maximum temperature 月绝对最高温度
absolute number 绝对值
absolute permeability 绝对磁导率
absolute power 绝对功率
absolute pressure 绝对压力
absolute pressure controller 绝对压力控制器
absolute pressure gauge 绝对压力表
absolute pressure regulator 绝对压力调节器
absolute pressure sensor 绝对压力传感器
absolute pressure transducer 绝对压力转换器；绝对压力传感器
absolute pressure vacuum gauge 绝对压力真空表
absolute roughness 绝对粗糙度
absolute scale of temperature 绝对温标
absolute similarity 完全相似性
absolute spectrum function 绝对频谱函数
absolute stability 绝对稳定性
absolute strength 绝对强度
absolute temperature 绝对温度；热力学温度
absolute temperature scale 绝对温标
absolute thermodynamic scale 绝对热力学标度
absolute thermodynamic temperature scale 绝对热力学温标
absolute thermometer 绝对温度表
absolute topography 绝对地形
absolute unit 绝对单位
absolute unit system 绝对单位制

absolute vaccum 绝对真空
absolute vaccum gauge 绝对真空计
absolute value 绝对值
absolute velocity 绝对速度
absolute viscosity 绝对动力黏度
absolute volume 绝对容积；绝对体积
absolute vorticity 绝对涡量
absolute weight 绝对重量
absolute zero 绝对零度
absolute zero of temperature 绝对零度
absorbability 吸收；可吸收性
absorbed 被吸收的
absorbed dose 吸收剂量
absorbed dose commitment 吸收剂量负担
absorbed dose index 吸收剂量指标
absorbed moisture 吸收的水分
absorbed power 吸收能量；吸收功率
absorbed radiation dose 吸收的辐射剂量
absorbed-in-fracture energy 冲击韧性；冲击强度；冲击功
absorbent 吸收剂
absorbent material 吸收剂
absorber 减震器；吸收剂；吸收体
absorber cooler 吸收冷却器
absorbing 吸收
absorbing capacity 吸收能力
absorption 吸收
absorption band 吸收谱带；吸收频带
absorption coefficient 吸收系数
absorption column 吸收柱
absorption line 吸收线
absorption liquid chiller 吸收式液体冷却器
absorption liquid chilling system 吸收式液体冷却系统
absorption of heat 吸热

absorption of shock 减振
absorption spectrum 吸收谱
absorption tower 吸收塔
absorptivity 吸收率；吸收系数
abstract 转移
abstract heat 散热
AC (alternating current) 交流电
AC (altocumulus cloud) 高积云
AC ammeter 交流电流表
AC cast (altocumulus castellanus) 堡状高积云
AC cug (altocumulus cumulogenitus) 积云性高积云
AC flo (altocumulus floccus) 絮状高积云
AC lent (altocumulus lenticularis) 荚状高积云
AC motor 交流电动机
AC op (altocumulus opacus) 蔽光高积云
AC power loss 交流功率损耗
AC tra (altocumulus translucidus) 透光高积云
AC voltage converter 交流电压变流器
AC voltage transformer 交流电压互感器；变压器
AC voltmeter 交流电压表
AC welder 交流焊接机
accelerate 加速；促进
accelerated 加速的；促进的
accelerated stall 过载失速；加速失速
accelerated test 加速试验
accelerating 加速
accelerating grade 加速坡
accelerating pressure gradient 顺压梯度
acceleration 加速；加速度；加速作用
acceleration amplitude 加速度幅值

acceleration boundary 加速度界限
acceleration in pitch 俯仰加速度
acceleration in roll 滚转加速度；侧倾加速度
acceleration in spectrum 加速度谱
acceleration in transducer 加速度传感器
acceleration in yaw 偏航加速度
acceleration of gravity 重力加速度
acceleration response 加速度响应
accelerometer 加速度传感器
acceptability 可接受性
acceptability criteria 可接受标准
acceptable 合格的；容许的；可接受的；验收的
acceptable condition 验收条件；合格条件
acceptable dose 容许剂量
acceptable environment limit 容许环境极限
acceptable level 容许水平
acceptable life 有效使用寿命
acceptable limit 容许极限
acceptable noise level 容许噪声级
acceptable product 合格产品
acceptable quality level 验收质量标准
acceptable standard 验收标准
acceptable test 验收试验
acceptable value 容许值
acceptable wind speed 容许风速
acceptance 验收；认可；接受
acceptance and transfer (A&T) 验收与移交
acceptance certificate 验收合格证；验收证书
acceptance check 验收
acceptance of work 工程验收
acceptance quality level (AQL) 可接受的质量水平
acceptance specification 验收规范
acceptance test (AT) 验收试验
acceptance value 合格判定值

accepted 接受的
accepted bid 认可的投标书；已接受的投标书
access 入口；人孔；检修孔
access cover 检修盖
access door 检修门
access hatch 观察孔；检查孔
access hole 检查孔
access of air 通气孔
access opening (ACS-O) 检修孔；人孔
access panel 观察板；观测台
access to nacelle cabin 入机舱通道
access to tower 入塔通道
accessory 附件；附属设备
accessory equipment 附属设备
accident 事故；偶然事件
accident analysis 事故分析
accident brake 紧急制动器
accident condition 事故状态
accident defect 事故损坏
accident discharge 紧急排放口
accident error 偶然误差
accident insurance 事故保险
accident prevention 事故预防；安全措施
accidental 意外的；偶然的；附带的
accidental alarm 事故报警
accidental error 偶然误差
accidental exposure 偶然曝光；偶然曝射；偶然辐射
accidental failure 偶然故障
accidental loading 偶然负荷
accidental maintenance 事故维修
accidental release 事故排放
acclimatization 环境适应性
accommodate 适应；容纳；调节
accommodation 调节；调节作用；适应

accommodation coefficient 调节系数，适应系数
accompanying 陪伴的；附随的
accompanying diagram 附图
accordant 一致的；调和的
accordant connection 匹配连接
accretion 添加
accumulate 累积
accumulate deformation 累积变形
accumulate error 累积误差
accumulated 积累的
accumulated error 累积误差
accumulated filth 积垢
accumulated horizon 堆积层
accumulated snowdrift area 吹雪堆积区
accumulating 累加
accumulating capacity 储备能力
accumulation 累积；堆积
accumulation dose 堆积剂量
accumulation factor 累积因子
accumulation pattern 堆积模式
accumulator 储压罐；蓄能器
accumulator metal 蓄电池极板合金
accumulator plate 蓄电板；蓄电池极板
accumulator rectifier 蓄电池整流器
accuracy 精度；准确度
accuracy class 精度等级
accuracy control 精确控制
accuracy in computation 计算精确度
accuracy in instrument 仪表的精确度
accuracy in measurement 测量精度
accuracy of adjustment 调节精度
accuracy of instrument 仪表精度
accuracy of manufacture 制造精度
accuracy of measurement 测量精度

accuracy of reading 读数精确度
accuracy of scale 刻度精确度
accuracy to size 尺寸准确度
accurate adjustment 精调
accurate grinding 精磨
AC-DC converter 交直流转换器
AC-DC relay 交直流继电器
ACE (Aylesbury collaborative experiment) 艾尔兹伯里（风载荷）合作实验
ACES (automatic checkout and evaluation system) 自动检测和估算系统
acetone 丙酮
acetylene 乙炔；电石气
acetylene welding 乙炔焊接
acid 酸；酸性的
acid corrosion 酸腐蚀
acid degree 酸度
acid fog 酸雾
acid gas 酸气
acid rain 酸雨
acid resistance 耐酸性
acid test 酸性试验
acid value 酸值
acidification 酸化；氧化
acidimeter 酸比重计；酸量计
acidity 酸度；酸性
acidproof 耐酸性；抗酸性
acidproof material 耐酸材料
acid-resistent 耐酸的
acid-resistent steel 耐酸钢
acknowledgement 确认
ACOE (automatic checkout equipment) 自动检测装置
ACORE (automatic checkout and recording equipment) 自动检测记录设备

acorn 整流罩
acoustic 声学的；音响的
acoustic absorption 声吸收
acoustic absorptivity 吸声率；吸声系数
acoustic baffle 吸声板
acoustic barrier 吸声层
acoustic board 吸声板
acoustic ceiling board 吸声天花板
acoustic celotex board 吸声纤维板
acoustic conductivity 声导率
acoustic coupling 声耦合
acoustic damper 消声器
acoustic damping 声阻尼
acoustic damping effect 声阻尼效应
acoustic environment 声环境
acoustic fiber board 吸声纤维板
acoustic filter 声过滤器
acoustic frequency 声频率
acoustic insulation 隔音
acoustic isolation 隔声
acoustic level 声级
acoustic material 声学材料；隔音材料
acoustic meter 测声计
acoustic noise 声学噪音
acoustic pollution 声污染
acoustic pressure level 声压级
acoustic radar 声雷达
acoustic radiation pressure 声辐射压力
acoustic reference wind speed 声的基准风速；声参考风速
acoustic resistance 声阻
acoustic stiffness 声刚度
acoustic thermometer 声学温度计
acoustic velocity 声速度
acoustic(al) board 吸声板

acoustical detector 声波探测器
acoustics 声学
ACP (auxiliary control panel) 辅助控制仪表板
across 穿过
across corner 对角
across cutting 横向切割
across the grain 横纹
across-flow (cross-flow) 横向流；交叉流；错流
across-wind (cross-wind) 横风向的
across-wind correlation 横风向相关
across-wind cross-correlation 横风向互相关
across-wind response 横风向响应
across-wind test 横风向试验
acrotorque 最大扭力
ACS-O (access opening) 检修孔；人孔
ACSR (aluminium conductor steel reinforced) 钢芯铝绞线
act 法则；规程
ACT (auxiliary current transformer) 辅助电流互感器
ACTE (actuate) 致动；开动；使动作
ACTG (actuating) 致动；启动；开动；驱使；激励；作用
acting 代理的；起作用的
acting force 作用力
acting surface 作用面
actinography 光能测定仪；辐射仪
actinometer （自记）曝光计，曝光表
action 作用；影响；效应；操作
action center 作用中心
action roller 活动滚轮；动辊
action time 作用时间
action wheel 主动轮
activate 刺激；激活

activate button 启动按钮
activate key 启动键
activated charcoal filter 活性炭过滤器
activation 活动；赋活；激活；活化；激励；启用
activation analysis 活化分析
activation detector 放射性探测器
activation power 临界功率
activation power for wind turbines 风力机临界功率
activation rotational speed 临界转速
active 积极的；主动的
active appendage 主动附件
active area 有效面积
active circuit elements 主动电路元件
active component 有功分量
active control 主动控制
active cooling surface 有效冷却表面积
active current 有功电流
active deposit 活性沉积；放射性沉积；放射性沉降物
active drag 正阻力
active dust 放射性尘埃
active face 刃面
active fall-out 放射性沉降物
active filter 有源滤波器
active force 有效力；主动力
active height 有效高度
active in respect to 相对……呈阻性
active load 有源负载
active loss 有功损耗
active material 活性物质；活性材料；激活材料；放射材料
active pollution 放射性污染物
active power 有功功率
active region 作用区

active safety 主动安全性
active solar system 主动太阳能系统
active source 放射源
active stall 主动失速
active stall power control 主动失速控制
active stall regulated wind turbine 主动失速调节风电机组
active volt-ampere 有功伏安
active yaw mechanism 主动偏航机构
active yawing 主动偏航
activity 放射性；活度
activity coefficient 功率因数
activity duration 作业持续时间；有效期限
activity median aerodynamic diameter (AMAD) 放射性气动力中位直径；活性中值空气动力学直径
activity sampling 工作抽样检查
actual 真实的；实际的
actual air 实际空气量
actual capacity 实际容量
actual cooling surface 有效冷却表面积
actual cycle 实际循环
actual density 实际密度；真实密度
actual distance 实际距离
actual efficiency 实际效率
actual enthalpy drop 实际焓降
actual enthalpy rise 实际焓增
actual error 实际误差
actual gas 实际气体；真实气体
actual head 实际扬程
actual internal area 实际流通截面
actual life 实际寿命
actual lift 实际升力
actual load 实际载荷
actual loading test 实际负载试验

actual loss 实际损失

actual measurement 实际测量

actual output 实际输出

actual parameter 实在参数；实际参数

actual power 实际功率

actual pressure 实际压力

actual pressure lapse rate 实际气力递减率

actual pump head 实际泵扬程

actual service life 实际使用寿命

actual size 实际尺寸

actual source 实际（污染）源

actual stack height 实际烟囱高度

actual state 实际状态

actual stress 实际应力

actual terrain height 实际地形高度

actual time 实际时间

actual torque 实际转矩

actual value 实际值

actual weight 实际重量

actual wind 实际风

actual wind energy 实际风能

actual wind power 实际风能

actual working pressure 实际工作压力

actual zero point 绝对零点，基点

actuate (ACTE) 致动；开动；使动作

actuating (ACTG) 致动；启动；开动；驱使；激励；作用

actuating arm 驱动杆

actuating cam 主动凸轮

actuating cylinder 动力气缸

actuating device 驱动装置

actuating element 执行元件

actuating force 驱动力

actuating lever 驱动杆

actuating mechanism 执行机构

actuating motor 启动电动机；伺服电动机

actuating pressure 作用压力；驱动压力；促动压力

actuating signal 启动信号

actuating system 传动系统；传动机构

actuating unit 传动装置；传动机构

actuation 开动；传动

actuation time 动作时间

actuator 执行元件；传动装置

actuator disc 促动盘；动盘；作用盘

actuator governor 调速控制器

acute angle 锐角

ACV (alternating current voltage) 交流电压

ACV (automatic control valve) 自动控制阀

acyclic 非周期性的

AD (aerodynamic decelerator) 气动减速器；气动减速装置；空气动力减速装置；空气动力减速器

A/D converter (analog-to-digital converter or analog-digital converter) 模数转换器；A/D转换器

adapt 使适用；使适应

adaptability 适应性；灵活性；可用性

adaptation 自适应；适配

adapter 管接头；连接器

adapter connector 连接器

adapter ring 接合环

adapter skirt 连接套

adapter sleeve 紧固套，连接套管

adaptive 适应的；合适的

adaptive control 自适应控制

adaptive control system 自适应控制系统

adaptive-wall wind tunnel 自适应壁风洞

add 附加；加

added 增加的；更多的

added enthalpy 焓增
added heat 附加热
added lift 附加升力
added loss 附加损失；附加损耗
added mass 附加质量
added mass coefficient 附加质量系数
added metal 添加金属
added resistance 附加电阻
addendum 齿轮
addendum angle 齿顶角
addendum circle 齿顶圆
addendum line 齿顶线
addendum modification on gear 齿轮的变位
ADDER (automatic digital data error recorder) 数字信息误差自动记录器
addition 添加；附加
addition agent 添加剂；合金元素
additional 附加的；额外的
additional drag 附加阻力
additional equipment 附加设备；辅助设备
additional heat loss 附加耗热量
additional load 附加载荷
additive 添加剂
additive for lubricating grease 润滑脂添加剂
additive property 可加性
additivity law 相加性定律
address 地址
adfreezing 冻结
ADG (air-driven generator) 风动发电机
adhere 黏附
adherence 坚持；依附
adhesion 黏着力；附着力
adhesion strength 黏着力
adhesive 带黏性的；胶黏；黏合剂
adhesive ability 黏着力

adhesive agent 黏附剂
adhesive bonding 胶结
adhesive coating 黏附层
adhesive force 黏附力；内聚力
adhesive material 黏附材料
adhesive power 黏附力；附着力
adhesive seal 胶泥密封
adhesive strength 黏结强度
adhesive tape 胶带
adhesive value 黏合值
adhesive wax 黏蜡；胶黏封
adiabatic 绝热的
adiabatic boundary layer 绝热边界层
adiabatic change 绝热变化
adiabatic compression 绝热压缩
adiabatic condensation 绝热凝结
adiabatic cooling 绝热冷却
adiabatic curve 绝热曲线
adiabatic energy storage 绝热贮能
adiabatic equation 绝热方程
adiabatic equilibrium 绝热平衡
adiabatic expansion 绝热膨胀
adiabatic flow 绝热流动
adiabatic gradient 绝热梯度
adiabatic heating 绝热升温
adiabatic index 绝热指数
adiabatic lapse rate 绝热递减率
adiabatic layer 绝热层
adiabatic line 绝热线
adiabatic process 绝热过程
adiabatic region 绝热区
adiabatic warming 绝热增温
adiabatical 绝热的
adjacent 邻近的；毗连的
adjacent accommodation 相邻的设备；厢房

adjacent building 邻近建筑
adjacent vortex 附着旋涡
adjoining 邻接的；毗连的
adjoining building 邻接建筑
adjust 调整；校正
adjustability 可调节性
adjustable 可调整的；可校准的
adjustable blade 可调叶片
adjustable bolt 可调螺栓
adjustable clamp 可调夹具
adjustable clearance 可调间隙
adjustable condenser 可调电容器
adjustable constant speed motor 可调恒速电动机
adjustable contact 可调触点
adjustable damper 可调减震器
adjustable die 活动扳牙；活动扳手
adjustable feet 可调活脚
adjustable grille 可调格栅
adjustable guide vane 可调导叶
adjustable mark 调整标记
adjustable mass balance 可调配重
adjustable nozzle 可调喷嘴
adjustable pliers 可调手钳
adjustable spanner 活扳，可调扳手
adjustable speed 调速
adjustable speed motor 调速电动机
adjustable surface 可调面
adjustable vane 可调叶片
adjustable varying speed motor 可调变速电动机
adjustable voltage stabilizer 可调稳压器
adjustable wall 可调（风洞）壁
adjustable wrench 活扳，可调扳手
adjustable-pitch 可调螺距，可调桨距；变螺距

adjustable-pitch propeller 变距螺旋桨
adjustage 辅助设备
adjuster bolt 调节螺栓
adjusting bolt 调节螺栓
adjusting coil 可调线圈
adjusting damper 调节风门
adjusting device 调节装置
adjusting gauge 整定仪表
adjusting instrument 调节仪器
adjusting pin 定位销
adjusting plate 调整板
adjusting range 调节范围
adjusting screw 调正螺丝钉；校正螺丝钉
adjusting valve 调节阀
adjustment 调整，调节，调节器
admissibility 许入；准许
admissible 可容许的；可采纳的
admissible concentration 容许浓度
admissible error 容许误差
admissible load 容许负荷
admissible parameter 导纳参数；容许参数
admittance 进入许可；导纳
admittance function 导纳函数；容许函数
admixture 混合；掺和；混合物；附加剂
ADP (automatic data processing) 自动数据处理
ADPE (automatic data processing equipment) 自动数据处理装置
ADPS (automatic data processing system) 自动数据处理系统
adrift 漂浮的
adsorption 吸附
adsorption affinity 吸附力
adsorption column 吸附柱
adsorption effect 吸附效应

advance 发展的；先行的

advance ball 滑动滚珠

advanced 先进的

advanced and innovative wind system 革新型风能系统

advanced research 远景研究，前沿研究

advanced science 尖端科学

advection 平流

advection fog 平流雾

advection inversion 平流逆温

advection layer 平流层

advection region 平流区

advection scale 平流尺度

advective cooling 平流冷却

advective region 平流区

advective term 平流项

adverse 相反的

adverse pressure gradient 逆压梯度

adverse wind 逆风

adversely 逆地，反对地

adze 扁斧

AE (absolute error) 绝对误差

AED (automated engineering design) 自动工程设计

aeolian deposit 风积物

aeolian erosion 风蚀作用

aeolian excitation 风成激励

aeolian feature 风成地貌；风成特征

aeolian vibration 风激振动；风成振动

aer 气压单位

aeration 换气；通风

aerial cable 架空电缆

aerial cableway 架空索道

aerial conductor 架空线；明线

aerial contamination 空气污染

aerial current 气流

aerial detection 空气检测；空中检测

aerial discharger 避雷器

aerial farming 航空农业

aerial growth 气生；地面生长；地上部分生长

aerial pollution 空气污染

aerial radiation thermometer 空气辐射温度计

aerial railway 高架铁道

aerial ropeway 架空索道

aerial surveillance 空中监测

aerial survey 航空测量；空中监测

aerial transmission 空中传播

aerial 航空的；生活在空气中的；空气的；高耸的；天线

aero 飞机的；航空的；飞行的

aerocurve 曲翼；曲翼飞机

aerodromometer 气流流速表

aerodynamic 空气动力学的；气体动力学的；气动的

aerodynamic (force) differential 气动力差

aerodynamic add-on device (AOA device) 气动附加装置

aerodynamic admittance 气动导纳

aerodynamic airfoil 气动翼型

aerodynamic analysis 空气动力分析

aerodynamic appendage 气动力附件

aerodynamic area 气动力面积

aerodynamic attachment 气动力附件

aerodynamic augmentation device 气动增强装置

aerodynamic balance 气动力平衡

aerodynamic bearing 气动轴承；空气动力轴承

aerodynamic behavior 气动性能

aerodynamic body 空气动力绕流体；流线型体

aerodynamic boundary layer 气动边界层

aerodynamic brake 气动制动；空气动力刹车；空气动力制动
aerodynamic braking 气动制动
aerodynamic capture 气动力捕获；空气动力捕获
aerodynamic center 气动中心；焦点
aerodynamic characteristics 气动特性
aerodynamic characteristics of rotor 风轮气动特性；风轮空气动力特性
aerodynamic chord of airfoil 翼型的气动弦线
aerodynamic coefficient 气动力系数；空气动力系数
aerodynamic compensation 气动补偿
aerodynamic configuration 气动构型
aerodynamic control 气动控制
aerodynamic controls 气动控制装置
aerodynamic coupling 气动耦合
aerodynamic criterion 气动判据；气动力准则
aerodynamic cross-coupling 气动交叉耦合
aerodynamic damping 气动阻尼
aerodynamic damping coefficient 气动阻尼系数
aerodynamic data 气动数据
aerodynamic decelerator (AD) 气动减速器；气动减速装置；空气动力减速装置；空气动力减速器
aerodynamic deflector 气动导流板
aerodynamic derivative 气动导数
aerodynamic derivative coefficient 气动导数系数
aerodynamic destabilizing 气动失稳
aerodynamic device 气动装置
aerodynamic directional instability 气动力方向不稳定性
aerodynamic dissipation 气动力耗散

aerodynamic disturbance 气动力扰动
aerodynamic downwash 气动力下洗
aerodynamic effect 气动力效应
aerodynamic effectiveness 气动力效能；空气动力效能
aerodynamic efficiency 气动效率
aerodynamic end effect 气动力端部效应
aerodynamic environment 气动环境
aerodynamic excitation 气动力激励
aerodynamic field 气动力场
aerodynamic force 气动力
aerodynamic force coefficient 气动力系数
aerodynamic force component 气动力分量；空气动力分量
aerodynamic force derivative 气动力导数
aerodynamic form 空气动力学形状；气动形状；流线型
aerodynamic handling force 气动操纵力
aerodynamic heating 气动加热
aerodynamic hinge moment 气动铰链力矩
aerodynamic hysteresis 气动力迟滞
aerodynamic influence coefficient 气动影响系数
aerodynamic instability 气动不稳定性
aerodynamic interaction 气动力相互作用
aerodynamic interference 气动力干扰
aerodynamic laboratory 空气动力实验室
aerodynamic lag 气动力滞后
aerodynamic layout 气动力外形
aerodynamic lift 气动升力
aerodynamic load 气动载荷
aerodynamic load distribution 气动力载荷分布
aerodynamic mass 气动质量
aerodynamic modification 气动修改；气动改

型

aerodynamic moment of inertia 气动力惯性矩

aerodynamic noise 气动噪声

aerodynamic nomogram 空气动力特性图

aerodynamic nonlinearity 空气动力特性的非线性；气动特性的非线性

aerodynamic optimization 空气动力特性最优化；气动特性最优化

aerodynamic parameter 气动参数

aerodynamic performance 气动性能

aerodynamic performance test 气动性能试验

aerodynamic piston theory 气动活塞理论

aerodynamic pitch 气动桨距

aerodynamic power controller 气动功率控制器

aerodynamic power regulation method 气动功率调节方法

aerodynamic profile 气动翼型

aerodynamic property 气动特性

aerodynamic quality 气动品质

aerodynamic radius 空气动力学半径；气动半径

aerodynamic reference center 气动力参考中心

aerodynamic refinement 气动改善；气动改型

aerodynamic resistance 气动阻力

aerodynamic roughness 气动粗糙度

aerodynamic roughness length 气动粗糙长度

aerodynamic roughness parameter 气动粗糙参数

aerodynamic self-excitation 气动力自激

aerodynamic self-starter 气动自动启动装置

aerodynamic shadow 气动阴影；气动尾迹

aerodynamic shape 气动形状

aerodynamic shape correction 气动外形修改

aerodynamic shape optimization 气动外形最优化

aerodynamic shaping 空气动力造型

aerodynamic shear 气力剪切力

aerodynamic shield 气动整流罩

aerodynamic shroud 整流罩

aerodynamic similarity 气动相似性

aerodynamic simulation 空气动力模拟

aerodynamic solidity 气动实度

aerodynamic sound 气动声音

aerodynamic source of sound 气动声源

aerodynamic speed control 气动速度控制

aerodynamic spoiler 气动扰流板

aerodynamic spoiler device 气动扰流装置

aerodynamic stability 气动稳定性

aerodynamic stack downwash 空气动力堆积下洗气流

aerodynamic stack height 空气动力堆积高度；气动高度

aerodynamic starting 气动力启动

aerodynamic stiffness 气动刚度

aerodynamic surface 气动力作用面

aerodynamic technology 空气动力工艺；气动工艺

aerodynamic term 空气动力项；气动项

aerodynamic test 气动力试验

aerodynamic theory 空气动力学理论；气动理论

aerodynamic torque 空气动力扭矩

aerodynamic trail 气动尾迹；航迹云

aerodynamic transfer function 气动力传递函数

aerodynamic tunnel 风洞

aerodynamic twist 气动扭转

aerodynamic wake 气动尾流
aerodynamic yaw 气动力偏航
aerodynamic(mean) chord 平均气动弦长
aerodynamically 空气动力学地
aerodynamically clean body 气动力流线型车身
aerodynamically fully rough 气动充分粗糙部分
aerodynamically induced vibration 气动力致振
aerodynamically shaped 流线型的；空气动力型的
aerodynamically smooth 气动光滑的
aerodynamically stable 气动力稳定的
aerodynamicist 空气动力学家；气动学家
aerodynamics (ARODYN) 空气动力学；气体动力学
aeroelastic 气动弹性的
aeroelastic building 气动弹性建筑物
aeroelastic characteristic 气动弹性特性
aeroelastic deformation 气动弹性变形
aeroelastic derivative 气动弹性导数
aeroelastic effect 气动弹性效应
aeroelastic eigenfunction 气动弹性特征函数
aeroelastic eigenvalue 气动弹性特征值
aeroelastic equation of motion 气动弹性运动方程
aeroelastic feedback 气动弹性反馈
aeroelastic instability 气动弹性不稳定性
aeroelastic interaction 气动力弹性干扰
aeroelastic load 气动弹性载荷
aeroelastic model 气动弹性模型
aeroelastic modeling 气动弹性模拟
aeroelastic oscillations 气动弹性振动
aeroelastic phenomena 气动弹性现象

aeroelastic response 气动弹性响应；气动弹性反应
aeroelastic section model 气动弹性节段模型
aeroelastic simulation 气动弹性计算
aeroelastic stability 气动弹性稳定性
aeroelastic stiffness 气动弹性刚度
aeroelastic structure 气动弹性结构物
aeroelastic transfer function 气动弹性传递函数
aeroelastic vibration 气动弹性振动
aeroelastic wing 气动弹性翼
aeroelastic wing theory 气动弹性翼理论
aeroelasticity 气动弹性力学
aeroelastics 空气弹性学
aerofoil (=airfoil) 翼型
aerofoil camber 翼型弯度
aerofoil design 翼型设计
aerofoil profile 翼型剖面；翼型
aerofoil theory 机翼理论；机翼理论
aerofoil thickness 翼型厚度
aerofoil surface finish 翼型表面光洁度
aerogenerator 风力发电机
aerography 大气图表；气象学
aerological analysis 高空分析
aerological ascent 高空观测
aerological instrument 高空仪器
aerological observatory 高空观象台
aerological sounding 高空观测
aerological station 高空观测站
aerological theodolite 高空经纬仪
aerology 高空气象学
aeromancy 大气预测
aerometeorograph 高空气象计
aerometer 气体密度计
aerometric network 高空气象网

aerometry 气体测量

aeromobile 气垫车

aeromotor 风力发动机

aeronautical experiment 航空实验

aeronautical meteorology 航空气象学

aeronautical type wind tunnel 航空风洞

aeronautics 航空学

aeronomy 高层大气物理学

aeropause 大气上界

aerophotography 航空摄影学

aeroradioactivity 大气放射性强度

aero-sensitive 气动力敏感的

aerosol 浮质；气溶胶；气雾剂；烟雾剂

aerosol generator 气溶胶发生器

aerosol mixing ratio 气溶胶混合比

aerosol particle 气溶胶粒子

aerosol sedimentation 气溶胶沉淀

aerosol source 气溶胶源

aerosol spectrometer 气溶胶谱仪

aerosphere 大气层

aerostat 气球；飞艇

aerostatics 空气静力学

aerothermodynamics 空气热力学；气体热力学

aeroturbine 空气涡轮机

aerovane 风向风速计

AES (American Electromechanical Society) 美国机电学会；美国电子学会

AES (American Engine Society) 美国发动机学会

AESC (American Engineering Standards Committee) 美国工程标准委员会

aesthetic constraint 美学制约

aesthetics 美学

AF (axial flow) 轴向流动

affine deformation 均匀变形

affinor 反对称张量，张量

affluent 支柱

AF-PMSG (axial flux permanent magnet synchronous generator) 轴向磁通永磁同步发电机

A-frame A型骨架

AFT (automatic fine tuning) 自动微调

AFT (automatic frequency tuner) 自动频率调谐器

aft bearing 后轴承

aft deck 后甲板

aft draft 尾吃水

after 在……之后

after cooler 后冷却器

after mast 后桅

after service 售后服务

after service center 售后服务中心

after body 后体；后车身

after body effect 后体效应

after cooling 再冷却

after drying 再干燥

after filter 再过滤器

after treatment 后处理

after sales service 售后服务

against-wind 逆风；迎风

ageing 时效；老化；变质

ageing behaviour 时效行为

ageing characteristic 老化特性

ageing of insulation 绝缘老化

ageing of materials 材料老化

ageing of valve 电子管老化

ageing process 老化过程

ageing test 老化试验

ageing time 老化时间

agglomeration 凝聚；结块

agglutination 凝聚作用；胶着作用
agglutination reaction 凝聚反应
aggregate capacity 总功率
aggregated momentum 总动量
agitation 搅拌；掺混
agricultural buildings 农业建筑物
agricultural chemicals 农业化学品；农药
agrotopoclimatology 农业地形气候学
AIEE (American Institute of Electrical Engineers) 美国电气工程师协会
AIEEE (American Institute of Electrical and Electronics Engineers) 美国电气和电子工程师协会
aiguille 钻孔器；钻头
AIIE (American Institute of Industrial Engineers) 美国工业工程师协会
aileron (AIL) 副翼
aileron actuator 副翼促动器
aileron cable 副翼操纵索
aileron chord 副翼弦
aileron control 副翼控制
aileron hinge 副翼铰链
aileron horn 副翼杆；副翼操纵杆
aileron yaw 副翼偏航
aim 目标；目的
aiming （风轮的）对风，调向
air 空气；气流；气动的
air at rest 静止空气
air bearing 空气轴承
air bells pollution 大气污染
air blast 强气流
air blast transformer 风冷式变压器
air bleed 空气泄放；放气；排气；通气；通气器
air blower 吹风机；散热风扇
air blowing 吹气；空气吹污
air brake 气闸；风闸；空气制动器
air braking system 空气制动系统
air breather 通风装置；通风孔
air bubble 气泡
air bubble collapse 气泡破碎
air buffer 空气缓冲器；空气隔层
air bumper 空气减震(缓冲、阻尼)器
air change loss 空气交换损失
air change rate 换气率
air chimney 通风烟囱
air circulation 空气环流；空气循环
air cleaning 空气净化
air clutch 气动离合器
air condenser 空气冷凝器
air conditioned building 带空调的建筑
air conditioner 空气调节器
air conditioning 空气调节
air conductivity 空气电导率；大气导电性
air control 气动控制
air control equipment 气动控制装置
air control motor 气动控制电动机
air control system 气动控制系统
air control valve 气动控制阀
air controlled direction valve 气动控制换向阀
air cool 空气冷却
air coolant 冷却空气；空气冷却剂
air cooled oil cooler (air-cooled-oil cooler) 风冷式油冷却器
air cooler 空气冷却器；风冷装置
air cooling 气冷；风冷
air cooling fin 散热片
air course 通风巷道；风道
air current 气流
air curtain 气幕
air cushion 气垫

air cushioning 空气减震
air dam 前扰流板
air damper 空气阻尼器
air damping 空气阻尼
air dashpot 空气阻尼器
air deflector 导风板
air density 空气密度
air diffusion 空气扩散
air direction 风向
air discharge 排气；空中放电
air door 通风孔；通风门
air dose 空气剂量
air draft 气流；空气通风
air drag 空气阻力
air drain 空气泄流；通风管
air-driven generation (ADG) 风动发电机
air dry weight 风干重
air duct 风道
air eddy 空气涡流
air exhaust 排气
air exit hole 排气孔
air fan 风扇
air filter 空气过滤器
air flow 气流
air flow breakaway 气流分离
air flow characteristic 气流特性
air flow condition 空气流动条件
air flow direction 气流方向
air flow guide 导流罩；导流板
air flow indicator 空气流量表；气流指示表
air flow instrument 气流仪
air flow line 流线
air flow meter 气流计
air flow noise 气流噪声
air flow rate 气流流速
air flow resistance 气流阻力
air flow structure 气流结构
air flow velocity 气流速度
air flow wattmeter 气流瓦特计
air flue 风道；烟囱
air force 风力
air friction 空气摩阻；空气摩擦
air gap 气隙；空气间隙
air gap flux 气隙磁通
air gauge 气压表；气压计
air hatch 通风口
air header 集气管；空气总管
air humidity 空气湿度
air impermeability 不透气性；气密性
air inclusion 空气杂质
air infiltration 空气渗透；漏风
air injection 空气喷射；喷气
air inlet 通风口
air inlet hub 进气整流罩
air input 进气量
air intake 进风口；进气孔；吸气管；进气流
air jacket 气套
air jet 空气喷射；喷气口
air leak 漏气
air line 送气管路；飞行航线
air load 气动荷载
air load testing 空气动力载荷试验
air management 大气管理
air mass 气团
air measurement 空气测量
air measuring and recording instrument 空气测录仪
air mechanics 空气力学；空气动力学
air meter 风速仪；气流计
air metering 空气测量

air mixture 空气混合物
air moisture 空气湿度
air moisture content 空气含湿量
air monitor 大气监测器
air monitoring 大气监测
air monitoring car 大气监测车
air monitoring network 大气监测网
air monitoring procedure 大气监测规程
air monitoring station 大气监测站
air noise 空气噪音
air oil cooler (AOC) 空气油冷却器
air outlet 空气出口；放气口
air parameter 空气参数
air particle monitor 空气粒子监测器
air passage 气路；气道；空气道
air patenting 空气淬火；风冷
air permeability 透气性
air pocket 气袋
air pollution 空气污染
air pollution alert 空气污染警报
air pollution code 空气污染法规
air pollution complaint 空气污染诉讼
air pollution concentration 空气污染物浓度
air pollution control district 空气污染管制区
air pollution control law 空气污染防治法
air pollution control regulation 空气污染防治条例
air pollution control system 空气污染防治系统
air pollution emission 空气污染排放物
air pollution episode 空气污染事件
air pollution forecasting 大气污染预报
air pollution index 空气污染指数
air pollution level 空气污染水平
air pollution meteorology 空气污染气象学

air pollution model 空气污染模式
air pollution modelling 空气污染模拟
air pollution monitoring system(APMS) 空气污染监测系统
air pollution potential 空气污染潜势
air pollution prediction 大气污染预报
air pollution region 空气污染地区
air pollution sensor 空气污染传感器
air pollution source 空气污染源
air pollution standard 空气污染标准
air pollution surveillance 空气污染监视
air pollution survey 空气污染调查
air pollution transport 空气污染物运输
air pollution trend 空气污染趋势
air pollution zone 空气污染区
air pressure probe 气压探头
air quality 大气质量
air quality control region 大气质量控制区
air quality criteria 大气质量判据
air quality impact analysis 大气质量影响分析
air quality index 大气质量指数
air quality model 大气质量模式
air quality standard 大气质量标准
air quenching 空气冷却；空气淬冷
air resistance 空气阻力；气动阻力
air resistance brake 空气阻力制动器
air resistance of permeability 空气渗透阻力
air resonance 空气共振
air return 回风
air return method 回风方式
air sampling 空气取样
air scoop 进气喇叭口；进风口
air screw 空气螺旋桨
air screw balance 空气螺旋桨平衡
air screw boss 空气螺旋桨毂

air screw pitch 空气螺旋桨距
air seal 气密；气封
air set 空气中凝固；常温自硬；自然硬化
air shaft 通风井
air shed 空气污染区域
air sink 气穴；气汇
air sizing 风力分级
air space cable 空气绝缘电缆
air space paper core cable (ASPC) 空气纸绝缘电缆
air spaced coax 空气绝缘同轴电缆
air speed 风速
air speed gauge 风速表
air speed tube 空气流速管；皮托管
air speedometer 气流速度计
air spoiler 扰流板
air spring suspension 空气弹簧悬挂；气垫吊架
air stagnation 大气滞止
air stream 气流
air stream contamination 气流污染
air stream deflector 导风板
air supported roof 充气屋顶；空气支撑屋顶
air switch 空气开关
air temperature (AT) 空气温度
air tight 气密；不漏气
air torrent 空气湍流
air tube cooler 空气管冷却器
air tunnel 风洞
air turbine alternator (ATA) 空气涡轮交流发电机
air turbulence 空气湍流
air vane 风标
air vane guide 气流导向叶片
air velocity vector 气流速度矢量
air vent 排气口；通风孔

air ventilation 通风；换气
air ventilation window 通风窗
air volume 风量
air wire 架空线
air acetylene welding 气焊；乙炔焊
airborne 空运的；飞机载的；航空的；空气所带的；空气传播的
airborne activity 机载活动
airborne concentration 气载浓度
airborne contaminant 气载污染物
airborne dust 气载尘埃
airborne noise 空气噪声
airborne particle 气载粒子
airborne pollutant 气载污染物
airborne pollution 大气污染；空气污染
airborne radioactivity 气载放射性；航空放射性
airborne radioactivity survey 航空放射性测量
airborne waste 气载废物
air-brake dynamometer 空气制动测力计
air-cooled 风冷的；空气冷却的
air-cooled condenser 空冷式冷凝器
air-cooled generator 空冷式发电机
air-cooled-oil cooler (air cooled oil cooler) 风冷式油冷却器
air-cushion 空气垫
air-damped 空气减震器的
air-driven generator (ADG) 风动发电机
airflow 气流；流线型的
airflow line 流线
airflow pattern 流谱
airflow pulsation 气流脉动
airflow rate 空气流量
airflow stability 气流稳定性

airfoil (aerofoil) 翼型
airfoil camber 翼型弯度
airfoil center of pressure 翼型压力中心
airfoil characteristic 翼型特性
airfoil chord 翼弦
airfoil contour 翼型廓线
airfoil drag 翼型气动阻力；翼型阻力
airfoil pressure distribution 沿翼型压力分布
airfoil profile 翼型；翼型剖面
airfoil section 翼型；翼剖面
airfoil shaped blade 翼型叶片
airfoil surface 翼面
airfoil theory 翼型理论；机翼理论
airfoil thickness 翼型厚度
airfoil wind machine 翼型板风力机
airfoil with reverse camber 反弯度翼型
airfoil-shaped vane 翼型叶片
air-gap diameter 气隙直径
air-gap field 气隙磁场
air-gap flux 气隙磁通
air-gap flux density 气隙磁通密度
air-gap flux distribution 气隙磁通分布
air-gap line 气隙磁化线
air-gap torque 气隙转矩
air-gap transformer 气隙变压器
air-gap voltage 气隙电压
air-gap width 气隙长度
air-gap winding 气隙绕组
air-hole 气孔
airiness 通风
air-jacketed condenser 空气套冷凝器
air-mass 气团
air-mass fog 气团雾
air-mass modification 气团变性
air-operated 气动的

air-operated tool 气动工具
air-over-hydraulic brake 空气液压刹车装置
air-powered servo 气动伺服机构
air pressure 大气压
air-release valve 放气阀（门）
air-sea interaction 气海作用
air-sea interface 气海界面
airspace 大气层；大气空间
air space paper core cable (ASPC) 空气纸绝缘电缆
air-speed head 空速管；气压感受器
air-speed tube 空速管
air-spring 气垫；空气弹簧
airstream deflector 导风板
airstream quality 气流品质
air-supported structure 充气结构物
air-termination system 接闪器
air-water interface 气水界面
alarm 告警装置
alarm and protection system 报警保护系统
alarm annunciator 警报信号器
alarm device 报警装置
alarm gauge 警报器
alarm indicator 报警指示器
alarm light 报警灯
alarm signal 报警信号
alary 翼状的；翼形
albedo 反照率；反射率
alclad 铝衣合金；覆铝层；包铝
alclad sheet 包铝板
alcohol manometer 酒精压力计
alerting signal 报警信号
algam 锡；铁皮
algebraic 代数的
algebraic average error 代数平均误差

algebraic equation 代数方程
algorithmic 算法的
aliasing 混淆
align 对准；校直
aligner 校准器；前轴定位器
alignment 对准；定位调整
alignment coil 校正线圈
aligment function 对准功能
aligment test 准直精度检查
alkali metal coolant 碱金属冷却剂
alkaline 碱性的
alkaline battery 碱性蓄电池
alkaline cleaning 化学除油；碱清洗
alkali-sensitive 碱性感；碱敏
all purpose engine oil 通用机油
all range 全量程
allen key 艾伦内六角扳手
allen wrench 六方扳手
all-gear(ed) 全齿轮的
alligator 皱裂；裂开
alligator crack 龟裂
alligator wrench 管扳手
alligator-hide crack 龟裂
all-metal construction 全金属结构
allobaric wind 变压风
allowable 容许的；许可的
allowable concentration 容许浓度
allowable environment condition 容许环境条件
allowable error 容许误差
allowable exposure 容许照射
allowable limit 容许极限
allowable load 容许负荷
allowable maximum current density 容许最大电流密度

allowable noise 容许噪声
allowable peak pressure 容许最高压力
allowable pressure 容许压力
allowable resistance 容许阻力
allowable stress 容许应力
allowable tolerance 容许公差
allowable value 容许值
allowable working pressure 容许工作压力
allowance 公差
allowance error 容许误差
allowance for finish 加工余量；精加工余量
allowance for machining 机械加工余量
allowance test 公差配合实验；容差试验
allowed band 公差带
alloy 合金
alloy steel (AS) 合金钢
alloy steel protective plating (ASPP) 合金钢保护电镀
almost-periodic 近周期性的
along-flow 顺流向的
along-wind 顺风向的
along-wind acceleration 顺风向加速度
along-wind buffeting 顺风向抖振
along-wind correlation 顺风向相关
along-wind cross-correlation 顺风向互相关
along-wind deflection 顺风向挠度
along-wind fluctuation 顺风向脉动
along-wind galloping 顺风向驰振
along-wind oscillation 顺风向振动
along-wind response 顺风向响应，顺风向反应
alpha decay α衰变
alpha disintegration α蜕变
alpha radiation α辐射
alpha ray α射线
altazimuth 地平经纬仪

alternate determinant 交错行列式
alternately spaced vortex 交替分布涡
alternating 交替的
alternating bending 交变弯曲
alternating bending moment 交替弯矩
alternating component 交流成分
alternating current (AC) 交流电
alternating current generator 交流发电机
alternating current machine 交流电机
alternating current motor 交流电动机
alternating current rectifier 交流电流整流器
alternating current voltage (ACV) 交流电压
alternating energy tax law 替代能源税收法
alternating load 交变负载
alternating motion 往复运动
alternating pressure 交变压力
alternating stress 交变应力
alternating voltage 交流电压
alternating vortex shedding 交替漩涡脱落
alternative energy source 替代能源
altigraph 气压计
altimeter 高度表
altimetry 测高学；高度测量法
altitude 海拔
altitude angle 仰角
altitude gauge 测高仪
altitude tunnel 高空模拟风洞
altitude wind tunnel 高空模拟风洞
altocumulus castellanus (AC cast) 堡状高积云
altocumulus cloud (AC) 高积云
altocumulus cumulogenitus (AC cug) 积云性高积云
altocumulus duplicatus 重叠高积云；复高积云

altocumulus floccus (AC flo) 絮状高积云
altocumulus lenticularis (AC lent) 荚状高积云
altocumulus opacus (AC op) 蔽光高积云
altocumulus stratiformis 成层高积云；层状高积云
altocumulus translucidus (AC tra) 透光高积云
aluminium 铝
aluminium alloy 铝合金
aluminium bar 铝棒
aluminium brass 铝黄铜
aluminium cable 铝芯电缆
aluminium coating 热镀铝法
aluminium conductor steel reinforced (ACSR) 钢芯铝绞线
aluminium foil 铝箔
aluminium oxide 氧化铝
aluminium sheet 铝片
aluminium steel 渗铝钢
aluminium-cell arrester 铝避雷针
aluminum continuous melting & holding furnaces (ATC system) 自动换刀系统；刀库
AM (amplitude modulation) 调幅
AMAD (activity median aerodynamic diameter) 放射性气动力中位直径；活性中值空气动力学直径
amalgamation 混合
ambient 环境
ambient air 周围空气
ambient atmosphere 环境大气
ambient concentration 环境浓度
ambient condition 环境条件
ambient density 环境密度
ambient environment 周围环境

ambient humidity 环境湿度；周围湿度
ambient noise 环境噪声
ambient pollution burden 环境污染负荷
ambient power density 周围风能密度
ambient pressure 环境压力；周围压力
ambient temperature 环境温度；室温
ambient turbulence 周围湍流；环流湍流
ambient vibration 环境振动
ambient wind 周围风
ambient wind angle 周围迎风角
ambient wind speed 周围风速
ambient wind stream 周围气流
AMC (automatic modulation control) 自动调制控制
AME (angle measuring equipment) 测角装置
Amenican Welding Institute (AWI) 美国焊接协会
AMER STD (American Standard) 美国标准
American Academy of Sciences (AAS) 美国科学院
American Association for the Advancement of Science (AAAS) 美国科学发展协会
American Association of Engineers (AAE) 美国工程师协会
American Automatic Control Council (AACC) 美国自动控制委员会
American Electromechanical Society (AES) 美国机电学会；美国电子学会
American Engine Society (AES) 美国发动机学会
American Engineering Standards Committee (AESC) 美国工程标准委员会
American National Standards Institute (ANSI) 美国国家标准学会
American Society for Testing and Materials (ASTM) 美国材料与试验协会
American Society of Mechanical Engineer (ASME) 美国机械工程师学会
American Standard (AMER STD) 美国标准
American Standard Thread (AST) 美国标准螺纹
American Standards Association (ASA) 美国标准协会
American Standards Committee (ASC) 美国标准委员会
American Standards of Test Manual (ASTM) 美国标准实验手册
American Welding Institute (AWI) 美国焊接协会
AMM canalog monitor modu(e) 类比监视器模组；模拟监视器模块
ammeter 电流表
amorphous metal transformer 非晶合金变压器
amort winding 制动线圈；阻尼线圈
amortisseur 减震器；消音器；缓冲器；阻尼器；阻尼绕组
amount of daily solar radiation 太阳日辐射量
amount of direct solar radiation 太阳直射辐射量
amount of radiation 辐射量
amount of theoretical air 理论空气量
amp out 放大器输出端
Ampere 安培
Ampere density 电流密度
Ampere gauge 电流表
Ampere meter 安培计
Ampere turn 安培匝数；安匝数；安匝
Ampere-hour 安培小时
Ampere's law 安培定律
ampere-turns 安匝（数）

ampersand 表示"&"符号；和
amplidyne 电机放大机
amplification 放大；放大率
amplified action 放大作用
amplifier 放大器
amplifier panel 放大器盘
amplitude (AMPTD) 振幅；幅度
amplitude distortion 波幅畸变
amplitude domain 幅域
amplitude frequency characteristic 幅频特性
amplitude growth 增幅
amplitude histogram of wind-speed 风速较差直方图
amplitude limiter 限幅器
amplitude modulation (AM) 调幅
amplitude of vibration 振幅
AMPTD (amplitude) 振幅；幅度
anabatic 上升的
anabatic flow 上升气流
anabatic wind 上升风；坡风
anacom 分析计算机
anaerobic 没有空气而能生活的；厌氧性的
anaflow 上升气流
anafront 上升锋面
analog 模拟；类似
analog computation 模拟计算
analog control 模拟控制
analog control system 模拟控制系统
analog controller 模拟控制器
analog data 模拟数据
analog digital converter 模拟数字转换器
analog input terminals 模拟量输入端子
analog signal 模拟信号
analog-to-digital converter 模数转换器；A/D 转换器

analogue 类似物
analogue board 模拟盘
analogue computer 模拟计算机
analogue computing system 模拟装置
analogue control 模拟控制
analogue filter 模拟滤波器
analogue simulation 相似模拟；类比模拟；仿真模拟
analogue study 模拟研究
analogue signal 模拟信号
analogue-digital converter or analogue-to-digital converter (A/D converter) 模—数转换器；A/D 转换器
analogy 类比
analogy switch 模拟开关
analysis 分析
analysis of variance 方差分析
analytical 分析的
analytical error 分析误差
analytical function 解析函数
analytical method 解析法
analytical model 解析模型
analytical prediction 解析预估
analyzer 分析器
anchor 锚；抛锚；锚定
anchor bar 锚筋
anchor beam 固定梁
anchor bolt 底座螺栓
anchor ring 环面
anchorage 固定支座
anchorage bolt 固定螺栓；地脚螺栓
anchored suspension bridge 锚定式悬索桥
anchoring system 锚定系统
ancillary attachment 特殊附件
ancillary equipment 辅助设备

andhi 对流性尘暴
anelasticity 内摩擦力；滞弹性
anelectric 不能因摩擦而生电的物质；非电化体
anemometer cup 风速计转杯
anemobarometer 风速风压计
anemobia(o)graph 风速风压记录器
anemochore 风播植物
anemocinemograph 电动风速计
anemoclinograph 铅直风速计
anemoclinometer 铅直风速仪；垂直风速表；风斜表
anemodispersibility 风力分散率
anemogram 风力自记曲线
anemograph 风速记录表，风力记录计；自动记风仪
anemology 测风学
anemometer 风速计；风速仪；流速表
anemometer factor 风速计系数
anemometer mast 测风杆
anemometer tower 测风塔
anemometer with stop-watch 停表风速表
anemometric 测定风力的
anemometrograph 记风仪；风向风速记录仪
anemometry 风速和风向测定法；风力测定
anemorumbometer 风向风速表
anemoscope 风速仪
anemovane 接触式风向风速器
aneroid 真空气压盒
aneroid manometer 无液压力计
aneroid-barometer 无液气压计
aneroidograph 空盒气压计；无液气压器
angle 角；角度
angle beam 角钢梁
angle bend 角形接头；角形弯管
angle encoder 角度编码器

angle gear 锥齿轮（传动）；斜交轴伞齿轮
angle grinder 角锉
angle iron 角钢
angle joint 角接头
angle measuring equipment (AME) 测角装置
angle of approach 渐进角
angle of attack (AOA) 迎角；攻角
angle of attack convergence 迎角减小
angle of attack of blade 叶片攻角
angle of attack of maximum lift 最大升力攻角；临界攻角
angle of attack vane 迎角风标
angle of bank (AOB) 倾斜角
angle of cant 喷管摆角
angle of curve 曲度
angle of declination 偏斜角
angle of deflection 偏移角；挠度
angle of delivery 出手角度；投掷角度
angle of departure 分离角；发射角；发射离地角；偏离角
angle of depression 俯角
angle of descent 下降角；坡道角
angle of elevation 离地角；仰角
angle of fall 落下角
angle of fire 射角
angle of friction 摩擦角
angle of impact 命中角
angle of incidence 入射角；迎角；安装角
angle of inclination 倾角
angle of lag 滞后角
angle of leeway 横漂角；偏航角
angle of oscillation 摆动角
angle of pitch 俯仰角；桨距角
angle of rack 倾角
angle of refraction 折射角

angle of release	出手角
angle of repose	安息角；休止角
angle of rest	休止角
angle of roll	滚转角；倾侧角
angle of roof	屋面坡度
angle of rotor shaft	风轮仰角
angle of scattering	散射角；散布角
angle of shear	剪切角
angle of shot	投掷角
angle of sideslip	侧滑角
angle of sight	视线角；瞄准角；高低角
angle of spread	展开角；散布角
angle of trim	纵倾角；配平角
angle of twist	扭转角
angle of view	视角
angle of wind approach	迎风角；来流迎角
angle of yaw	偏航角
angle off	偏角
angle pedestal bearing	斜轴承；斜托架轴承
angle plate	角盘
angle steel	角钢
angle table	角钢托座；角撑架
angular	有角的
angular accuracy	角精确度
angular bevel gear	斜交伞齿轮
angular contact ball bearing	向心推力球轴承；角面接触滚珠轴承；径向止推滚珠轴承
angular coordinate	角坐标
angular correlation	角相关
angular deflection	角偏转
angular discrepancy	角度偏差
angular displacement	角位移
angular distribution	角分布
angular force	角向力；旋转力
angular freedom	角自由度
angular frequency	角频率
angular kinetic energy	旋转动能
angular misalignment	角偏差
angular momentum	角动量
angular momentum vector	角动量矢量
angular motion transducer	角运动传感器
angular movement	角运动
angular oscillation	角振荡
angular rate	角速率
angular rate component	角速度分量
angular rotation	角位移；角旋转
angular speed	角速度
angular spin rate	自旋角速率；角旋转速率
angular surface	斜面
angular thread	三角螺纹
angular torsional displacement	扭转角位移
angular velocity	角速度
angular velocity vector	角速度矢量
anisotropic	各向异性的
anisotropic diffusion	各向异性扩散
anisotropic distortion	各向异性变形
anisotropic scattering	各向异性散射
anisotropic turbulence	各向异性湍流
anisotropic turbulence scale	各向异性紊流度
annealing	退火；韧化；退火处理
annual	每年的；历年的；年度的
annual available wind energy	年有效风能
annual average	年平均
annual average wind power density	年平均风功率密度
annual capacity factor	年利用系数；年利用率
annual coldest month	历年最冷月
annual consumption	年消耗量
annual cost	年费用
annual discharge rate	年排放率

annual dose 年剂量
annual energy production 年发电量
annual equation 年差；周年差
annual extreme daily mean of temperature 年最高日平均温度
annual extreme-mile wind speed 年极端里程风速
annual flow 年径流；年流量
annual hottest month 历年最热月
annual integrated activity concentration 年度综合放射性浓度
annual load curve 年负荷曲线
annual load factor 年负荷系数
annual maximum 年最高
annual mean temperature 年平均温度
annual overhaul 年度检修
annual precipitation 年降水量
annual rainfall 年降水量
annual range 年较差
annual range of temperature 年温度范围
annual repair 全年需修量
annual report 年度报告
annual utility 年利用率
annual utility factor 年利用率系数
annual value 历年值
annual variation 年变化
annual wind regime 年风况
annual wind speed frequency distribution 年风速频率分布
annular 环形的
annular ball bearing 径向轴承
annular cowling 环形整流罩
annular flow 环状流
annular knurl 滚花
annular return-pressure wind tunnel 环形回压风洞
annular seal 环状密封
annular tube 套管
annular valve 环形阀
annular wheel 内齿轮
annulus 环形物；环；圆环域
annulus gear 内齿圈
annum 年
annunciator 信号器；电铃指示装置；报警器
anode 阳极；正极
anodization 阳极氧化
ANSI (American National Standards Institute) 美国国家标准学会
antenna 天线；触角
anthropogenic emission 人为排放
anthropogenic heat 废弃热
anti- 反对；逆；防；非；减；耐
anti-air-pollution system 大气污染防治系统
antiattrition 减少磨损
anti backlash spring 消隙弹簧
anticipated life 预期寿命
anti collision light 防撞灯
anti condensation heater 防潮加热器；防结露加热机
anticorrosin 防蚀；防锈；耐蚀
anticorrosin and antifouling paint 防蚀防污涂层
anticorrosin coating 防蚀层
anticorrosin composition 防蚀化合物；防蚀油漆
anticorrosin insulation 防蚀层
anticorrosive 防蚀的；防腐蚀的
anticorrosive coating 防腐涂层
anticorrosive insulation 防腐绝缘层；防蚀层
anticorrosive paint 防腐漆；防锈漆

anticorrosive treatment 防蚀处理
anticyclone 反气旋；高气压
anticyclone circulation 反气旋环流
anticyclone curvature 反气旋曲率
anticyclone eddy 反气旋涡旋
anticyclonic curvature 反气旋曲率
anticyclonic eddy 反气旋涡旋
anticyclonic wind 反气旋风
antifluctuator 缓冲器
antiflutter device 抗颤振装置
antifreeze solution 防冻液
antifreezer 防冻剂
antifreezing lubricant 抗冻润滑剂
antifriction 减摩剂；润滑剂
antifriction bearing 减摩轴承；抗摩轴承
antifriction bearing grease 减摩轴承润滑脂
antifriction bearing pillow 减摩轴承垫座
antifriction grease 减摩润滑脂
antifriction material 润滑剂
antifriction metal 减摩金属
antigalloping damper 抗驰振阻尼器
antihunt action 阻尼作用；防震作用
anti-icing 防冰
antileak 防泄漏
antiparallel 逆平行的；反平行的
antiparallel thyristor 反平行晶闸管；反并联晶闸管
antipollution standard 抗污染标准
antiresonance 防共振；反共振
anti-return valve 止回阀
antiroll bar 防滚杆；防侧摆杆
antirolling device 消摇装置；减摇装置
antirust 防锈的
antirust action 防锈作用
antirust coat 防锈层

antirust compound 防锈混合剂
antirust paint 防锈漆
antirusting paint 防锈涂层
antiscale 防垢剂
anti-slip mat 防滑垫
antisplash guard 挡泥板
antisymmetrical mode 反对称模态；反对称振型
antitorque moment 反扭力矩；抗扭力矩
antitrade wind 反信风
anti-trigonometric function 反三角函数
antitriptic wind 摩擦风
antivibration 防震
antivibration bases 防震基础
antivibration device 减震装置
antivibrator 防震器；减震器
antiwear 抗磨损的
antiwear property 耐磨特性
antiwind grade 抗风等级
anvil 铁砧
AO (assembly order) 装配指令
AO (assembly outline) 装配大纲
AOA (angle of attack) 迎角；攻角
AOA device (aerodynamic add-on device) 气动附加装置
AOB (angle of bank) 倾斜角
AOC (air oil cooler) 空气油冷却器
AO/OSP (assembly order/operational sequence planning) 装配大纲；操作顺序计划
AORN (assembly outline requirement notification) 装配工艺规程要求通知单
AO-SR (assembly outline-ships record) 装配大纲—架次记录
AO-SRS (assembly outline-ship recording system) 装配大纲—架次记录系统

AOTCR (assembly outline tracing control report) 装配大纲跟踪控制记录
aperiodic 非周期的
aperiodic curve 非周期曲线
aperiodic motion 非周期运动
aperiodic phenomenon 非周期现象
apex angle 顶角
apex of roof 尖屋顶
aphelion 远日点
APMS (air pollution monitoring system) 空气污染监测系统
apocynthion 远月点
apogee 远地点
A-post 前立柱；A支柱
apparatus 装置；设备；仪表；仪器
apparent 表面的；表观的；视在的
apparent angle of attack 视迎角
apparent density 表观密度；视在密度
apparent energy 表观能量
apparent gravity 视重力
apparent mass 表观质量
apparent power 表观功率；视功率
apparent reluctance 视在磁阻；表现磁阻
apparent sound power level 表观声功率级；视在声功率级
apparent specific gravity 表观比重
apparent statistical property 表观统计性质
apparent temperature difference 表观温差；视在温差
apparent velocity 视速度
apparent wind 表观风
apparent wind angle 表观风向角
apparent wind direction 表观风向
apparent wind velocity 表观风速
appendage 配件；附属设备

appendix 附录；附属物
appentice 厢房；窗前的雨篷
appliance 器具；器械；装置
applicability 适用性
application 应用
application drawing 操作图；应用图
applied 应用；实施
applied aerodynamic 应用气体动力学
applied climatology 应用气候学
applied fluid dynamics 应用流体动力学
applied force 外加力
applied load 外加负载
applied mechanics 应用力学
applied meteorology 应用气象学
applied stress 作用力；外加应力
apply 申请
apply force 施加力
apply oil 上油
apply work 做功
approach 方法；途径
approach fetch 来流风区气流吹程
approach flow 迎风气流
approach velocity 来流速度；驶近速度；行近流速
approaching 接近
approaching airstream 来流；迎面气流
approaching flow 来流；迎面气流
approaching wind 迎面风；来流
approval 批准；认可
approval test 验收试验
approximate 近似的；大概的
approximate calculation 近似计算
approximate expression 近似式
approximate formula 近似公式
approximate method 近似方法

approximate similarity 近似相似性
approximate value 近似值
approximation 近似；逼近
appurtenance 附属设备；配件；附件
apron wall 前护墙
APS (auxiliary power supply) 辅助电源；自备供电设备
APSA (automatic particle size analyzer) 自动粒度分析仪
apsacline 斜倾型
APU (auxiliary power unit) 附带电源设备；辅助动力装置；辅助电源装置
AQL (acceptance quality level) 可接受的质量水平
aquadag 导电敷层；胶体石墨
aquagraph 导电敷层
aqueous vapour 水汽；水蒸气；蒸汽
AR (aspect ratio) 叶片展弦比
araldite 环氧树脂；合成树脂粘结剂
aramid fibre 芳族聚酰胺纤维
arbitrary function 任意函数
arbor 主轴；柄轴；心轴
arc 弧；弓形；拱；电弧
arc control device 灭弧装置
arc cutting 电弧切割
arc gouging 电弧刨削
arc horn 角形避雷器
arc line 弧线
arc profile 圆弧翼型
arc welding （电）弧焊
arc welding electrode 电焊条
arcade 拱廊；拱形建筑物
arc-control device 灭弧装置
arch 弓形
arched roof 拱形屋顶；拱形车顶

Archimedean 阿基米德的
Archimedean principle 阿基米德原理
Archimedean screw 阿基米德螺线；螺旋升水泵；阿基米德螺旋泵
Archimedean spiral 阿基米德螺线
architect 建筑师
architectural aerodynamics 建筑空气动力学
architectural complex 建筑群
architectural details 建筑细部；建筑详图
architectural meteorology 建筑气象学
architectural model 建筑模型
architectural planning 建筑规划
architectural wind tunnel 建筑风洞
arch-limb 拱翼；顶翼；穹翼
arcing 电弧作用
arcing ground 电弧接地
arcing horn 角形避雷器
area 面积；区域
area density 面密度
area forecast 区域预报
area map 区域图
area mean pressure 面积平均压力
area moment 面积矩
area moment of inertia 面积惯性矩
area of safe operation (ASO) 安全工作区
area pollution 区域污染
area ruling 面积律
area source 面源；区域源
area swept 扫掠面积
areal 地区的
areal coordinates 重心坐标
areal deformation 表面变形
areal velocity 掠面速度；面积速度
ARG (argument) 幅角
argon arc welding 氩弧焊

argument (ARG) 幅角
arid 干旱的
arid climate 干燥气候
arithmetic 算术；算法；计算
arithmetic average 算术平均
arithmetic average efficiency 算数平均效率
arithmetic average temperature difference 算数平均温差
arithmetic mean 算术平均值
arm 臂；支架；指针
arm crane 悬臂式起重机
arm of wheel 轮辐
armature 电枢；衔铁
armature circuit 电枢电路
armature coil 电枢线圈
armature current 电枢电流
armature m.m.f. wave 电枢磁势波
armature winding 电枢绕组
armoured cable 有金属套皮的电缆；铠装电缆
armoured layer 护面层
ARODYN (aerodynamics) 空气动力学
arrangement 布置
arranger 传动装置
array 数组；阵列
array effect 排列效应
array of wind turbines 风轮机阵
array pattern 阵列型式；阵方向图
array spacing 阵内间距；排列间距
arrest point 临界点；驻点；转变点
arrested dune 固定沙丘；稳定沙丘
arrester 制动器；避雷器
arresting device 制动装置
arresting gear 制动器
arris 尖脊；边棱
arrises 棱；棱角线

arrises of joint 接缝圆角
articulated 接合；链接；有关节的；铰接式的
articulated blade 铰接叶片
articulation 铰接
artificial climate 人工气候
artificial cloud 人造云
artificial contamination 人为污染
artificial disturbance 人为扰动
artificial draft 人工通风
artificial precipitation 人工降雨
artificial precipitation stimulation 人工催雨
artificial radioactivity 人工放射性
artificial radioelement 人工放射性元素
artificial radionuclide 人造放射性核素
artificial rainfall 人工降雨
artificial velocity gradient 人工速度梯度
artificial ventilation 人工通风
artificial wind 人造风
AS (alloy steel) 合金钢
ASA (American Standards Association) 美国标准协会
ASC (American Standards Committee) 美国标准委员会
ascendant 上升的
ascendant air current 上升气流
ascending 上升的
ascending air 上升空气
ascending line of flight 向上飞路线
ash 灰；灰分
ash air 含灰空气
ash analysis 灰分分析
ash content 含灰量
ash fall 灰尘沉降；烟尘降落
ash-laden gas 含尘气；富尘气体
askania 液压自动控制装置

ASL (above sea level) 海拔高度
ASL (Atmospheric Sciences Laboratory) 大气科学实验室
ASME (American Society of Mechanical Engineer) 美国机械工程师学会
ASO (area of safe operation) 安全工作区
ASPC (air space paper core cable) 空气纸绝缘电缆
aspect 方面
aspect ratio (AR) 叶片展弦比
aspect ratio of blade 叶片展弦比
aspirated psychrometer 通风式干湿表
aspirated thermometer 通风式温度计
aspiration condenser 吸气式冷凝器；通风电容器
ASPP (alloy steel protective plating) 合金钢保护电镀
assemblage 装配；安装
assemble 装配
assemble and test (A&T) 装配与测试
assembling 装配
assembling bolt 装配螺栓
assembly 装配
assembly adhesive 装配胶黏剂
assembly drawing 装配图
assembly inspection 装配检查
assembly order (AO) 装配指令
assembly order/operational sequence planning (AO/OSP) 装配大纲；操作顺序计划
assembly outline (AO) 装配大纲
assembly outline requirement notification (AORN) 装配工艺规程要求通知单
assembly outline-ships record (AO-SR) 装配大纲—架次记录
assembly outline-ship recording system (AO-SRS) 装配大纲—架次记录系统
assembly outline tracing control report (AOTCR) 装配大纲跟踪控制记录
assembly parts 组装部件
assembly set 组合件
assist 帮助
assisted 协助的；辅助的
assisted stall 协助失速；辅助失速
associated mass 附连质量；附加质量
association 联合；协会
assumed load 假设载荷
assurance 保证
assurance period 保险期
AST (American Standard Thread) 美国标准螺纹
ASTM (American Society for Testing and Materials) 美国材料与试验协会
ASTM (American Standards of Test Manual) 美国标准实验手册
ASYM (asymmetry) 不对称的；不对称
asymmeter 非对称计
asymmetric 不对称的
asymmetric force 非对称力
asymmetric mode 非对称模态
asymmetric plane system 不对称平面系
asymmetry (ASYM) 不对称的；不对称
asymptote 渐近线
asymptotic expression 渐近式
asymptotic(al) 渐进的
asymptotic(al) curve 渐近线
asymptotic(al) distribution 渐进分布
asymptotic(al) method 渐进法
asymptotic(al) stability 渐进稳定性
asynchronous 异步的
asynchronous adaptation 异步适应；异步适配

asynchronous generator 异步发电机
asynchronous machine 异步电机
asynchronous motor 异步电动机
AT (acceptance test) 验收试验
AT (air temperature) 空气温度
A&T (acceptance and transfer) 验收与移交
A&T (assemble and test) 装配与测试
ata (absolute atmosphere) 绝对大气压单位 (阿泰)
ATA (air turbine alternator) 空气涡轮交流发电机
ATC system (aluminum continuous melting & holding furnaces) 自动换刀系统；刀库
athletic ground 运动场地
atm press 大气压；大气压力
atmosphere 大气；大气层
atmosphere aerosol 大气气溶胶
atmosphere circulation 大气环流；大气循环
atmosphere constituent 大气成分
atmosphere contamination 大气污染
atmosphere data 大气数据
atmosphere density 大气密度
atmosphere dispersion factor 大气弥散因子
atmosphere gauge 气压表；气压计
atmosphere layer 大气层
atmosphere moisture 大气湿度；大气水分
atmosphere pollutant 大气污染物
atmosphere pollution 大气污染
atmosphere pollution burden 大气污染负荷
atmosphere standard 标准大气压；标准大气
atmosphere-ocean interaction 气海相互作用
atmospheric 大气的
atmospheric advection 大气平流
atmospheric baseline observing station 大气基准观测站
atmospheric boundary layer 大气边界层
atmospheric carcinogen 大气致癌物
atmospheric circulation 大气环流
atmospheric condition 大气条件
atmospheric constituent 大气组分
atmospheric contaminant 大气污染物
atmospheric contamination 大气污染
atmospheric convection 大气对流
atmospheric corrosion 大气腐蚀
atmospheric damping 大气阻尼
atmospheric density 空气密度
atmospheric depression 低气压
atmospheric depth 大气深度；大气厚度
atmospheric diffusion 大气扩散
atmospheric diffusion equation 大气扩散方程
atmospheric diffusion process 大气扩散过程
atmospheric diffusion test 大气扩散试验
atmospheric dilution 大气稀释
atmospheric dispersion 大气弥散；大气分散
atmospheric disturbance 大气扰动
atmospheric dust 大气尘埃
atmospheric emission standard 大气排放标准
atmospheric environment 大气环境
atmospheric extinction 大气消光
atmospheric flow 大气流动
atmospheric heat balance 大气热平衡
atmospheric humidity 大气湿度
atmospheric hydromechanics 大气流体力学
atmospheric impurities 大气杂质
atmospheric inversion 大气逆温
atmospheric mass 大气团；大气质量
atmospheric mass transport 大气质量运输
atmospheric model 大气模式；大气模型
atmospheric modelling technique 大气模拟

技术；大气建模技术

atmospheric momentum transport 大气动量运输

atmospheric monitoring 大气监测

atmospheric noise 大气噪声

atmospheric optical thickness 大气光学厚度

atmospheric ozone layer 大气臭氧层

atmospheric parameter 大气参数

atmospheric photochemistry 大气光化学

atmospheric physics 大气物理学

atmospheric pollution 大气污染

atmospheric precipitation 大气降水

atmospheric pressure 大气压；大气压力

atmospheric process 大气过程；大气作用

atmospheric radiation 大气辐射

atmospheric reaction 大气反应

atmospheric refraction 大气折射

atmospheric regional station 大气地区观测站

atmospheric sampling 空气取样

atmospheric scale 大气尺度

atmospheric scavenging 大气净化

atmospheric science 大气科学

Atmospheric Sciences Laboratory (ASL) 大气科学实验室

atmospheric shear flow 大气剪切流动

atmospheric similarity 大气相似性

atmospheric singularity 大气奇异现象

atmospheric stability 大气稳定度

atmospheric stability constant 大气稳定度常数

atmospheric stability length 大气稳定度长度

atmospheric stagnation 大气滞止；大气停滞

atmospheric stagnation pressure 大气驻压

atmospheric stagnation temperature 大气驻温

atmospheric static conditions 标准大气条件

atmospheric stratification 大气层结

atmospheric structure 大气结构

atmospheric surface layer 大气表面层

atmospheric temperature 大气温度

atmospheric thermal 大气热泡

atmospheric thermodynamics 大气热力学

atmospheric tide 大气潮

atmospheric trace gas 大气痕量气体

atmospheric transmission 大气透射；大气传递

atmospheric transmissivity 大气透射率

atmospheric transparency 大气透明度

atmospheric transport 大气运输

atmospheric transport model 大气运输模式

atmospheric turbulence 大气湍流

atmospheric turbulence model 大气湍流模型

atmospheric turbulence spectrum 大气湍流谱

atmospheric turbulence structure 大气湍流结构

atmospheric vortex 大气旋涡

atmospheric whirl 大气旋风

atmospheric wind 大气风

atmospheric wind tunnel 大气压风洞

atomic 原子的

atomic absorption method 原子吸收法

atomic energy 原子能

atomic hydrogen welding 原子氢焊

atomic power station 核电站；原子能电站

atomic surface burst 地面核爆炸

atrium 天井

ATT (attachment) 辅助设备

attached vortex 附着旋涡

attachment 附件

attachment (ATT) 辅助设备
attachment flow 附着流；附壁流
attachment hardware 连接用的金属件；五金附件
attachment point 附着点
attack angle 迎角；攻角
attainable precision 可达精确度
attemperator 温度控制器；保温装置；保温器
attendance 看护；值班；保养；维护
attention 维修
attenuate 衰减
attenuation 衰减；稀释
attenuation coefficient 衰减系数
attenuation factor 衰减因子
attenuation rule 衰减法则
attenuator 阻尼器；分压器；衰减器
attitude 方位；姿态
attitude angle 姿态角；方位角
attitude of the joint 焊缝特性
attrition 摩擦
attrition rate 损耗率
attrition resistance 抗磨耗性
ATU (auxiliary test unit) 辅助测试装置
audible 听得见的
audible alarm 音响报警
auditory 听觉的
auditory fatigue 听觉疲乏
augmentation 增强
augmentation device 增强装置
augmentation horizontal-axis wind turbine 增力型水平轴风轮机
augmenter 增压器
austemper 奥氏体回火
austemper case hardening 等温淬火表面硬化
austemper stressing 等温淬火表面应力

authorized pressure 容许压力
auto- 自动的；自动装置的；自己的
autoalarm 自动报警器
autoanalyzer 自动分析器
autobalance 自动平衡器
autobias 自动偏置
autobrake 自动制动器
autocompensation 自动补偿
auto control 自动控制
autoconvection 自动对流；自动修正
autocorrelation 自相关
autocorrelation coefficient 自相关系数
autocorrelation function 自相关函数
autocovariance 自协方差
autocovariance technique 自协方差技术
auto-draft 自动制图
autographic recording apparatus 自动图示记录仪
autographic apparatus 自动记录仪
autographic data acquisition system 自动数据采集系统
autographic data exchange system 自动数据交换系统
autographic data processing 自动数据处理
autographic drag spoiler 自动阻力板；自动阻力扰流板
autographometer 自动图示仪；自动地形仪
autoinduction 自感；自动感应
autoinhibition 自动抑制作用
auto-manual switch 自动—手动开关；半自动开关
automated engineering design (AED) 自动工程设计
automatic 自动的；自动装置的
automatic and hand operated changeover

switch 自动—手动转换开关
automatic assembly 自动装配
automatic bypass valve 自动旁通阀
automatic call device 自动报警器；自动呼叫装置
automatic check valve 自动止回阀
automatic checkout and evaluation system (ACES) 自动检测和估算系统
automatic checkout and recording equipment (ACORE) 自动检测记录设备
automatic checkout equipment (ACOE) 自动检测装置
automatic checkout system 自动检测系统
automatic checkout tester 自动检测仪
automatic circuit breaker 自动断路器
automatic compensation 自动补偿
automatic control 自动控制
automatic control cutout valve 自动切断阀
automatic control device 自动控制装置
automatic control engineering 自动控制工程
automatic control equipment 自动控制设备
automatic control program 自动控制程序
automatic control system 自动控制系统
automatic control valve (ACV) 自动控制阀
automatic data processing (ADP) 自动数据处理
automatic data processing equipment (ADPE) 自动数据处理装置
automatic data processing system (ADPS) 自动数据处理系统
automatic data reduction tunnel 实验数据自动处理风洞
automatic detecting system 自动检测系统
automatic detector 自动检测器
automatic device 自动装置；自动仪表

automatic digital data error recorder (ADDER) 数字信息误差自动记录器
automatic driver 自动传动装置
automatic fine tuning (AFT) 自动微调
automatic flap 自动襟翼
automatic following 自动跟踪；自动对风
automatic frequency tuner (AFT) 自动频率调谐器
automatic governing 自动调节
automatic indicator 自动指示器
automatic logger 自动记录仪
automatic lubrication installation 自动润滑装置
automatic lubrication system 自动润滑系统
automatic measurement 自动测量
automatic measuring device 自动测量器
automatic modulation control (AMC) 自动调制控制
automatic monitor 自动监测仪
automatic monitoring system 自动监测系统
automatic nonreturn valve 自动逆止阀；自动止回阀
automatic operating 自动操作；自动运行
automatic operating system 自动操作系统
automatic oscillograph 自动示波器
automatic particle size analyzer (APSA) 自动粒度分析仪
automatic phase compensation 自动相位补偿
automatic power control 自动功率控制
automatic pressure-reduction valve 自动减压阀
automatic recorder 自动记录器
automatic reducing valve 自动调节阀；自动减压阀
automatic regulating device 自动调节装置

automatic regulating system 自动调节系统
automatic regulating valve 自动调节阀
automatic regulation 自动调整
automatic regulator 自动调节器
automatic sample handling system 样品自动处理系统
automatic sampler 自动取样器
automatic sampling system 自动取样系统
automatic sequence control 自动程序控制
automatic shut-off valve 自动断流阀
automatic shut valve 自动关闭阀
automatic signal 自动信号
automatic sizing 自动测量
automatic smoke sampler 自动烟尘取样器
automatic starting control 自动起动控制
automatic starting device 自动起动装置
automatic starting relay 自动起动继电器
automatic station 无人值守电站
automatic stop valve 自动停止阀;自动断流阀
automatic stopping device 自动停止装置
automatic switch 自动开关;自动断路器
automatic temperature recorder 温度自动记录器
automatic test equipment 自动测试设备
automatic throttle valve 自动节流阀
automatic timing control 自动定时控制
automatic voltage regulator (AVR) 自动电压调整器
automaticity 自动化程度
autopurification 自动净化
autoreclosing cycle 自动接通周期
autoreclosure 自适应重合闸;自动重合闸
autoregression method 自回归法
autorestart 自动再启动
autorotation rate 自旋速率

autospectral analysis 自谱分析仪
autospectral density 自谱密度
autospectrum 自谱
autoswitch 自动开关
autotransformer 自耦变压器
auxiliary 附件;辅助设备;辅助的;副的;附加的
auxiliary aerodynamic surface 辅助气动作用面
auxiliary apparatus 辅助设备
auxiliary battery power supply 辅助动力供应电池
auxiliary circuit 辅助电路
auxiliary contact 辅助触头
auxiliary control panel (ACP) 辅助的控制仪表板
auxiliary current transformer (ACT) 辅助电流互感器
auxiliary device 辅助装置
auxiliary electric power supply 配套供电工程
auxiliary energy source 辅助能源
auxiliary engine 辅助动力机
auxiliary flap 辅助襟翼
auxiliary motor 辅助电动机
auxiliary power supply (APS) 辅助电源;自备供电设备
auxiliary power unit (APU) 附带电源设备;辅助动力装置;辅助电源装置
auxiliary rotor 辅助风轮
auxiliary switch 辅助接点
auxiliary switching power supply 机内辅助开关电源
auxiliary test unit (ATU) 辅助测试装置
auxiliary windshield wiper 副刮水器;辅助雨刷

availability 可利用率；有效性
availability factor 可利用系数；可利用率
available 可用的；有效的
available accuracy 实际精确度
available capacity 有效容量
available cross section 有效面积
available deposition velocity 有效沉积速度
available diffusion time 有效扩散时间
available factor 可利用系数
available heat transfer 有效传热
available heat transfer coefficient 有效传热系数
available horsepower 有效功率；可用功率
available input power 有效输入功率
available life 可用期；有效寿命
available load 有效负荷；可用负荷
available output 有效输出
available power 有效功率；可用功率
available power curve 有效功率曲线
available power efficiency 有效功率效率
available power loss 有效功率损耗；有功损耗
available pressure 可用压力
available time 有效工作时间；开机时间
available velocity 有效速度
available wind 有效风
available wind energy 有效风能；可用风能
available wind power 有效风能；可用风能
available work 可用功
availably 有效地
average (AVG) 平均；平均的
average absorbing power 平均吸收功率
average absorbing coefficient 平均吸收系数
average annual damage index 年平均损失指数
average capacity factor 平均利用率

average departure 平均偏差
average deviation 平均偏差
average diameter (AVGDIA) 平均直径
average error 平均误差
average flow 平均流量
average flow rate 平均流量率；平均流量
average flow velocity 平均流速
average full-load power 平均全负荷功率
average life 平均寿命
average life time 平均使用期限
average load 平均负荷
average load curve 平均负荷曲线
average loss of energy 平均能量损失
average monthly wind speed 月平均风速
average noise level 平均噪声级
average of the daily load curve 平均日负荷曲线
average operation time 平均运行时间
average output 平均输出；平均产量
average output power 平均输出功率
average temperature 平均温度
average thickness 平均厚度
average useful life 平均使用期限；平均使用寿命
average value 平均值
average velocity 平均速度
average wind direction 平均风向
average wind speed 平均风速
average work load 平均工作负荷
averaging time 平均时间
AVG (average) 平均
AVGDIA (average diameter) 平均直径
aviation industry 航空工业
avometer 万用表
AVR (automatic voltage regulator) 自动电压

调整器

A-weighted sound pressure level A-加权声级；A声级

AWI (American Welding Institute) 美国焊接协会

awl 锥子

axes system 坐标轴系

axial 轴的；轴向的

axial airflow 轴向气流

axial centrifugal 轴向离心式

axial clearance 轴向间隙

axial compression force 轴向压力

axial concentration 轴线浓度

axial deformation 轴向变形

axial dimension 轴向尺寸

axial distance 轴向距离

axial eddy 轴向涡流

axial elongation 轴向伸长

axial expansion 轴向膨胀

axial flow (AF) 轴向流动；轴流式

axial flow compressor 轴流式压缩机

axial flow fan 轴流式风扇

axial flux permanent magnet synchronous generator (AF-PMSG) 轴向磁通永磁同步发电机

axial force 轴向力

axial length 轴长

axial load 轴向负荷

axial moment theory 轴向力矩理论

axial mount 轴向装配

axial movement 轴向运动；轴向移动

axial of rotation 旋转轴

axial of twist 扭转轴

axial pitch 轴向齿距；轴向节距；轴向螺距

axial pressure 轴向压力

axial pressure gradient 轴向压力梯度

axial seal 轴向密封

axial strain 轴向应变

axial symmetry 轴对称

axial thrust bearing 轴向止推轴承

axial vector 轴矢量

axial velocity 轴向速度

axial volute 轴向蜗壳

axial vortex 轴向涡流

axial-compressor-driven wind tunnel 轴流式压气机驱动风洞

axially symmetric bending 轴对称弯曲

axially symmetric shape 轴对称形状

axis 轴；轴线；坐标轴

axis distance 轴距

axis of abscissa 横坐标轴；横轴

axis of inertia 惯性轴

axis of ordinates 纵坐标轴

axis of reference 参考轴

axis of rotation 旋转轴

axis of symmetry 对称轴

axisymmetric 轴对称的

axisymmetric (wind) tunnel 轴对称风洞

axisymmetric body 轴对称体

axisymmetric determinant 轴对称行列式

axisymmetric flow 轴对称流动

axisymmetric plume 轴对称羽流

axisymmetric puff 轴对称喷团

axisymmetric thermal 轴对称热

axle 轮轴；车轴

axle arm 轴臂；驱动桥定位臂

axle cap 轴罩；轴帽

axle elongation 轴伸长

axle fracture 轴断裂

axle grease 轴用脂

axle pin 轴销

axle-base 轴距

axle-neck 轴颈

Aylesbury collaborative experiment (ACE) 埃尔兹伯里（风荷载）合作实验

Aylesbury experimental building 埃尔兹伯里实验建筑物

azimuth 方位

azimuth angle 方位角

azimuth drive 方位角驱动

azimuthal angle 方位角

azimuthal position 方位；方位位置

现代英汉风力发电工程

A Modern English-Chinese Dictionary of Wind Power Engineering

B

back 支座；背面；反面
back action 反作用
back EMF 反电动势；逆电动势
back gear 反齿轮
back pressure 反压力；背压
back pull 反拉力
back saw 夹背锯；脊锯；镶边手锯
back sealing weld 封底焊
back seat gasket 底座垫圈
back streaming 反流
back substitution 倒转代换；回代
back suction 反吸
back surface 背面
back-and-forth movement 往复运动
backflow 回流
background activity 本底放射性
background air 本底空气
background air pollution 本底空气污染
background concentration 本底浓度
background excitation factor 本底激励因子
background level 本底水平；自然本底值
background pollution 本底污染
background radiation 本底辐射
background response 本底响应
background turbulence 背景湍流

background wind load 本底风载；背景风荷载
backing 后背；衬背
backing weld 底焊；封底焊
backing wind 逆转风
backlash 反冲；无效行程；间隙；偏移
backlash eliminator 齿隙消除装置
back-lash spring 消隙弹簧
backplate 背面板；背面电极
backscatter 反向散射
backset 涡流
backshaft 后轴
backslope angle 反坡角度
backup 备份
backup energy storage 后备储能
backup mechanical brake 备用机械制动器
backup plate 垫片
backup power system 备用动力系统
backup ring 垫圈；支撑环
backup roll 支撑轧辊
backup washer 支撑垫圈
backward 向后的
backward curved vane 后弯曲叶片
backward flow 逆流；反向流
backward jet 反向射流；反向喷射

backward running 反向运动；倒转
backward visibility 后视度
backwash 后涡流；回流
backwind 背后风
back run 反转
bad conductor 不良导体
bad contact 接触不良
bad earth 接地不良
baffle (BAF) 挡板；折流板；阻力板
baffling wind 无定向风；迎面风
bag 袋子
Bagnold's threshold parameter 拜格诺阈值参数
balance 平衡
balance axes system 平衡轴系
balance element 平衡元件
balance equation 平衡方程
balance pressure-reducing valve 平衡减压阀
balance strut 平衡支柱
balance support 平衡支架
balance test 平衡试验
balance unit 平衡装置
balance weight 平衡锤；平衡块；配重
balanced aileron 平衡副翼
balanced flap 平衡襟翼
balancer 平衡器
balancing 平衡
balancing adjustment 平衡调整
balancing check 平衡检查
balancing equipment 平衡设备
balancing torque 平衡力矩
balancing weight 平衡重
balcony 阳台
balk 障碍
balk board 隔板；障碍板；防护板

balk ring 摩擦环；阻环
ball 球
ball and roller bearing 滚珠和滚柱轴承
ball and socket coupling 球窝连接器；球窝耦合
ball bearing (BB) 滚珠轴承
ball bearing (BBRG) 滚珠轴承
ball bearing torque (BBT) 滚珠轴承扭矩
ball bond 球焊接头
ball hardness 钢球硬度；布氏球测硬度；布氏球印硬度
ball lighting 球状闪电
ball peen hammer 圆头锤；圆头手锤
ball race 滚珠座圈
ball saddle 滚珠支撑
ball-eye 球头挂环
ball-hook 球头挂钩
ballistics 弹道学
balloon sounding 气球探测
balsa wood model 软木模型；轻木模型
balustrade 栏杆
band 区；带；波段
band compensation 频带补偿
band elimination filter 带阻滤波器
band saw 带锯
bandpass 带通
bandpass filter (BF) 带通滤波器
bandpass tuner 带通调谐器
bandwidth 带宽
bandwidth of vortex shedding 旋涡脱落带宽
bandwidth parameter 带宽参数
bandwidth technique 带宽技术
banked battery 并联电池组
banking 外缘超高
banking angle 外缘超高角；侧倾角

BAP (basic assembler program) 基本汇编语言
bar 巴（气压单位）
bar feed lock 进给杆锁
bar magnet 磁棒
bar pressure 大气压力
barb 风羽
barbed arrow 风矢
barchan dune 新月型沙丘
bare electrode 无药焊条
baric gradient 压力梯度
baroclinic atmosphere 斜压大气
baroclinic flow 斜压流动
baroclinic model 斜压模式
baroclinity 斜压性
barograph 自计气压计
barometer (BRM) 气压计
barometric fluctuation 气压变动
barometric gradient 气压梯度
barometric low 低气压
barometric pressure (BP) 气压；大气压；大气压力
barometric pressure sensor (BP sensor) 大气压力传感器
barometry gradient 气压梯度
baromil 毫巴（气压单位）
barosphere 气压层
barotropic atmosphere 正压大气
barotropic flow 正压流动
barotropic fluid 正压流体
barrel (bbl.) 桶；筒（容量或重量单位，1 bbl.=42加仑）
barrel cam 筒形凸轮
barrel roller 鼓形滚柱
barrel roller bearing 鼓形滚柱轴承
barrel roof 筒形薄壳屋顶
barretter 镇流电阻器
barretter resistance 镇流电阻
barrier 挡板；栅栏；障碍物
base 基础；基地
base bending moment 底部弯矩
base board 基线板；踢脚板
base cavity 底部空穴；尾部
base coat 底漆
base data 原始数据
base display unit (BDU) 主显示器
base drag 底部阻力
base frame 基础构架
base lacquer 底漆
base level 基准面
base material 基底材料
base moment 基底力矩
base pad 基底
base pressure 底压
base suction 底部吸力
baseline 基线
baseline design 基准设计；基线设计
basic 基本的
Basic Assembler Program (BAP) 基本汇编语言
basic data 基本数据；原始资料
basic design feature 基本设计要素
basic dimension 基准尺寸
basic element 基本元件
basic equation 基本方程
basic error 基准误差
basic instrumentation 基本检测仪表
basic internal pressure 基本内压
basic load 基本负荷；主要负荷
basic mechanical design feature 主要技术性能

basic operation condition 基本运行工况；主要工作状况
basic regional wind velocity 地区基本风速
basic requirement 基本要求
basic size 基本规格
basic specifications 主要技术规格；基本参数
basic wind speed 基本风速
basin 盆地
BAT (battery) 蓄电池；电池
battery charger 蓄电池充电器
batch 一批
batch produce stage 批量生产阶段
batch soldering 批量焊接
batch soldering equipment 批量焊接设备
Batchelor constant 白切勒常数
battery (BAT) 蓄电池；电池
battery backup (BBU) 电池(组)备用
battery bank 蓄电池组
battery charging system 蓄电池充电系统
battery energy storage 蓄电池贮能
battery in quantity 并联电池组
battery input 蓄电池充电
battery pack 电池包；电池组件
battery power drill 电池钻
battery solution 电池电解液；电池溶液
Baud 波特
Baud rate 波特率
bay 海湾
bayonet 卡口；接合销钉
BB (ball bearing) 滚珠轴承
bbl. (barrel) 桶；筒(容量或重量单位，1 bbl.=42 加仑)
BBRG (ball bearing) 滚珠轴承
BBT (ball bearing torque) 滚珠轴承扭矩
BBU (battery backup) 电池(组)备用

BDIG (brushless doubly-fed induction generator) 无刷双馈异步发电机
BDU (base display unit) 主显示器
BDV (breakdown voltage) 击穿电压
beach drifting 沿滩漂移；海滩移动
beacon 灯塔
beacon lamp 标向灯
beacon light tower 探照灯塔
bead weld 堆焊
beaded covering 串珠覆盖
beam 横梁；光线
beam balance 杠杆式天平
beam bending stress 悬梁弯曲应力
beam bridge 梁式桥
beam callipers 卡尺
beam channel 槽型梁
beam idler gear 惰性轮齿
beam spar 腹板式翼梁
beam trammel 骨架
beam wind 横风；侧风
beam with compression steel 双筋梁
bearer 支架；托架；支座；载体
bearing 轴承
bearing adjustment 轴承调整
bearing alignment 方位对准
bearing area 支承面
bearing bracket 轴承座
bearing cage 轴承保持架；轴承罩
bearing capacity 容许负荷；承载能力
bearing fittings 轴承配件
bearing friction loss 轴承摩擦损失
bearing housing 轴承座；轴承箱；轴承套
bearing point 支承点
bearing processing equipment 轴承加工机
bearing seal 轴承密封

bearing surface	轴承面；承载面；支承面
bearing bush	轴承衬
beat	迎风航行；差拍；搏动
beat frequency	拍频；差频
beat phenomenon	差拍现象
Beaufort	蒲福
Beaufort force	蒲福风力
Beaufort number	蒲福（级）数
Beaufort Scale	蒲福风级
bed load	推移质；底沙；底负载
bedding area	承载面
bedplate	底座
before the wind	顺风
begin block	开始分程序；开始区块；起始块
behavio(u)r	工况；运转状态；特性动态
behaviour characteristics	性能特性
behaviour in service	运转性能；使用性能
behaviour of cross section	截面的变化曲线
behaviour of fluids dynamic	流体动力特性
bell crank	曲柄
bell housing	外罩；屏蔽套；钟形罩；外壳
bell roof	钟形屋顶
bellow type	波纹管式
bell-shaped hill model	钟形山丘模型
belt	传送带；传动带
belt drive	带传动
belt pitching mechanism	皮带调桨机构
belted	束带的；装甲的
belted cable	铠装电缆
belting	制带的材料；带类；调带装置
BEM (blade element momentum)	叶素动量
BEM theory (blade element momentum theory)	BEM理论；叶素动量理论
bench	长凳；工作台
bench insulator	绝缘座
bend	使弯曲
bend pipe	弯管
bend radius	弯弧内径
bend strength	抗弯强度
bend test	弯曲试验
bending	弯曲度
bending beam	抗弯梁
bending deflection	挠度
bending displacement	弯曲位移
bending frequency	弯曲频率
bending load	弯曲负荷
bending mode	弯曲模态
bending moment	弯曲力矩
bending oscillation	弯曲振动
bending radius	弯曲半径
bending response	弯曲响应；弯曲反应
bending stiffness	抗弯刚度
bending strain	弯曲应变
bending strength	抗弯强度
bending test	弯曲试验
beneficial	有益的
beneficial wind effect	有利风效应
bent-over jet	弯曲射流
bent-over plume	弯曲羽流
Bernoulli	伯努利
Bernoulli binomial distribution	伯努利二项式分布
Bernoulli constant	伯努利常数
Bernoulli equation	伯努利方程
Bernoulli formula	伯努利公式
Bernoulli surface	伯努利面
Bernoulli test	伯努利实验
Bernoulli trail	伯努利尾迹
Bernoulli vector	伯努利矢量
Bernoulli's law	伯努利定律

Bernoulli's theorem 伯努利定理
best fit-line 最佳拟合线
best lift drag ratio 最佳升阻比
beta background β本底
beta decay β衰变
Betz 贝茨
Betz coefficient 贝茨系数
Betz equation 贝茨等式；贝茨效率公式
Betz limit (0.593) 贝茨极限 (0.593)
Betz manometer 贝茨压力计
Betz' law 贝茨定律；贝茨理论
bevel 斜角；斜面
bevel angle 斜角
bevel drive 伞齿轮传动
bevel gear 斜齿轮
bevel wheel 斜齿轮
bevelled washer 斜垫圈
BF(bandpass filter) 带通滤波器
BF (blind flange) 盖板；闷头法兰
BFR(buffer) 缓冲器
biased estimator 有偏估计量
biconvex airfoil 双凸形翼面
bicubic interpolation 双三次插值
bicycle-type multi-bladed wind machine 自行车轮型多叶片风力机
bicycle-type windmill 自行车轮型风车
bidimensional flow 二维流动
bidirectional power electronic converter 双向电力电子变流器
bidirectional power flow 双向功率流
bidirectional triode thyristor 双向晶闸管；双向三端可控硅
bidirectional vane 双向风标
bifurcation plume 分岔型羽流
big repair 大修

bilateral circuit 双向电路
bill board 广告牌
bimotored 双马达的
binary flutter 二维颤振；二元颤振
binder 黏合剂
binding 装订；捆绑
binding beam 联梁
binding bolt 连接螺钉
binding course 黏结层
binding face 结合面
binding head screw 圆顶宽边接头螺钉
binding post 接线柱
binding screw 接线螺钉；紧固螺钉
binomial distribution 二项式分布
binomial equation 二项式方程
binomial expansion 二项式展开
binomial expression 二项式
bioaccumulation factor 生物累积因子
bioassay 生物鉴定法
bioclimate 生物气候
biological effect 生物效应
biological half-life 生物半衰期
biological hazard 生物公害
biological wind indicator 植物风力指示器
biosphere 生物圈
biospheric contamination 生物圈污染
Biot-Savart law 毕奥-萨伐尔定律
biparted hyperboloid 双叶双曲面
biphase 双相的
biphase rectification 双相整流；全波整流
biphase rectifier 全波整流器
bipolar junction transistor (BJT) 双极性晶体管；双极面结型晶体管
biquadratic equation 四次方程；双二次方程
Biram's wind meter 翼型风速仪

bit 位；比特
bitumen 沥青
bitumen cable 沥青电缆；沥青绝缘电缆
bitumen sheet 油毛毡
bituminized paper 沥青纸
bitumen-sheathed paper cable 沥青绝缘纸电缆
bivane 双向风标
BJT (bipolar junction transistor) 双极性晶体管；双极面结型晶体管
black blizzard 黑尘暴
black body 黑体
black buran 黑风暴
black ice 雨凇
black wind 黑风
black-body radiation 黑体辐射
blade 叶片；桨叶；旋翼
blade accumulator 叶片蓄能器
blade activity factor 桨叶功率因子
blade aerial 刀形天线
blade aerodynamics 桨叶空气动力(学)
blade aerofoil 叶片翼型
blade angle 叶片角；桨叶角
blade angle change 桨叶角的变化
blade angle of attack 叶片迎角
blade antenna 刀形天线
blade articulation 桨叶关节连接
blade aspect ratio 叶片翼弦比；叶片展弦比
blade azimuth angle 叶片方位角
blade back 叶背
blade balance 桨叶平衡
blade bearing 叶片轴承
blade butt 桨叶根部
blade coning angle 叶片锥角
blade count 叶片数

blade cuff 桨叶根套
blade damper 桨叶减摆器；桨叶减震器
blade edge 叶片缘；桨叶缘
blade element 叶素
blade element theory 叶素理论；桨叶剖面理论
blade face 叶面
blade feathering 叶片顺桨
blade for iron saw 剧刃；盘刃
blade geometric twist 叶片几何扭转
blade hub 叶片轮毂
blade interference 桨叶干涉
blade length 叶片长度
blade lift 桨叶升力
blade lift coefficient 桨叶升力系数
blade load 叶片负载
blade losses 叶片损失
blade mass factor 桨叶质量因数
blade neck 叶颈
blade orbital angle 叶片运行角
blade passing frequency 叶片通过频率
blade passing noise 越桨噪音
blade pitch 桨距；叶片节距
blade pitch angle 桨距角
blade pitch change 桨距变化
blade pitching motion 桨叶俯仰运动
blade planform taper ratio 叶片平面形状的锥度比
blade profile characteristic 叶片轮廓特性
blade root 叶根
blade root attachment 叶根连接
blade root diameter 叶根直径
blade rotating angle 叶片旋转角度
blade section pitch 叶片剖面桨距
blade segment 叶段
blade setting 叶片安装角

blade setting angle 叶片安装角
blade shank 叶柄
blade shedding of ice 叶片甩冰
blade skin 叶片蒙皮
blade socket 叶片座
blade spacing 叶片间隔
blade span axis 桨叶轴线
blade spar 桨叶梁
blade twist test 叶片扭转度试验
blade stall 叶片失速
blade stall condition 叶片气流分离状态
blade stall regime 叶片气流分离状态
blade station 桨叶站位
blade strut 叶片撑杆
blade taper 叶片楔面；叶片锥度
blade thickness 叶片厚度
blade tilt 桨叶倾角
blade tip 叶尖
blade tip eddy 叶尖旋涡
blade tip loss 叶尖损失
blade tip speed 叶尖速度
blade tip spoiler 叶尖扰流器
blade tip stall 叶顶失速
blade tip velocity 叶尖速度
blade tip vortex 叶尖涡
blade tip-root loss factor 叶端损失系数
blade twist 叶片扭转；桨叶扭转
blade vibration 叶片振动
blade vibration frequency 叶片振动频率
blade wake 桨叶尾流
blade wheel 叶轮
blade width ratio 叶片宽度比
blade-section chord 叶片剖面弦长
blade-to-jet speed ratio 叶片—气流速度比
blading 叶片组

blank 空白的
blank bolt 光螺栓；非切制螺栓
blank flange 死法兰；盲法兰
blank run 空转
blank wall 无窗墙；闷墙
blanket 覆盖；掩盖
blanket area 敷层面积
blanketed area 遮蔽面积；气动阴影面
blanketing effect 气动阴影效应
blanking die 冲裁模；下料模；切口冲模
blankoff flange 盲法兰
blankoff pressure 极限低压强
blast cooling 鼓风冷却
blast draft 强制通风；压力通风
blast fan 鼓风机；风扇
blast furnace stack 高炉烟囱；高炉炉身
blasting air 喷气
BLC (boundary layer control) 边界层控制
BLCS (boundary layer control system) 边界层控制系统
bleacher 露天看台
bleed of boundary layer 边界层抽吸
bleed orifice 放气孔
BLF (building-life factor) 建筑寿命因子
blind 挡板
blind door 百叶门
blind wall 无窗墙；闷墙
blinder 遮阳板
blinding plate 盲板
blizzard 暴雪
block 块体；阻塞块
block and tackle 滑轮组
block building 块体建筑物
block coefficient 方形系数
block curve 连续曲线；实线曲线

block diagram 方框图
block mount 组合装配
block up 封闭；堵塞
block work 块体结构
blockage 阻塞；阻塞度
blockage correction 阻塞修正
blockage correction factor 阻塞修正因子
blockage effect 阻塞效应
blockage ratio 阻塞比
blocking 锁定
blocking anticyclone 阻塞反气旋
blocking effect 阻塞效应
blocking for wind turbines 锁定风力机
blocking interference 阻塞干扰
blocking voltage 闭锁电压；阻塞电压
blowdown test 风洞吹风试验
blower snow fence 导雪栏栅
blowing dust 吹尘；高吹尘
blowing sand 飞沙；流沙
blowing snow 吹雪；风雪流
blowing soil 土壤吹失
blowpipe 吹管
blowpipe analysis 吹管分析
blowpipe system （阵风模拟用的）吹管系统
bluff 钝形的
bluff body 钝体
bluff body aerodynamic 钝体空气动力学
bluff building 钝体型建筑
bluff cylinder 钝形柱体
bluff section 钝形截面
bluff structure 钝体型结构
bluffness 钝度
blunt base 钝底；钝尾
blunt body 钝头体
blunt cowling 钝头整流罩
blunt nose 钝头
blunted (blunt wing) tip 钝翼尖
BM (breakdown maintenance) 故障维修
board 木板
boat tailing 钝尾；船形尾部
bobtail 截尾
boil 沸腾
boiling water reactor 沸水反应堆
BOL (boundary layer) 边界层
bolster 支架；鞍座
bolt 螺栓
bolt blank 螺栓坯件
bolt callouts 螺栓名称；螺栓标注
bolt damage sensor 螺栓损坏传感器
bolt head 螺栓头
bolt head retaining slots 螺栓头固定槽
bolt nut 螺母
bolt symboles 螺栓符号
bolt threads in bearing 支撑部分的螺栓螺纹；轴承螺栓螺纹
bolt washer 螺栓垫圈
bolted connection 螺栓连接
bolting 用螺栓固定
bolts interchangeable hole patterns 孔型可互换的螺栓
bolts wrench clearance 螺栓扳手间隙
Boltzmann constant 玻耳兹曼常数
bond 结合
bond flux 焊剂
bond line 黏合层；黏合剂
bond master 环氧树脂类黏合剂
bond strength 结合强度
bonded 结合
bonded in bolts 用螺栓连接的
bonding 黏合

bonding bar 等电位连接带
(equipotential) bonding conductor 等电位连接导体
(equipotential) bonding electrical 电器元件的固定
bonding mental to mental 金属与金属的胶结
bonding point 接合点
bond-meter (bond-tester) 胶接检验仪
bonnet 烟囱罩；发动机罩
boost 增压
boost-buck 升压去磁
booster mill 增压提水风车
boot lid 引导盖
bore 钻孔；钻
boring 钻孔
boring heads 搪孔头
boring machines 镗床
Bose-Einstein statistics 玻色-爱因斯坦统计法
boss bolt 轮毂螺栓
both 两个的
both sides welding 双面焊接
bottom 底部
bottom deflecting snow fence 下导防雪栅栏
bottom die 底模
bottom die base 底模基础
bottom effect 底部效应
bottom flange 底部法兰；下阀盖；下翼缘；底凸缘
bottom layer 底层
bottom panel 底板；（风洞）底壁
bottom pressure 底压
bottom surface 底面
bottom surface camber 下弧；下曲面
bottoming tap 平底螺丝攻；平底丝锥
boulder 大圆石；巨砾

bouncing 跳动
bound 束缚；范围；限制
bound circulation 附着环量
bound vortex 附着涡；约束涡
bound vortex filament 附着涡丝
bound vortex system 附着涡系
bound vorticity 附着涡量
boundary 边界；界线
boundary conditions 边界条件
boundary constraint 边界约束；（风洞）洞壁约束
boundary dimension 外形尺寸
boundary effect 边界效应；（风洞）洞壁效应
boundary layer (BOL) 边界层
boundary layer accumulation 边界层堆积；边界层增厚
boundary layer bleed 边界层抽吸
boundary layer blowing 边界层吹除
boundary layer buildup 边界层形成
boundary layer configuration 边界层形状
boundary layer control (BLC) 边界层控制
boundary layer control system (BLCS) 边界层控制系统
boundary layer correction 边界层修正
boundary layer depth 边界层厚度；边界层深度
boundary layer development 边界层扩展
boundary layer equation of motion 边界层运动方程
boundary layer flow 边界层流动
boundary layer flux 边界层通量
boundary layer friction 边界层摩擦
boundary layer friction loss 边界层摩擦损失
boundary layer function 边界层函数
boundary layer geometry 边界层几何形状
boundary layer growth 边界层增长

boundary layer height 边界层高度；边界层厚度

boundary layer immersion ratio 边界层浸没比

boundary layer measurement 边界层测量

boundary layer meteorology 边界层气象

boundary layer phenomenon 边界层现象

boundary layer probe 边界层探针

boundary layer profile 边界层廓线

boundary layer removal 边界层的去除

boundary layer reversal 边界层回流

boundary layer separation 边界层分离

boundary layer separation loss 边界层分离损失

boundary layer skin friction 边界层表面摩擦

boundary layer structure 边界层结构

boundary layer suction 边界层抽吸

boundary layer temperature 边界层温度；边界温度

boundary layer theory 边界层理论

boundary layer thickness 边界层厚度

boundary layer transition 边界层转捩

boundary layer transition strip 边界层转捩绊线；边界层过渡地带

boundary layer vorticity 边界层涡量

boundary layer type wind 边界层型风

boundary layer type wind tunnel 边界层型风洞

boundary lubrication 边界润滑

boundary resistance 边界层阻力

boundary science 边界层科学；边缘科学

boundary spar 边缘梁；边缘翼梁

boundary surface 界面；边界曲面

boundary value 界限值；监界品位；边值

boundary value problem 边界值问题

boundary wave 边界波；界面波

bounded function 有界函数

Boussinesq approximation 布辛涅斯克近似

bow 弓形

bow collector 集电弓

bowl 旋转

bowl shaped structure 凹面形结构

Bowman formula 波曼公式

box 箱子

box beam 箱型梁

box fan 箱式风扇；鸿运扇

box model 箱模式

box spanner 管钳子

box spanner inset 插入式套筒扳手

box spar 箱式叶梁

box girder bridge 箱梁桥

BP (barometric pressure) 气压；大气压；大气压力

BP Sensor (barometric pressure sensor) 大气压力传感器

B-pillar B支柱

brace 支柱；带子

brace bolt and nut 拉条螺栓及螺母

brace nut 拉条螺母

brace screw 撑柱螺丝

brace summer 双重梁；支撑梁

brace wrench 曲柄头扳手

bracing 支撑；支柱

bracing piece 加劲撑杆，斜梁

bracing stiffness 支撑刚度

bracket 托架；括弧；支架

bracket crane 悬臂式起重机

brad 曲头钉

bradawl 小锥

braid 编织物

brake action 制动作用
brake assist 刹车辅助；制动力辅助
brake assist system 制动力辅助系统
brake disc 刹车盘
brake disk 制动盘
brake drag 制动阻力
brake dressing 制动器润滑脂
brake efficiency 制动效率
brake electromagnet 制动电磁铁
brake equalizer 制动平衡器
brake fluid 刹车油
brake horsepower 轴功率；轴马力
brake lining 闸衬片；制动衬片
brake mechanism 制动机构
brake off 松开制动
brake on 制动
brake on-off switch 制动开关
brake pad 闸垫
brake pad thickness 制动块厚度
brake power 制动力；制动功率
brake release 制动释放；松开制动器
brake setting 制动器闭合
brake shoe 制动瓦；闸瓦；制动蹄片；制动块；煞车块；制动靴
brake spring 制动弹簧
braker (BRK.) 制动器
braking 制动系统
braking deceleration 制动减速率
braking device 制动制置
braking disc 制动盘
braking effect 制动作用
braking efficiency 制动效能；制动效率
braking effort 制动作用力
braking energy 制动能
braking mechanism 制动机构

braking moment 制动力矩
braking path 刹车滑行距离
braking period 制动时间；制动周期
braking power 制动功率
braking releasing 制动器释放
braking surface 制动面
braking system 制动系统
braking type 刹车形式
braking vane 刹车尾舵；刹车板
branch 分支
branch connection 分支接续
branch line 支线
branch of joint 连接分支
braze 铜焊；用黄铜镀；用黄铜镀制造
braze welding 钎焊；硬焊；铜焊
brazed joint 硬钎焊接
brazier 扁头螺钉
breakaway coupling 断开式联轴节
breakaway device 安全分离装置；安全脱钩装置；保险装置
breakaway force 起步阻力
breakaway starting current of an A.C. 交流电动机的最初启动电流
breakdown 击穿
breakdown current 击穿电流
breakdown field strength 绝缘强度
breakdown maintenance (BM) 故障维修
breakdown point 屈服点；击穿点
breakdown test 断裂实验
breakdown torque 停转力矩；极限转矩；崩溃转矩；损坏力矩
breakdown voltage (BDV) 击穿电压
breaker bolt 安全螺栓
breaker (BRKR) 断路器
breaking 破坏；阻断

breaking capacity 破坏能力
breaking coefficient 断裂系数
breaking current 断路电流
breaking device 断路器
breaking down point 破坏点
breaking down test 耐压实验；耐破坏试验；击穿试验
breaking effort 制动力
breaking elongation 断裂伸长；破断伸长
breaking elongation rate 断裂伸长率
breaking load 破坏负载
breaking moment 断裂力矩
breaking point 强度极限
breaking strength 破坏强度
breaking stress 破坏应力
breaking test 破坏试验；断裂试验
breakwind 防风林；风障
breast drill 胸压手摇钻；胸压钻；曲柄钻
breather （风洞）换气装置
breathing rate 呼吸率
breeze 微风
bricklayer's hammer 瓦工锤
bridge 桥梁
bridge abutment 桥台；桥肩
bridge aerodynamics 桥梁空气动力学
bridge aeroelasticity 桥梁气动弹性
bridge beam 横梁
bridge buffeting 桥梁抖振
bridge flutter 桥梁颤振
brightness 亮度
brine pumping 盐水提抽
Brinell hardness 布氏硬度
Brinell hardness number 布氏硬度数
Brinell hardness test 布氏硬度试验
Brinell tester 布氏硬度计

briquettability 压塑性；压制性
brittleness 脆性；脆度
BRK (brake) 制动器
BRKR (breaker) 断路器
BRM (barometer) 气压计
broad band 宽波段
broad band excitation 宽带激励
broad band response 宽带响应
broad band turbulence 宽带湍流
broad scale flow 大尺度流动
broken sky 裂云天空
bronze 青铜
Brookfield viscometer 布氏黏度计
brown cloud 棕色烟云
brown fume 棕色烟雾
Brownian diffusion 布朗扩散
Brunt-Vaisala frequency 布伦特-维塞拉频率
brush 电刷
brush discharge 电晕放电；刷形放电
brush contact 电刷触点
brush holder 电刷座
brush plating 刷镀
brush wastage 碳刷磨损
brushless 无刷的
brushless doubly fed induction generator (BDIG) 无刷双馈异步发电机
brushless exciter 无刷励磁机
brushless power electronic circuit 无刷电力电子电路
brute force 强力
bubble 磁泡；水泡；气泡
bubble suction 空泡吸力
BUC (buffer controller) 缓冲控制器
Buckingham's pi theorem 白金汉 π 定理
buckle 扣住；使变弯

buckling 屈曲；变形；弯曲
buckling load 屈曲荷载
buckling strain 弯曲应变
buckling strength 抗弯强度
buckling stress 弯曲应力
buffer action 缓冲作用
buffer amplifier 缓冲放大器
buffer beam 缓冲梁；缓冲杆
buffer circuit 缓冲电路
buffer controller (BUC) 缓冲控制器
buffer spring 缓冲弹簧
buffet 抖振
buffeting characteristic 颤振特性
buffeting deflection 抖振挠度
buffeting excitation 抖振激励
buffeting factor 抖振因子
buffeting response 抖振响应
buffing wheel 抛光轮
build down 降低；衰减
build in terrain 建成地形
build up 组合
build up area 建成区
build up factor 累积因子
building 建筑；建筑物
building block 积木；组成部件；标准组件
building code 建筑规范
building complex 建筑群；综合建筑
building coverage 建筑覆盖率
building density 建筑密度
building depth 建筑进深；建筑深度
building frame 建筑框架
building life factor (BLF) 建筑寿命因子
building line 建筑线；房屋界线
building permit 建设工程规划许可证；建筑施工执照；建筑许可

building serviceability 建筑物舒适性
building site 建筑场址
building standard 建筑标准
building wake effect 建筑物尾流效应
building wake region 建筑物尾流区
built-in oscillation 固有振荡
built-up die 组合模具
built-up environment 建成环境
built-up gear 组合齿轮
bulb 电灯泡
bulb resistance 灯泡电阻
bulding-out 补偿值
bulding-out circut 补偿电路
bulk assembly 大部件装配
bulk buoyancy 总体浮力
bulk density 单位容积容重；松密度
bulk heat transfer coefficient 整体传热系数
bulk mass transfer coefficient 整体质量运输系数
bulk modulus 体积模量；体积弹性模量；体积压缩性
bulk modulus of elasticity 体积弹性模量
bulk moulding compound 预制整体模塑料；块状模塑料
bulk Richardson number 总体理查森数
bulk stability parameter 总体稳定度参数
bulk transfer coefficient 整体运输系数
bull 大型的
bull horn 手提式扩音器
bull wheel 大齿轮
bump 冲撞；隆起物
bumper 缓冲器；防撞器；防撞杠；缓冲器；减震器
bumper angle 缓冲角铁
bumping 碰撞

bumping collision 弹性碰撞
bumpy 颠簸的；气流不稳的；崎岖的
bumpy flow 涡流
bundle of vortex filaments 涡丝束
bundle power line 集束输电线
bundled cables 集束电缆
bundled conductor 集束电线
buoyancy 浮力
buoyancy correction 浮力修正
buoyancy flux 浮力通量
buoyancy length scale 浮力长度尺度
buoyancy lift 浮力；浮升
buoyancy term 浮力项
buoyant acceleration 浮力加速度
buoyant convection 浮力对流
buoyant jet 浮升射流
buoyant lift 浮力
buoyant plume 浮升羽流
buoyant puff 浮升喷团
buoyant rise 浮力抬升
buoyant source 浮力源
buoyant thermal 浮升热泡
burble 漩涡
burble angle 失速角
burbling 气泡分离；气流分离；流体起旋；旋涡
burbling point 失速点；气流分离点
burn-in 老化
burn-off 烧化；熔化焊穿；雾消
burnt gas 废气；燃烧过的气体
bursting 爆炸；猛然打开；突然开始
bus 母线
bus coupler 总线耦合器
bus duct 母线槽
bus regulator 母线电抗器；母线电压蝶器
bus voltage loss 母线电压损失

bus voltage regulator 母线电压调整器；母线电压调节器
busbar 母线
busbar expansion joint 母线伸缩节
busbar price 母线价格
busbar separator 母线间隔垫
bush （绝缘）套管；加（金属）衬套于……
bush hammer 凿石锤；气动凿毛机
bushing 轴衬；套管
bushing for arm assembly boss 臂组件轮毂用衬套
butt contacts 对接触点
butt muff coupling 刚性联轴器
butt weld 对接焊缝
butt welding 对接焊；对焊
butte 孤山；小尖山
butterfly damper 蝶形阀
Buys-Ballot's law 白贝罗定律
buzzing 嗡鸣
by the wind 顺风航行
bypass 旁路；旁通；分流
bypass switch 旁通开关
byte 字节

现代英汉风力发电工程

A Modern English-Chinese Dictionary of Wind Power Engineering

C

CAB (cable) 电缆
cabane 翼间支架
cabinet converter 变频器柜
cabinet door 柜门
cabinet nacelle 机舱机柜
cabinet nacelle transformer 机舱变压器柜
cabinet tower 塔基机柜
cable (CAB) 电缆
cable armor 电缆铠装
cable bent 缆绳垂度
cable bond 电缆接头
cable breakdown 电缆击穿
cable bridge 缆索桥
cable bundle 束；光纤束；捆；卷
cable coupling capacitor 电缆耦合电容器
cable cutter 电缆剪
cable fitting 电缆配件
cable gland 电缆衬垫
cable guide 拉线导向块；电缆引导管；缆索导向器
cable laying 电缆敷设
cable making tools 造线机
cable net 索网
cable network 电缆网；索网
cable net wall 索网幕墙

cable rack 电缆架
cable reel 电缆盘
cable roof 拉索屋顶
cable routing 电缆路由选择
cable sag 缆索垂度
cable shear 电缆剪
cable sheath 电缆包皮层
cable shielding 电缆屏蔽
cable shoes 电缆靴
cable stayed bridge 斜拉桥
cable stayed grider bridge 斜梁桥
cable supported membrane roof 拉索薄膜屋顶
cable tie 电缆带
cable to cable tie 拉索扣带
cable tray 电缆盘
cable trunk 电缆管道
cable twist 扭缆
cable twist counter 电缆扭转计数器
cable twist sensor 电缆扭绞传感器
cable twisting 电缆扭曲
cable untwisting 电缆解扭
cabling diagram 电缆敷设图
cabtyre sheathing 硬橡胶电缆包皮
cacaerometer 空气污染检查器

CACD (computer aided circuit design) 计算机辅助电路设计
CAD (computer aided design) 计算机辅助设计
CADD (computer aided design and drafting) 计算机辅助设计制图
CADE (computer aided design and engineering) 计算机辅助设计与工程
cage 笼型
cage assembly 升降台
cage bar 鼠笼条
cage lifter 升降机
cage motor 鼠笼式电机
cage rotor 鼠笼式转子
calathiform 杯形的
calcination temperature (CT) 煅烧温度
calculate 计算；核算；评价
calculated height 计算高度
calculated instruction 计算说明书
calculated load 设计负荷
calculated risk 计算危险值
calculated value 计算值
calculating machine 计算机
calculation 计算；估计
calculation error 计算误差
calculation of loading 负荷计算
calculation sheet 计算书
calculus 微积分学
calibrating constant 校准常数
calibrating device 校准装置
calibrating instrument 校准装置
calibration 校准
calibration accuracy 校准精度
calibration chart 校准图表
calibration coefficient 校准系数
calibration curve 校准曲线
calibration factor 校准因素
calibration model 检验模型；标准模型
calibration model test 标模试验
calibration tunnel 标准风洞
calibration value 校准值
calibration wind tunnel 校准风洞
caliper 卡钳
calliper brake 线闸；卡钳制动器
calliper rule 卡尺
calliper square clasp 游标卡尺的滑尺
calm 无风；零级风
calm belt 无风带
calm breeze 静风
calm central eye 无风眼
calm layer 无风层
calm smog 无风烟雾
calm night 静夜
calm plume 无风烟羽
calm sea 无浪海面
calm spell 无风期
calm zone 无风带
calorific value 热值；卡值
cam 凸轮
CAM (computer aided manufacturing) 计算机辅助制造
cam adapter 凸轮联轴器
cam bowl 凸轮滚子
cam break 凸轮制动器
CAMA (computer aided mathematics analysis) 计算机辅助数学分析
camber 弯度
camber curvature （翼型）中弧线弯度
camber distribution 延翼展的弯度分布
camber effect 弯度效应；前轮外倾效应

camber line 翼型中心线
camber(ed) beam 曲梁
cambered aerofoil 曲翼面
cambered airfoil 弯曲翼面
cambered axle 弯轴
cambered blade （弦向）弯曲叶片
cam-ring chuck 三爪卡盘
camshaft 凸轮轴
canalization 开凿运河；运河网造管术；渠化
canard 鸭翼；前翼
cancelled structure 空腹结构
candidate site 候选场址
canopy 天蓬；华盖；顶棚
canopy flow 冠层流
canopy hatch 舱口
canopy layer 冠层；林冠覆盖
canopy sublayer 冠次层
canopy top 林冠顶
cant 斜面
cant deficiency angle 欠超高角
cant of curve 曲线超高
cant of rail 轨道超高
cant of sleeper 轨枕倾斜
canted blade 斜面叶片
cantilever 伸臂，悬臂；悬臂梁
cantilever beam 悬臂梁
cantilever bridge 悬臂桥
cantilever force 悬臂受力；悬臂力
cantilever girder 悬臂梁
cantilever rotor 悬臂风轮
cantilever structure 悬臂梁结构
cantilever tower 悬臂塔架；独立塔架
canvas bag 帆布垒包
canyon 峡谷；街谷
canyon wall 峡谷壁

canyon wind 下吹风；下降风
cap 盖；帽子
cap bolt 倒角螺栓；盖螺栓
cap of thermal 热泡冠；帽热
capacitance 电容
capacitance effect 电容效应
capacitance type sensor 电容式传感器
capacitive 电容性的
capacitive reactance 容抗
capacitive susceptance 容纳
capacitor bank 电容器组
capacitor bank switchgear 电容器组开关设备
capacitor depressing voltage 电容降压
capacitor for voltage protection 保护电容器
capacitor initial voltage 电容初始电压值
capacitor section 电容区
capacitor switch 电容器开关
capacitor switching event 电容器开关事件；电容器切换事件
capacity 容量
capacity factor 容量因数；利用率
capacity fall off 电容漏电；电容量减退
capacity for work 做功能力
capacity loading 满载
capacity of the wind 风卷挟力
capacity value 功率
capillary 毛细管
capital cost 投资费
capital repair 大修
capsizing moment 倾覆力矩
capsule building 盒式建筑
captive balloon 系留气球
capture 捕获；获风
capture area 捕获面积；获风面积
capture cross section 捕获截面；获风截面

capture efficient 捕获效率

capture rate 捕获率

captured air bubble craft 封闭气泡式气垫艇；气囊船

carbide bit 硬质合金刀头

carbide blade 硬质合金刀片

carbon 碳

carbon body 电刷

carbon brush 碳刷

carbon carburizing steel 碳素渗碳钢

carbon constructional quality steel 优质碳素结构钢

carbon credit 碳信用；碳信用额

carbon felt blanket 含碳毡垫

carbon fiber 碳纤维

carbon fiber material 碳纤维材料

carbon fiber reinforced plastic 碳素纤维增强塑料

carbon filament lamp 碳丝灯泡

carbon glass reinforcement 碳—玻璃加强件

carbon penetration 渗碳

carbon reduction 碳还原法

carbonization 碳化

carbonization zone 碳化层

carbonsteel 碳钢

carborundum grain 金刚砂

carburator 渗碳器

carburetted iron 碳化铁

carburization 渗碳

carburization material 渗碳剂

carburization zone 碳化层

carburized layer 渗碳层

carburized structure 渗碳组织

carburizing agent 渗碳剂

carburizing box 渗碳箱

carburizing by molten salts 液体渗碳

carburizing by solid matters 固体渗碳

carburizing furnace 渗碳炉

carburizing gas 气体渗碳剂

carburizing liquid 渗碳液

cardan shaft 万向轴

cardinal winds 主要风向

care 关怀；照顾

care and maintenance party 维护保养组

cargotainer 集装箱

carpeting 地毯状（Ⅶ级植物风力指示）

carrier 载体；载波

carrier gas 载气

carrier liquid 载液

carrier material 载体

carrier mobility 载体迁移率

carrying 运输的

carrying area 升力面

carrying bolt 支撑螺栓

carrying capacity 载带能力；装载量

carrying current 极限电流

carrying idler 空载

carrying plane 支承面

carrying power 承载力

Cartesian coordinate system 笛卡儿坐标系

cartridge fuse 保险丝管

cascade connection 串联

cascade impactor 多级碰撞取样器

cascade tunnel 叶栅风洞

case 情况

case hardening 表面硬化

case-carbonizing 表面渗碳

case-carburizing 表面渗碳

cased-muff coupling (solid coupling or butt muff coupling) 刚性联轴器

case-hardened casting 冷硬铸件
case-hardened glass 钢化玻璃
case-hardening 表面硬化
case-hardening carburizer 表面硬化渗碳剂
case-hardening steel 渗碳钢
casing 外壳；发电机舱
cast 投；掷；铸件
cast alloyiron 合金铸铁
cast aluminium 铸铝
cast aluminium alloy 铸造铝合金
cast aluminium rotor 铸铝转子
cast blade 铸造叶片
cast iron 铸铁
cast jacket 整铸套箱
cast joint 铸焊；浇铸连接
cast steel 铸钢
cast teeth standard 铸齿标准
cast temperature 铸造温度
cast to shape 精密铸造
cast unit 整件铸造
cast welding 铸焊
castability 铸造性；流动性
castellated barrier wall 城垛形挡板
castellated coupling 牙嵌式连接
caster 主销纵倾；铸工
casting 铸件；铸造
casting aluminium 铸铝
casting cooling system 铸件冷却装置
casting copper 铸铜
casting die 压铸模
casting gray iron 铸灰口铁
casting malleable iron 可锻铸铁
casting other 其他铸造
casting steel 铸钢
castle nut 槽形螺母

casual repairs 临时修理
CAT(clear air turbulence) 晴空湍流
cat's nose 猫鼻风
cat's paw 猫掌风
catamaran 双体船
cataract 大瀑布
catastrophic 灾难的
catastrophic failure 严重故障；灾难性故障；突变失效
catastrophic oscillation 灾难性振动
catch bolt 止动螺栓
catching load 制动力
catching of toothed wheels 齿轮啮合
catenarian 悬链线；悬索
catenary 悬链线；悬索
catenary angle 悬垂度
catenoid 悬链曲面；链状的
cathode 阴极
cathode inductance 阴极电感
cathode-ray tube (CRT) 阴极射线管
cathodic protection system 阴极保护系统
Cauchy type distribution 柯西分布
caulking 填……以防漏；堵缝；压紧
caulking metal 填隙合金；填隙合金金属材料
cavitation 气蚀现象
cavitation bubble 空泡
cavitation damage 空化损坏
cavitation effect 空化效应
cavitation erosion 空蚀
cavitation nucleus 空化核
cavitation number 空化数
cavitation parameter 空化参数
cavitation phenomenon 气蚀现象
cavitation shock 气蚀冲击
cavitation tunnel 空泡试验筒；空蚀试验槽；气

穴风洞
cavity 空穴
cavity boundary 空穴边界
cavity drag 空穴阻力
cavity flow 空穴流动
cavity region 空穴区
cavity wake 空穴尾流
CAVT (constant absolute vorticity trajectory) 等绝对涡度轨迹
CBC (comparator buffer) 比较器缓冲器
CB (contact breaker) 接触断路器
Cb (cumulo-nimbus) 积雨云
Cb cal (cumulo-nimbus calvue) 秃积雨云
Cb cap (cumulo-nimbus capiltatus) 鬃状积雨云
Cb inc (cumulo-nimbus incus) 砧状积雨云
Cb mam (cumulo-nimbus mammatus) 悬球状积雨云
CCB (convertible circuit breaker) 可变换断路器
CCW (counter clock wise) 逆时针
CDM (clean development mechanism) 清洁发展机制
ceiling 天花板；（风洞）顶壁；云幂
ceiling fan 吊扇
ceilometers 云幂仪
cell 环型；单体；（测力，测压）传感器
cellular 多孔的
cellular girder 空腹梁
cellular glass 泡沫玻璃
cellular insulant 多空隔热材料；泡沫保温材料
cellular material 多孔材料
cellular motion 环形运动
cellular plastics 泡沫塑料
cellular radiator 蜂窝式散热器
cellular rubber 泡沫橡胶
cellular structure 网格结构
cellular type radiator 孔式散热器
cellule 小细胞
celotex 隔声材料；纤维板
Celsius 摄氏
Celsius thermometer 摄氏温度计
cement 水泥；接合剂；接合；用水泥涂；巩固；黏牢
cement lined piping 水泥衬里
cement type binder 水泥型黏合剂
census data 户口普查资料
cent 百分
center 中心
center bit 中心钻；转柄钻；中心钻头
center calliper 测径中心卡尺
center calm 中心无风区
center chord 中心弦
center distance 中心距
center gear 中心轮
center line keel beam 中央龙骨
center of buoyancy 浮力中心
center of disturbance 扰动中心
center of lift 升力中心
center of mass 质量中心
center of oscillation 振动中心；摆动中心
center of pressure 压力中心
center of pressure distribution 压力中心分布
center of resistance 阻力中心
center of rotation 转动中心
center of span 跨度中心；桥跨中点
center of stiffness 刚度中心
center of turn 转向中心
center pillar 中立柱
center puncher 中心冲

center section 中翼段；中心截面
centering adjustment 对准中心
centering chuck 定心夹盘
centerline chord 中心弦
centi 厘
centigrade 百分度；摄氏度
centimeter 厘米
central 中心的
central airfoil 中部翼面
central board 中央操纵台
central calm 中心无风区
central conic 有心圆锥曲线
central control room 中央控制室
central difference method 中心差分法
central equilibrium 中心平衡
central limit theorem 中心极限定理
central movement 有心运动
central span 中跨
centralized 集中的
centralized control 集中控制
centralized smoked stack 集中吸烟栈
centralized wind energy systems 集中式风能转化系统；集中风能系统
centrifugal 离心的
centrifugal clutch 离心式离合器
centrifugal effort 离心力
centrifugal fan 离心式通风机
centrifugal force (CF) 离心力；向心力
centrifugal governor 离心式传感器；离心调速器
centrifugal hinge moment 离心铰链力矩
centrifugal load 离心载荷
centrifugal unit 离心单元
centripetal force 向心力
cepstrum 倒频谱

CER (certified emission reduction) 可认证减排量；经核证的减排量
ceramic 陶瓷
ceramic die 陶瓷模
ceramic heated tunnel 有陶瓷加热的风洞
certification 证明
certification rule 认证规则
certified 被证明的
certified emission reduction (CER) 可认证减排量；经核证的减排量
CF (centrifugal force) 离心力
CF (cooling fan) 冷却风扇
CF (correction factor) 修正系数
chain 链
chain bridge 链索桥
chain drive 链传动
chain making tools 造链机
chain pulley system 链—滑轮系统
chain vice 链式钳
chain wheel 滑轮
chamber pressure 室压
chamfer 切角；刻槽
change 改变
change gear 变速齿轮；交换齿轮
change gear set 齿轮变速组
change gear train 变换齿轮系
change over switching 换向开关
change-over circuit 转换
change-over switching 换接
changer 变换器
channel 海峡；狭管；风洞；水槽
channel beam 槽型梁
channel flow 狭管流；河道径流；河槽径流
channeled wind 夹道风
channeling effect 夹道效应；沟道效应

chaotic vorticity	混沌涡量
character	特性
character of service	工作状态
characteristic	特有的；表示特性的；特征；特色
characteristic area	特征面积
characteristic class	特征曲线类型
characteristic constant	特征值
characteristic correlation length	特征相关长度
characteristic curve	特性曲线
characteristic data	特性参数
characteristic dimension	特征尺寸
characteristic equation	特征方程
characteristic error	特性误差
characteristic frequency	特征频率
characteristic function	特性函数
characteristic length	特征长度
characteristic life	特征寿命
characteristic net	特性曲线；特性曲线网
characteristic parameter	特征参数
characteristic temperature	特征温度
characteristic value	特征值
characteristic velocity	特征速度
characteristic wavelength	特征波长
charge	充电；负载；费用；电荷；掌管；控告；命令
charge a battery	给蓄电池充电
charge of rupture	破坏荷载
charge of surety	容许荷载
charging inductance	充电电感
chart	图表
chassis	底盘
chatter	使咔哒咔哒作声
chatter proof bolt	防震螺栓；颤振防爆螺栓
check	检查
check against	检查；核对
check bolt	防松螺栓
check crack	细裂纹
check dam	拦沙坝；拦水坝
check nut	防松螺母
check ring	挡圈
check test	校核试验
check valve	止回阀
cheese mold	干酪压模
cheese-head screw	有槽凸圆柱头螺钉；圆头螺钉
chemical aerosol	化学气溶胶
chemical contamination	化学污染
chemical coolant	化学冷却剂
chemical corrosion	化学腐蚀
chemical effluent	化学排放物
chemical fume	化学烟雾
chemical pollutant	化学污染物
chemical vapor	化学压力灭菌法
chemosphere	光化层
CHF (criticle heat flux)	临界热通量
chill casting	冷铸
chill hardening	冷硬化
chilled	冷冻的
chilled cast iron	冷硬铸铁
chilled steel	冷钢
chimney	烟囱
chimney cap	烟囱风帽
chimney cloud	烟云
chimney draft	烟囱通风；烟囱抽力
chimney effect	烟囱效应；抽吸效应
chimney effluent	烟囱排放物
chimney emission	烟囱排放
chimney exit	烟囱（出）口
chimney flue	烟道

chimney fume 烟囱排烟；烟囱通风
chimney gas 烟气
chimney height 烟囱高度
chimney mouth 烟囱口
chimney neck 烟道
chimney plume 烟囱烟羽
chimney shaft 烟囱筒体
chimney spot 烟囱地点
chimney stack 组合烟囱
chimney superelevation 烟囱超高
chimney ventilation 烟囱通风
Chinese windmill 中国式竖轴风车
chinook wind 钦诺克风；奇努克风
chipping 修琢
chisel 凿子；砍凿
chocking effect 堵塞效应
choke 阻气门
choke valve 节流阀
choking 壅塞
choking action 阻塞作用
choking effect 壅塞效应
choking flow 壅塞流
chop stroke 削球
chopped 砍
chopped strand mat 短切原丝毡
chopper 断路器；斩波器
chopper circuit 斩波电路
choppy wind 不定向风；疾风
chord 弦；翼弦；弦度
chord axis （叶片）弦轴
chord deflection offset 弦线偏距
chord direction 弦向
chord force 弦向分力
chord length 弦长
chord line 弦线

chord taper ratio 翼弦锥度比
chordal addendum 弦齿高
chordal thickness 弦齿厚
chordwise 弦向
chordwise load change 弦向载荷变化
chordwise slotted 弦向开缝的
chordwise term 弦向项
chromatographic analysis 色谱分析
chromatographic column 色谱柱
chronometer 精密计时表
chronotron 瞬间计时器；摆线管
chuck 夹盘；轴承座
churning 旋动；搅动
churning losses 搅动损失
CI (convective instability) 对流不稳定性
circle 圆
circle theorem of hydrodynamics 流体动力学圆柱绕流定理
circling motion 圆周运动
circlip 环形；弹性挡圈
CI-cu (cirro-cumulus) 卷积云
CI mot (cirrus nothus) 伪卷云
CI ve (cirrus vertebratus) 脊状卷云
circuit 电路
circuit airfoil 圆弧翼型
circuit board 电路板
circuit board layout 电路板布局
circuit branch 支路
circuit breaker (CIR. BKR) 断路器；保护断路器；断路开关
circuit components 电路元件
circuit diagram 电路图
circuit parameters 电路参数
circuitation 旋转矢量；旋转；旋度
circular 圆形的

circular airfoil 圆弧翼型	cirrostratus fibratus (CS fib) 毛卷层云
circular arc section 圆弧翼型	cirrostratus filosus (CS fil) 毛卷层云
circular building 圆弧形建筑	cirrostratus nebulosus (CS neb) 薄暮卷层云
circular cylindrical coordinate 圆柱坐标	cirrus 卷云
circular flow 环行流动	cirrus nothus (CI not) 伪卷云
circular frequency 角速度；圆频率；角频率	cirrus vertebratus (CI ve) 脊状卷云
circular streamline 圆形流线	city climate 城市气候
circular wind tunnel 圆形试验段风洞	city complex 城市建筑群
circular working section 圆形截面试验段	city dwellers 城市居民
circulating 循环的；流通的	city engineering 土木工程；城市工程
circulating planetary wind 环流行星风	city environment 城市环境
circulating water 循环冷却水	city fog 城市烟雾
circulation 流通；传播	city layout 城市布局
circulation around circuit 封闭环流	city planning 城市规划
circulation cell 环流圈	city pollution 城市污染
circulation control 环量控制	city street canyon 城市街谷
circulation control airfoil 环量控制翼型	city structure 城市结构
circulation controlled rotor 环量控制型风轮	city ventilation 城市通风
circulation flow 环流	city wall 城墙
circulation index 环流指数	CI ve (cirrus vertebratus) 脊状卷云
circulation layer 环流层	civil building 围护建筑物；民用建筑
circulation lubricating 循环润滑	civil engineering 土木工程
circulation pattern 环流型；循环型	civil structure 土木结构
circulation regime 环流状态	CL (crane load) 吊车
circulatory flow 环流	cladding 围护结构；饰面物；熔覆；包层
circumcircle 外接圆	cladding load 围护结构荷载
circumferential backlash 圆周侧隙	cladding material 镀层
circumferential force 切向力	cladding panel 围护墙板
circumferential joint 圆圈接缝	cladding stiffness 围护结构刚度
circumferential velocity 圆周速度	cladding structure 围护结构
circumpolar whirl 环极旋风	clamp 加紧；固定住
circumradius 外接圆半径	clamp dog 制块
circumscribe circle 外接圆	clamp securing bolt 卡箍紧固螺栓
cirro stratus (CS) 卷层云	clamping apparatus 夹具
cirro-cumulus (CI-cu) 卷积云	clamping disk 夹紧盘

clamping/holding systems 夹具/支持系统
Clark Y airfoil 克拉克 Y 翼型
class 层
class of damage 破坏等级
class of minor damage 轻度破坏等级
class of pollution 污染等级
classical 经典的
classical airfoil flutter 经典翼型颤振
classical bending torsion flutter 经典弯扭颤振
classical blockage correction 经典阻塞修正
classical buffeting 经典抖振
classical fluid mechanism 经典流体力学
classical flutter 经典颤振
classical setting 规则地形；经典地形布置
classical strip theory 经典片条理论
classical Wagner function 经典瓦格纳函数
classification of solar energy resource 太阳能资源等级
claw 爪
claw coupling 爪形连接器
claw hammer 拔钉锤
clay model 油泥模型
CLD (cloud) 云
clean air 洁净大气
clean air act 空气洁净法
clean air legislation 空气洁净法规
clean development mechanism (CDM) 清洁发展机制
clean energy 清洁能源
clean energy source 无污染能源；清洁能源
clean flow 无分离流动；无旋流动
cleanliness 良流线性
clear 清楚的；清澈的
clear air turbulence (CAT) 晴空湍流

clear area 有效截面
clear span 净跨；大跨度
clear test section 风洞试验段
clear width 净宽
clearance 裕度；排除故障；清除
clearance fit 间隙配合
clearance for expansion 膨胀间隙
clevis U形夹
clevis drawbar 牵引环；联结钩；环卡组合式牵引装置
clevis joint 拖钩；脚架接头
cliff 悬崖
climate 气候
climate data 气候资料
climate effect 气候影响
climate element 气候要素
climate environment 气候环境
climate fluctuation 气候变动
climate modification 人工影响气候
climate zones 气候带
climatic chart 气候图
climatic classification 气候分类
climatic control 气候控制
climatic data 气候资料
climatic effect 气候效应；气候影响
climatic element 气候要素
climatic environment 气候环境
climatic fluctuation 气候变动
climatic noise 气候噪声
climatic stability 气候稳定度
climatic wind speed 气候风速
climatic wind tunnel 全天候风洞
climatological characteristics 气候学特性
climatology 气候学
clockwise (CW) 顺时针；右旋

clockwise rotation 顺时针的轮转；顺时针方向转动

clockwise spin 顺旋

clod 中砾块

cloddy structure 块状结构

close fit 紧密配合

close nipple 螺纹接口；全螺纹短节；螺纹接套

close structure 紧密结构

closed circuit 闭合电路

closed-circuit wind tunnel 回路式风洞

closed-cycle cooling 闭合循环式冷却

closed hydraulic loop circuit 闭式液压回路

closed isobar 闭合等压线

closed jet return flow tunnel 闭式工作段回流风洞

closed jet wind tunnel 闭口风洞

closed-section(-throat) wind tunnel 封闭试验段风洞

closed streamline 闭合流线

closed test section （风洞）封闭式试验段

closed-throat wind tunnel 闭口风洞

closed vortex line 闭合涡线

cloth covering 布制蒙皮

cloud (CLD) 云

cloud amount 云量

cloud base 云底

cloud cluster 云团

cloud column 云柱

cloud cover 云量

cloud drop 云滴

cloud droplet 小云滴

cloud element 云元素

cloudiness 云量

cloud layer 云层

cloud nuclei 云核

cloud resulting from industry 工业污染云

cloud seeding 云催化

cloud street 云街

cloud tower 云塔

cloud tube 云管

clout 猛击

clout nail 大帽钉

clover leaf body 三叶草图形

club hammer 锤子；榔头

cluster 颗粒团；建筑组群；群聚；丛生

cluster of wind turbines 风轮机群

cluster weld 丛聚焊缝

clutch 离合器；联轴器；凸轮；扳手

clutch disc facing 离合器盘衬片；离合器摩擦片

clutch disk 离合器摩擦片

clutch drag 离合器阻力

clutch driving disc 离合器主动盘

clutch facing 离合器衬片

clutch facing rivets 离合器面片铆钉

clutch gear 离合器齿轮

CMS (condition monitoring system) 状态监测系统

CNC (computer numerical control) 计算机数字控制机床；数控机床

CNC bending presses 电脑数控弯折机

CNC boring machines 电脑数控镗床

CNC drilling machines 电脑数控钻床

CNC EDM wire-cutting machines 电脑数控电火花线切削机

CNC electric discharge machines 电脑数控电火花机

CNC engraving machines 电脑数控雕刻机

CNC grinding machines 电脑数控磨床

CNC lathes 电脑数控车床

CNC machine tool fittings 电脑数控机床配件

CNC milling machines 电脑数控铣床
CNC shearing machines 电脑数控剪切机
CNC toolings 电脑数控刀杆
CNC wire-cutting machines 电脑数控线切削机
CNT(counter) 计数器
co(-)ordinate 坐标
coagulation 凝集
coal dust 煤尘
coal fired power plant 燃煤火电场
coal smoke pollution 煤烟污染
coalescence 并合；聚结
Coanda effect 柯恩达效应；附壁效应
coarsé estimation 粗估
coarse grade coefficient 粗粒径级系数
coarse grained snow 粗粒雪
coarse granular 粗团粒
coarse gravel 粗砾石
coarse grid 粗网眼
coarse sand 粗粝；粗砂
coarse topography 粗切地形
coarseness 粗糙度；粒度
coastal area 海岸区
coastal climate 沿海气候
coastal environment 沿海环境
coastal feature 海岸地形
coastal structure 沿海结构
coaster 单向联轴节；沿海航船；近海船
coasting 海岸线
coasting body 惯性体
coaxial 共轴的；同轴的
coaxial cable 同轴电缆
coaxial configuration 同轴结构
coaxial electrical cable 同轴电缆
cock 吊车；塞门；旋塞

cocking 压簧杆
cocking handel 机柄
cockloft 顶层；顶楼
code 代码
code of practice 实用规范
coefficient 系数
coefficient of convection 对流系数
coefficient of correction 修正系数
coefficient of cubical elasticity 体积弹性系数
coefficient of cubical expansion 体积膨胀系数
coefficient of damping 阻尼系数
coefficient of diffusion 扩散系数
coefficient of dilatation 膨胀系数
coefficient of discharge 流量系数
coefficient of dynamic viscosity 动力黏度系数
coefficient of fineness 船型系数；丰满系数
coefficient of flow velocity 流速系数
coefficient of friction 摩擦系数
coefficient of frictional resistance 摩擦阻力系数
coefficient of haze (COH) 霾系数
coefficient of kinematic viscosity 运动黏度系数
coefficient of local resistance 局部阻力系数
coefficient of performance (COP) 性能系数；（风轮）效率
coefficient of porosity 孔隙度
coefficient of resilience 弹性
coefficient of resistance 阻力系数
coefficient of restitution 恢复系数
coefficient of roughness 粗糙系数
coefficient of safety 安全系数
coefficient of torsional 扭转刚度系数

coefficient of torsional rigidity 扭转刚度系数
coefficient of turbulence flow 紊流系数
coefficient of uniformity 均匀系数
coefficient of viscosity 动力黏度；黏滞系数
coercive force 矫顽磁力；矫顽力；抗磁力；保磁力
cogging 齿；轮齿；嵌齿
cogging torque 齿槽转矩；顿转扭矩
cog wheel 齿轮
cogwheel 嵌齿轮
cogwheel coupling 齿形联轴器
cogwheel gearing 齿轮传动装置
COH (coefficient of haze) 霾系数
coherence 相干性；凝聚
coherence function 相干函数；凝聚函数
coherence vortex shedding 相干涡脱落
cohesive force 内聚力；黏合力
coil 线圈
coil inductance 线圈电感
coil spring 弹圈
coil winding 线圈绕组
coincide in phase with 与…同相
cold 寒冷的
cold deformation 冷变形
cold front 冷锋
cold gas plume 冷气羽流
cold hardening 冷加工硬化
cold quenching 冷（介质）淬火
cold rolled grain oriented transformer 冷轧晶粒取向变压器
cold strain 冷变形
cold temperature brittleness 低温脆性
cold temperature flexibility 低温韧性
cold temperature resistance 耐低温性
cold test 低温试验

cold tolerance 耐寒性
cold weld 冷焊
cold welding 冷压焊
cold work hardening 冷加工硬化
collapse 破坏；(涡) 破碎
collapse critical deflection 临界破坏挠度
collapse load 破坏荷载
collapsing 崩溃；塌陷
collapsing force 破坏力
collapsing load 破坏荷载
collapsing pressure 破坏压力
collapsing stress 破坏应力
collar 联轴节
collar bearing 环形止推轴承
collar step hearing 环状阶式轴承
collar stop 环形挡块
collar thrust bearing 环形推力轴承
collecting main 母线
collecting ring 集流环
collection efficiency 收集率；捕获率
collection of buildings 建筑群
collective dose 集体剂量
collector 集电极；(开口风洞) 喇叭口集气环；集热器
collector ring 集电环
collet chuck 套爪夹头；套爪卡盘；弹簧夹头；筒夹
collision efficiency 碰撞效率
collision stopping power 碰撞阻止能力
colloid 胶体；胶态
colloidal dispersion 胶态分散体；胶体分散系
colloidal particle 胶粒
colloidal suspension 胶态悬浮
colorimeter 比色计
colorimetric analysis 比色分析

coloured filament of water 染色水丝
coloured pigment 染料
comb 排管；电梳
combe 狭谷；冲沟
combination 结合；联合；合并；化合；化合物
combination callipers 内外卡钳
combination pliers 台钳；剪钳；钢丝钳；组合钳
combined action 联合行动
combined environment 综合环境；综合条件
comb pitot 梳状皮托管
combustion driven tube 燃气风洞；燃烧驱动管
comfort criteria 舒适判据
comfort factor 舒适因子
comfort index 舒适指数
comfort parameter 舒适参数
comfort zone 舒适区
comfortable wind environment 舒适风环境
commanding apparatus 操纵设备
commissioning 试运转；试车
commissioning test 投运试验
common breakdown 常见故障
common difference 公差
common divisor 公约数
common earthing system 共用接地系统
communication 通信
communication aerials 通信天线
communication cable 通信电缆
communication failure 通信故障
community atmosphere 城市大气；居民区大气
community pollution 城市污染；居民区污染
commutation 换向；整流；配电
commutation condenser 换向电容器
commutation condition 换向状况
commutation cycle 换向周期；整流周期
commutation voltage 换向电压
commutator 换向器；整流器；转向器；整流子
commutator bar 整流条
commutator change over switch 换向器；切换开关
commutator segment 换向片
commutator-brush combination 换向器—电刷总线
compactness 紧凑性
comparator buffer (CB) 比较器缓冲器
compartment mode 库室模式
compensating 补偿；修正
compensating circuit 补偿电路
compensation 补偿
competence 能力
competence of the wind 风运能力
complementary energy method 余能法
complete flagging 完全旗状（Ⅳ级植物风力指示）
complete overhaul 全面翻修
complete stall 气流完全分离；完全滞止
complex 复杂的
complex damping 复合阻尼
complex differentiation 复变微分
complex impedance 复数阻抗
complex number 复数
complex periodic 复周期的
complex potential 复位势
complex setting 复合地形布置
complex terrain 复杂地形带
compliance 柔度
component 组成的；构成的
component of force 分力；力分量
component of turbulence 湍流分量

component of velocity 分速度
component vector 矢量的分量
composite blade 复合材料叶片
composite construction 复合材料结构
composite cooling 复式冷却
composite determinant 合成行列式
composite laminate structure 复合夹层结构
composite material 复合材料
composite metal 复合金属
composite plume 复合羽流
composite stress-strain relation 复合材料应力应变关系
composition backing 焊接垫板
compound 混合物；化合物；复合的；混合；配合
compound flap 组合襟翼
compound generator 复励发电机
compound lateral scale 复横向尺度
compound locomotive 复式机车
compound motion 复合运动
compound oscillation 复合振动
compounded 复励；混合的；复合
compressed 被压缩的
compressed air manometer 压缩空气式压力表
compressed air storage 压缩空气贮能
compressed air wind tunnel 增压风洞；压缩空气风洞
compressed state 压缩状态
compressibility 可压缩性
compressibility coefficient 压缩系数
compressibility correction 压缩性修正
compressibility effect 压缩性效应
compressibility factor 压缩系数；压缩因子
compressibility stall 激波失速
compressible 可压缩性；可压缩的

compressible aerodynamic 可压缩空气动力学
compressible airflow 可压缩气流
compressible flow 可压缩流动
compressible fluid 可压缩流体
compressible fluid flow 可压缩流体流动
compression 压缩；浓缩
compression fracture 压缩断裂
compression load 压缩负荷
compressive 压缩的
compressive strain 压缩应变
compressive strength 抗压强度
compressor driven tunnel 压气机驱动风洞
computational 计算的
computational aerodynamic 计算空气动力学
computational mesh 计算网格
computer 计算机
computer aided circuit design (CACD) 计算机辅助电路设计
computer aided design (CAD) 计算机辅助设计
computer aided design and drafting (CADD) 计算机辅助设计制图
computer aided design and engineering (CADE) 计算机辅助设计与工程
computer aided manufacturing (CAM) 计算机辅助制造
computer aided mathematics analysis (CAMA) 计算机辅助数学分析
computer code 计算机代码
computer dynamics 计算机动力学
computer numerical control (CNC) 计算机数字控制机床；数控机床
computer simulation technique 计算机模拟技术
concave airfoil surface 叶凹面

concentrated	集中的
concentrated load	集中荷载
concentrated smoke stack	烟突丛
concentrated vortex	集中旋涡；合成旋涡
concentrated vorticity	集中涡量
concentration	浓度
concentration boundary layer thickness	浓度边界层厚度
concentration contour	浓度分布廓线
concentration diffusion	浓差扩散
concentration factor	浓集因子
concentration gradient	浓度梯度
concentration index	浓度指数
concentration intensity	浓度
concentration isopleths	等浓度线
concentration limit	浓度极限
concentration method	浓集法
concentration of floating dust	飘尘浓度
concentration of pollutant	污染物浓度
concentration on ground level	地面浓度
concentration profile	浓度分布廓线
concentrator	集风装置；浓缩器；聚光器；聚能器；集热器
concentric chuck	同心卡盘
conceptual design	方案设计
concrete	混凝土；具体的
concrete drill	混凝土钻
concurrent	一致的；并发的
concurrent forces	共点力；汇交力
condensate drain	冷凝排水
condensate drain orifice	冷凝水排水孔
condensation	凝结；冷凝
condensation cloud	凝结云
condensation coefficient	凝结系数
condensation level	凝结面；凝结高度
condensation nuclei	凝结核
condensation trail	航迹云；凝结尾迹
condensed	浓缩的
condensed steam	冷凝蒸汽
condenser	凝汽器；冷凝器；电容器
condition	环境条件；工况
condition monitoring system (CMS)	状态监测系统
conditional equilibrium	条件平衡
conditional sampling	条件取样
conductance	电导
conducting ring	导电环；集电环
conducting wire	导线
conduction	传导
conduction band	导电区
conduction current	传导电流
conduction current density	传导电流密度
conduction loss	传导损耗
conductive earth	接地
conductivity	导电性
conductor	导体
conductor clamp	卡线钳
conductor holder	夹线器
conductor sway	电线摆动
conduit	管道，导管；沟渠
conduit box	导管分线匣
conduit entry	导管引入装置
conduit outlet	电线引出口
cone	圆锥；风向袋；（风轮）锥角
cone angle	锥角
cone bearing	锥形轴承
cone coupling	锥形联轴节
confidence	置信度
confidence interval	置信区间
confidence level	置信水平

configuration 构型；外形
confined 有限制的
confined vortex 约束涡
confined vortex wind machine 约束涡型风力机
confluence 汇合；群集
conformal 保角的
conformal maping 保角映射
conformal transformation 保角变换
conformance 顺应；一致
conformity 一致；符合
conformity testing 合格试验
confriction 摩擦力
congested area 人口稠密区
congestus 浓云
conical 圆锥的
conical nozzle 圆锥形喷嘴
conical roof 圆锥形屋顶
conical steel 锥形钢筒
conical vortex 锥形涡
coning 锥形；形成圆锥形；锥度
coning angle 锥度角
coning damping 锥旋阻尼
coning dihedral effect 圆锥二面角效应
coning hinge 锥旋铰链
coning plume 圆锥形羽流
conjugate power law 共轭幂定律
connecting 连接
connecting cable 接线电缆
connecting diagram 接线图
connecting flange 连接法兰
connection 连接
connection in parallel 并联
connection in series 串联
connector 连接器；连接头

consecutive points 相邻点
conservation 守恒
conservation area 自然保护区
conservation equation 守恒方程
conservation field of force 保守力场
conservation law 守恒定律
conservation of angular momentum 角动量守恒
conservation of charge 电荷守恒
conservation of circulation 环量守恒
conservation of energy 能量守恒
conservation of energy theorem 能量守恒定理
conservation of mass 物质守恒
conservation of mass energy 质量能量守恒
conservation of mass theorem 质量守恒定理
conservation of matter theorem 物质守恒定理
conservation of momentum 动量守恒
conservation of momentum theorem 动量守恒定理
conservation of Reynolds stress 雷诺应力守恒
conservation of vorticity 涡量守恒
conservation parameter 保守参数
conservation relation 守恒关系
consistency 结特性；一致性；稳定性
consistent grease 润滑脂
constant 常数
constant absolute vorticity trajectory (CAVT) 等绝对涡度轨迹
constant amplitude fluctuation 等幅脉动
constant chord blade 等截面叶片
constant current 直流
constant current wire anemometer 恒电流热

线风速计
constant flux layer 等通量层
constant frequency generator 恒频发电机
constant lapse rate layer 等逆减率层；不断递减率层
constant level balloon 定高气球
constant pitching moment point 等俯仰力矩点
constant power 恒定功率
constant power output region 恒定功率输出区
constant speed electrical generator 恒速发电机
constant speed fixed pitch blade 恒速定桨距叶片
constant speed operation 恒速运行
constant speed squirrel cage induction generator 恒速鼠笼式异步发电机
constant speed synchronous generator 恒速同步电机
constant speed wind turbine 恒速风电机组
constant term 常数项
constant temperature (CT) 等温
constant temperature wire anemometer 恒温热线风速计
constant torque 恒定转矩
constant TSR operation region 恒定TSR运行区
constant speed rotor 恒速风轮；恒速转子
constant stress layer 等应力层
constant velocity 定速；等速
constant voltage 恒压
constituent 成分
constitutional 本质的
constitutional detail 结构零件

constrained 拘泥的
constrained body 约束体
constrained condition 约束条件
constrained oscillation method 约束振动法
constrained oscillations 强迫振动
constrained vortex 约束涡
constraint 约束
constraint effect 约束效应；（风洞）边界效应
construction 施工；建筑物
construction details 建筑细部
construction work 施工工程
consumption penalty 消耗量增大
contact 触头；触点
contact(ing) area 接触面（积）
contact breaker (CB) 接触断路器
contact load 接触荷载
contact terminal 接触端点；触头
contact type cup anemometer 接触式风杯风速计
contact welding 接触焊
contactor 触头；接触器；触点；开关；断续器；电流接触器
container 集装箱；容器
container truck 集装箱货车
contaminant 污染物
contaminant loading 污染物负荷
contaminanted exhaust system 污染气体排放系统
contamination 污染；玷染
contamination accident 污染事故
contamination control 污染控制
contamination dose 污染剂量
contamination factor 污染系数
contamination hazard 污染危险
contamination zone 污染带

content 内容
continental 大陆的
continental air 大陆空气
continental climate 大陆性气候
continental wind 大陆风
continuity 连续性
continuity condition 连续条件
continuity equation 连续方程
continuity test 连续性实验
continuous blocks 连续块
continuous current 恒电流；直流
continuous diffusion 连续扩散
continuous dynamic fumigation 连续动态熏蒸
continuous flow 连续流
continuous function 连续函数
continuous hypersonic tunnel 持续高超音速风洞
continuous line source 连续点源
continuous loop 连续循环
continuous medium 连续介质
continuous model 连续介质模型
continuous operation 持续运行
continuous oscillation 等幅振荡
continuous power （最坏风况下）持续功率；连续功率
continuous rating 持续功率；连续功率；长期运转的定额值
continuous rating power 持续功率；连续功率
continuous source 连续源
continuous spectrum 连续谱
continuous strand mat 连续毡；连续玻璃毡；连续原丝毡
continuous variable 连续变量
continuous vibration 连续振动；等幅振动
continuous weld 连续焊缝
continuous wind tunnel 连续式风洞
continuously recording sensor 连续记录的传感器
continuously variable gearbox 无级变速器；连续可变变速器
continuous sampling device 连续取样装置
continuum 连续介质
continuum flow 连续介质流动
continuum model 连续介质模型
contour 轮廓；外形；等值线
contour correction （风洞）边界影响修正
contour line 等高线
contour map 等高线图
contour of profile 翼型外形
contour parameter 外形参数
contour plate 压型板；仿形样板
contraction 收缩
contraction cone 风洞收缩段；收缩圆锥
contraction section 风洞收缩段
contraflexure 反弯曲；反挠曲
contrail 凝结尾迹；航迹云
contrary 相反的
contrary current 逆流
contrary winds 逆风
contrate gear 端面齿轮
contrate gear pair 端面齿轮副
contrate wheel 端面齿轮
control 控制
control accuracy 控制准确度
control and monitoring system 控制和监控系统
control apparatus 控制电器
control board 操纵板
control cabinet 控制柜

control cable 控制电缆
control circuit 控制电路
control console 控制台
control cubicle 控制柜
control desk 控制台
control device 控制装置
control gear 控制设备；控制器；控制机构；标准齿轮；检验用齿轮
control loop 控制回路
control monitoring equipment 监控装置
control of air pollution 空气污染控制
control of walking 行走控制
control of wind 风害防治
control panel 控制面板
control point 控制点
control point adjustment 调节器
control room 控制室
control signal 控制信号
control surface 控制面
control system (for wind turbine) 风力机控制系统
control system type 控制系统类型
control valve 控制阀
control valve actuator 阀控传动机构
control variable 控制变量
control volume 控制体
control wiring 控制线路
controllability 可控制性；可操纵性
controllable 可控制的；可管理的
controllable orifice 可控阀；可控孔
controllable reactance 可控电抗
controlled start up 可控起动
controlled temperature pressure range wind tunnel 温度压力可调式风洞
controller 控制器

convection 对流；运流
convection cell 对流单体
convection circulation 对流性环流
convection cloud 对流云
convection coefficient 对流系数
convection cooling 对流冷却
convection current 对流气流
convection layer 对流层
convection loss 对流损耗
convection of air 空气对流
convection section 对流区
convective 对流的；传递性的
convective boundary layer 对流边界层
convective circulation 对流性环流
convective diffusion 对流扩散
convective diffusion equation 对流扩散方程
convective eddy 对流涡旋
convective element 对流元
convective equilibrium 对流平衡
convective flux 对流通量
convective heat loss 对流热损失
convective heat transfer 对流传热
convective instability (CI) 对流不稳定性
convective intensity 对流强度
convective internal boundary layer 对流内边界层
convective mixed layer 对流混合层
convective parcel 对流气块
convective regime 对流流型
convective region 对流区
convective stability 对流稳定度
convective storm 对流风暴
convective transfer 对流输送
convective turbulence 对流性湍流
convenience 便利

convenience receptacle 电源插座
conventional 传统的
conventional energy 常规能源
conventional power plant 常规发电厂；常规电站
conventional windmill 常规风车
convergence 收敛；(风洞) 收缩段
convergence angle 收缩角；收敛角
convergence current 收缩流
convergence error 收敛误差
convergence iterative procedure 收敛迭代法
convergent oscillation 减幅振动
conversion 转换；变换
conversion coefficient 转换系数
conversion efficiency 转换效率
conversion factor 转换因子；换算因子
converter 变流器；变频器
convertible circuit breaker (CCB) 可变换断路器
convex airfoil surface 叶凸面
convex flange 凸面法兰
coolant 冷却液
coolant apparatus 冷却液装置
coolant channel 冷却剂通道
coolant charging system 冷却剂灌注系统
coolant circuit 冷却剂回路
coolant concentrate 冷却液添加剂（用以降低冰点，提高沸点及防锈等）
coolant conditions 冷却剂条件
coolant fluid 冷却液流体
coolant inlet 冷却液进口
coolant temperature 冷却液温度；冷却介质温度
cooler 冷却器
cooler casing 冷却器外壳

cooler condenser 冷凝器
cooler performance 冷却器性能
cooling 冷却的
cooling agency 冷却液；切削液
cooling agent 冷却剂
cooling air baffle 冷却空气导流板
cooling down 降温
cooling effect 冷却效果
cooling efficiency 冷却效率
cooling element 冷却元件
cooling equipment 冷却设备
cooling equipment system 冷却设备系统
cooling facility 冷却设备
cooling fan (CF) 冷却风扇
cooling filling 淋水装置；冷却填料
cooling fin 散热片
cooling installation 冷却装置
cooling medium 冷却介质；冷却剂
cooling range 冷却温降；冷却幅度
cooling rate 冷却速度
cooling rib 散热片
cooling speed 冷却速度
cooling system 制冷系统
cooling tower 冷却塔
cooling tower plume 冷却塔羽流
cooling tower shell 冷却塔壳体
COP (coefficient of performance) 性能系数；(风轮) 效率
cooper loss 铜耗
cooper wool filter 紫铜毛滤清器
coordinate adjustment 坐标调整
coordinate conversion 坐标变换
coordinate system 坐标系
copper 铜
copper bar 铜棒

copper fill factor 铜填充因数
copper loss 铜耗
copper winding 铜线绕组
cording diagram 接线图
core 核心
core burst 涡核猝发
core honeycomb 蜂窝型材；蜂窝填料
core inductance 铁心电感
core layer 夹心层
core loss 铁心损耗
core sandwich 构架夹芯结构
core splitting 涡核破碎
Coriolis acceleration 科里奥利加速度
Coriolis effect 科里奥利效应
Coriolis force 科里奥利力
Coriolis parameter 科里奥利参数
corking （风洞）滞塞
corner 角落拐角处
corner effect 拐角效应
corner lamp 角灯
corner stream 角区流
corner vane （风洞）拐角导流片
corner vane cascade （风洞）拐角导流片栅
cornering 拐弯；横偏
cornering capability 抗横偏能力；转弯能力
cornering force 拐弯力
cornice 挑檐；雨水槽；檐条
corona 电晕
corona current 电晕电流
corona effect 电晕放电效应
corona resistance 耐电晕放电击穿性
corona ring 电晕环
corona voltage 电晕电压
corpuscular radiation 微粒辐射
corrasion 风蚀，动力侵蚀

corrected 校正的；修正的
correction 修正
correction efficient 修正系数
correction factor (CF) 修正系数
correction for blockage 阻塞修正
correction for buoyancy 浮力修正
correction for compressibility 压缩性修正
correction for Reynolds number 雷诺数修正
correction for scale effect 缩尺效应修正；尺度效应修正
correction for tunnel wall 洞壁干扰修正
correction value 修正值
corrective 矫正物
corrective maintenance 故障检修；设备保养
corrective term 修正项
correlation 相关
correlation analysis 相关分析
correlation coefficient 相关系数
correlation factor (CF) 修正系数
correlation function 相关函数
correlation length 相关长度
correlation matrix 相关矩阵
correlation model 相关性试验模型；相关模型
correlation scale 相关尺度
correlogram 相关图
corridor 通路
corrode 腐蚀；侵蚀
corrodent 可腐蚀性；腐蚀剂；腐蚀性物质；腐蚀的
corrosion 腐蚀
corrosion allowance 允许腐蚀度
corrosion by gas 气体腐蚀
corrosion control 防腐控制
corrosion control equipment 防腐控制设备
corrosion inhibitor 缓蚀剂；防腐蚀剂

corrosion of metals 金属腐蚀
corrosion preventive 防腐；防蚀
corrosion proof 防腐蚀；耐腐蚀
corrosion proof type 防腐型；抗腐蚀型
corrosion protection 防腐保护；防腐蚀；锈蚀防护
corrosion resistance coating 耐腐蚀涂层
corrosion resistance tests 耐腐试验
corrosion resistivity 耐腐蚀性
corrosive environment 易腐蚀环境
corrosiveness 腐蚀性；腐蚀作用
corrugated covering 波纹蒙皮
corrugation 褶皱
cosine transform 余弦变换
cosmic dust 宇宙尘
cospectrum 余谱；同相谱；共谱
cost effectiveness 成本效率
cost feasibility 成本可行性
cost per kilowatt hour of the electricity generated by WTGS 风电度电成本
cotter pin 开口销
Couette flow 库埃特流动
Counihan type vortex generator 库尼汉式旋涡发生器
countdown 倒计数
countdown warning 计数报警
counter (CNT) 计数器
counter force 反作用力
counter gradient flux 逆梯度通量
counter pressure 反压力
counter stream 逆流
counter torque 反扭矩
counter weight 平衡锤；配重；平衡铊
counterclockwise (CCW) 逆时针
counter flow 逆流
counter jet 反向射流；反向喷射
counter-rotating bladed wind machine 对转叶轮风力机
counter type cup anemometer 计数式风杯风速计
counting apparatus 计数管
couple 对
coupled 耦合的；连接的
coupled bending-torsion oscillation 弯扭耦合振动
coupled degree of freedom 耦合自由度
coupled flutter 耦合颤振
coupled frequency 耦合频率
coupled instability 耦合不稳定性
coupled oscillation 耦合振动
coupled resonance 耦合共振
coupler 车钩；耦合器；联结器
coupling 联轴器
coupling capacitor 结合电容
coupling coefficient 耦合系数
coupling jaw 联轴器爪
coupling transformer 耦合变压器
course angle 航向角；行车方向角
course sensibility 航向灵敏度
courtyard 院子；楼群院场
covariance 协方差；协变性
covariance matrix 协方差矩阵
covariant 协变量
covariant differentiation 协变微分
covariant divergence 协变散度
covariant tensor 协变张量
covariant theory 协变理论
covariant vector 协变矢量
cover strip 整流带；蒙皮条；覆盖条；防蚀镶片
covering 蒙皮

cowl 整流罩

cowl former 整流罩框架

cowling 整流罩

cowling mount 整流罩架

C-pillar C支柱

crack 裂缝；开裂

crack detection 裂缝检查

crack detector 裂痕探测器

crack growth rate 裂缝扩展速度

crack initiation 裂纹萌生

cracking resistance 抗断裂能力

crane 起重机

crane load (CL) 吊车

crank 不稳定的；摇晃的；曲柄

crank case 曲柄轴箱

crank shaft 曲轴；机轴

crank shaft vibration damper 曲轴减震器

crank ship 易倾船

crash 撞碎；坠毁

crash pad 防震垫

crash value 峰值

crash voltage 峰值电压

crate 柳条箱

credible accident 可信事故

creep 蠕变

creep deformation 蠕变变形

creep of snow particle 雪粒蠕动

crescent dune 新月型沙丘

crest 波峰；脊顶；峰值

crest curve 凸形曲线

crest load 尖峰负荷

crest value 峰值；最大值

crevice corrosion 裂缝腐蚀

criterion 判据；准则

critical 临界的；关键的

critical angle of attack 临界攻角

critical area 关键部位

critical behavior 临界状态

critical clearing time 极限切除时间

critical coefficient 临界系数

critical concentration 临界浓度

critical condition 临界条件；临界状态

critical current 临界流

critical damping 临界阻尼

critical design case 临界设计情况

critical divergence wind speed 临界发散风速

critical emission rate 临界排放率

critical equation 临界方程

critical flow regime 临界流动状况

critical flutter wind speed 临界震颤风速

critical frequency 临界频率

critical galloping wind speed 临界驰振风速

critical group 关键居民组；关键人群组

critical heat flux (CHF) 临界热通量

critical instability 临界不稳定性

critical inversion layer 临界逆温层

critical length of grade 临界坡道长度

critical load distribution 临界载荷分布

critical nuclide 关键核素

critical overturning wind speed 临界倾覆风速

critical point 临界点

critical pressure gradient 临界压力梯度

critical regime 临界状态

critical Reynolds number 临界雷诺数

critical Richardson number 临界理查森数

critical slant angle 临界倾角

critical stall speed 临界失速速度

critical value 临界值

critical zone 危险地带；临界地带

criticality accident 临界事故
crop climate 作物气候
cross 交叉；十字
cross beam 横梁
cross bending 横向弯曲
cross cospectrum 交叉余谱
cross derivative 交叉导数
cross flow 横向流动；横流向；交叉流动
cross headwind 侧逆风
cross helical gear 螺旋齿轮
cross mark 十字标记
cross member reinforcement 横梁加强
cross product 矢量积
cross rib 横肋
cross slotted screw 十字槽螺钉
cross ventilation 对流通风；穿堂风
cross wind 横风；横风向的
cross wind displacement 横风位移
cross arm 横臂
cross correlation coefficient 互相关系数
cross correlation function 互相关函数
cross country ability 越野性能
cross coupling 交叉耦合
cross covariance 互协方差
cross flow wind turbine 贯流风机
cross peen hammer 横头锤
cross section 横断面；横切面；截面
cross section area 截面积
cross sectional shape 横截面形状
cross shear 横剪切
cross spectral density 交叉谱密度
cross stability derivative 交叉稳定导数
cross variance function 互方差函数
crosshead 小标题；子题；十字头；丁字头
cross wind 横风向的；横风

cross wind buffeting 横风抖振
cross wind diffusion 横风扩散
cross wind direction 横风向
cross wind galloping 横风驰振
cross wind gust 横向阵风
cross wind installation 横风装置
cross wind loading 横风荷载
cross wind paddles wind machine 横风桨板式风力机
cross wind stability 横风稳定性
cross wind test 横风试验
cross wind vibration 横风振动
cross wind wake force 横风尾流力
cross wind-axis wind machine 横风轴风力机
crosswise 斜地；成十字状地；交叉地
crowbar 撬棍；铁翘
crowbar protection 过压保护装置
crowbar protection circuit 过压保护电路
crown 树冠；路拱；拱顶
crown block 定滑轮
crown cornice 大屋檐
crown cover degree 树冠盖度
crown density 树冠郁闭度
CRT (cathode-ray tube) 阴极射线管
crude data 原始数据
cryogenic (wind) tunnel 低温风洞
cryogenic spill 低温溢出
cryptoclimate 室内小气候
crystal 结晶；晶体
crystal rectifier 晶体整流管
CS (cirrostratus) 卷层云
CS fib (cirrostratus fibratus) 毛卷层云
CS fil (cirrostratus filosus) 毛卷层云
CSI (current source inverter) 电流源逆变器
CS neb (cirrostratus nebulosus) 薄暮卷层云

CT (constant temperature) 等温
CT (critical temperature) 临界温度
CT (current transformer) 电流互感器
CU con (cumulus congestus) 浓积云
CU fra (cumulus fractus) 碎积云
CU hum (cumulus humilis) 淡积云
CU med (cumulus mediocris) 中积云
cube factor 风速立方因子
cubic building 立方体建筑
cubic equation 三次方程
cubic root mean wind speed 立方根平均风速
cubic strain 容积应变
cubic term 三次项
cubical content(s) 容积量
cubical dilatation 体积膨胀
CUF (cumuliform) 积状云
cumulative distribution function 累计分布函数
cumulative dose 累积剂量
cumulative error 累积误差
cumulative frequency distribution 累积频率分布
cumulative probability 累积概率
cumulative vorticity 累积涡量
cumulatively compounded motor 积复励电动机
cumuliform (CUF) 积状云
cumulo(-)nimbus (CB) 积雨云
cumulo(-)nimbus calvue (CB cal) 秃积雨云
cumulo(-)nimbus capillatus (CB cap) 鬃状积雨云
cumulo(-)nimbus incus (CB inc) 砧状积雨云
cumulo(-)nimbus mammatus (CB mam) 悬球状积雨云
cumulus 积云

cumulus congestus (CU con) 浓积云
cumulus fractus (CU fra) 碎积云
cumulus humilis (CU hum) 淡积云
cumulus mediocris (CU med) 中积云
cup 杯子
cup anemometer 转杯风速计
cup contact anemometer 电接风杯风速计
cup counter anemometer 计数风杯风速计
cup generator anemometer 转杯磁感风速计
cup type anemometer 杯状风速计
cup valve 杯形阀
cupped 杯形的
cupped wind machine 杯式风力机
curb 缘饰
Curie 居里
curl 旋度
curl up (使)卷
curling ball 旋转球
current 电流
current carrying capacity 载流容量
current density 电流密度
current flow 电流
current intensity 电流强度
current limiting 电流限制
current limiting reactor 限流电抗器
current pulse 电流脉冲
current rate 电流强度；气流强度；流速
current repair 小修
current rush 电流骤增
current source inverter (CSI) 电流源逆变器
current supply 电源
current transformer 电流互感器
current transient 暂态电流
current voltage diagram 伏安特性曲线
curtain wall 幕墙；护墙

curvature function of airfoil 翼型弯度函数
curvature of trajectory 轨道曲率
curve peak 曲线顶点
curve superelevation 曲线超高
curved 倒弧角；弯曲的
curved bridge 曲线桥
curved rail 弯轨
curved roof 曲面屋顶
curved stroke 曲线球
curved tooth bevel gear 曲面齿锥齿轮
curved tooth gear coupling 弧形齿轮联轴器
curved viaduct 曲线高架桥
curved wire gauze screen 弯线网筛
curvilinear equation 曲线方程
curvilinear tunnel 曲线隧道
cushion 垫子
cushion material 缓冲材料
cushioning 缓冲器
cushioning effect 减震作用
cushioning material 缓冲垫料
cusp 尖头；尖端
cusp of trailing edge 后缘尖端
custom 定制；习惯；风俗；海关
custom designed software 用户设计软件
customised 依照客户要求而具体制造的
customised wireless SCADA system 定制无线监控系统
cut 切割
cut grass 已割的草地
cut in 干预；插入；加塞
cut in wind speed 最小出功风速；切入风速
cut off frequency 截止频率
cut off plate 节流板；挡板
cut off wind speed 最大出功风速
cut out fuse 断流保险丝

cut out wind speed 最大出功风速；切出风速；截止风速
cutout line 切割线
cutout spar 切口（翼）梁
cutter 切削齿；切割机；切割者
cutter change factor 齿轮刀具变位系数
cutter compensation 刀具补偿
cutting face 切削面
cutting line 切割线
cutting performance 切割性能
CW (clockwise) 顺时针
cybernation 自动控制
cybernetics 控制论
cycle motion 周期运动
cycle of oscillation 振荡周期
cyclic 循环的
cyclic creep and stress rupture 循环蠕变和应力断裂
cyclic frequency 角频率
cyclic loading 交变载荷
cyclic pitch 周变桨距
cyclic strength 疲劳强度
cyclic stress 周期应力
cyclic stress limit 周期性应力极限
cyclic twist 循环扭曲
cyclic twist stress 循环扭曲应力
cyclic variation 周期性变化
cyclically 循环的
cyclically alternating vortex 周期性交替涡
cyclolysis 气旋的减弱或消失
cyclone 气旋
cyclone damage 气旋破坏
cyclone path 气旋路径
cyclone trace 气旋轨迹；气旋路径
cyclonic 气旋的；飓风的

cyclonic circulation 气旋式环流
cyclonic rain 气旋性雨
cyclonic storm 气旋(性)风暴
cyclonic vorticity 气旋涡量
cyclonic whirl 气旋型涡流
cyclonic wind 气旋风
cyclostrophic balance 旋衡
cyclostrophic wind 旋衡风
cylindrical 圆柱形的
cylindrical bearing 滚柱轴承
cylindrical coordinate 柱面坐标
cylindrical gear 圆柱齿轮

cylindrical nozzle 圆柱形喷嘴
cylindrical polar coordinate 柱面极坐标
cylindrical puff 圆柱形喷团
cylindrical roller bearing 圆柱滚子轴承
cylindrical roller thrust bearing 圆筒形滚柱推力轴承
cylindrical rotor 隐极转子
cylindrical rotor generator 圆柱形转子发电机；隐极发电机
cylindrical spiral 螺旋线
cylindrical wave 柱面波

现代英汉风力发电工程

A Modern English-Chinese Dictionary of Wind Power Engineering

D

DA (direct action) 直接作用
D'Alembert's paradox 达朗贝尔详谬
daily 日常的
daily amplitude 日变幅
daily extremes 日极端值
daily inspection 日常检验
daily load 日负荷；昼夜负荷
daily load curve 日负荷曲线
daily load factor 日负荷率
daily maintenance task 每日维修保养工作
daily maximum temperature 日最高温度
daily mean 日平均值
daily mean temperature 日平均温度
daily minimum temperature 日最低温度
daily operation costs 日运营成本
daily precipitation amount 日降水量
daily range 日较差
daily range of temperature 日温度变化范围
daily run 日运转；日运转期
daily task system of maintenance 日维修工作制度
daily temperature fluctuation 日温变化；日温波动
daily variation 日变动
daily work 每日工作

Dalton's law 道尔顿定律
dam 坝；挡风板
damage 损失
damage by dragging 被风吹坏
damage by storm 风暴损失
damage by wind 风灾破坏
damage capability 破坏能力
damage control 修复损害控制；破坏性控制
damage index 损失指数
damage oscillation 阻尼振动
damage ratio 损失比
damage repair 损坏修理
damaged 被损害的
damaged beyond repair 损坏难修
damaging 有破坏性的
damaging stress 破坏应力
damp 湿气；润湿；阻尼
damp air 潮湿空气
damper (DMPR) 阻尼器
damp proof 防潮；防湿
damp proof course 防潮层；防水层
damp proof material 防潮材料
damp proofing 防潮
damp proofness 耐湿性
damped 阻尼的；衰减的

damped oscillations 阻尼振动
damper (DMPR) 阻尼器；减振器；阻尼器；挡板；防振锤
damper brake 制动闸
damper by friction of liquid 液体摩擦减震
damper capacity 吸震能力
damper coefficient 阻尼系数
damper cylinder 减震筒
damper damping action 减震作用；阻尼作用
damper device 减震装置
damper of friction 摩擦式减震器
damper system 阻尼系统
damper valve 减震阀
damper winding 阻尼绕组；阻尼线圈
damping 阻尼；衰减；减震
damping apparatus 减震设备；阻尼器
damping arrangement 减震装置
damping baffle 缓冲隔板
damping blanket 防振垫
damping capacity 减震能力
damping channel 减震系统
damping characteristic 阻尼特性
damping coefficient 阻尼系数
damping constant 衰减常数；阻尼常数
damping controller 阻尼控制器
damping device 减震装置
damping dissipation 阻尼损耗
damping driven oscillation 阻尼驱动振动；负阻尼振动
damping effect 缓冲作用
damping factor 阻尼因子
damping material 减震材料；隔声材料
damping pad 减震垫
damping parameter 阻尼系数
damping ratio 阻尼系数

damping regime 阻尼状态
damping resistance 衰减阻力
damping rubber 橡皮减震器
damping screen 阻尼网
damping spring 减震弹簧
damping term 阻尼项
damping washer 减震垫圈
Danish concept 丹麦机型；AO型失速风电机组；定桨距齿轮驱动风电机组
Darboux vector 达布矢量
Darrieus machine 达里厄型风力机
Darrieus type rotor 达里厄型风轮
dashpot 减震器；阻尼器
dashpot relay 缓冲器
data 数据
data analysis (DATAN) 数据分析
data bank 资料储存系统；资料库
data base 数据库
data circuit 数据电路
data collection system 数据汇集系统
data correction 数据修正
data fetch 取数据
data flow 数据流
data flow chart 数据流程图
data flow detection 数据流检测
data flow machine 资料流程电脑
data form 数据记录表
data gathering 数据采集
data handling 数据处理
data integrity 数据完整性
data logger 数据记录器
data logging 数据记录
DATAN (data analysis) 数据分析
data plate 铭牌
data pool 数据库

data processing 数据处理
data reading 数据读出
data reduction 数据简化
data reliability 数据可靠性
data set for power performance measurement extrapolated 数据组功率特性测试
data smoothing 数据平滑；数据信号平滑；数据修匀
data terminal equipment 数据终端设备
data transmission 数据传输
date set 数据组
date set for power performance measurement 测试功率特性的数据组
datum 数据；已知条件；基点
datum face 基准面
datum level 基准；水准
datum line 基准线
datum plane 基准面
datum point 基准点
datum state 基准状态
datum surface 基准面
daughter activity 子体放射性
day breeze 日风
day-night effect 昼夜效应
daytime mixing layer 白昼混合
DC (direct current) 直流
DC field coil 直流磁场线圈
DC generator 直流发电机
DC link 直流链；直流母线
DC link bus 直流母线
DC link capacitor 直流链电容器
DC link converter 直流环节变流器；直流母线变流器
DC link voltage 直流母线电压
DC machine 直流电机

DC motor 直流电动机
DC power source 直流电源
DC supply 直流电源
DC-DC converter 直流—直流转换器
DCI (ductile cast iron) 球墨铸铁
DDC (direct digital control) 直接数字控制
DDC system 直接数字控制系统
DDRS (digital data recording system) 数字数据记录系统
dead air 闭塞空气；停滞空气；静止空气；(扰动气流的) 停滞区
dead air pocket 滞流区
dead band 死区
dead block 缓冲板
dead earth 完全接地；直通地；固定接地
dead load 静态载荷；固定负载
dead load weight 固定负载重量
dead main 空载线
dead region 死区
dead sounding 隔声层
dead space 死空间；死角；死舱位；静区；无信号区；阴影区
dead time 延迟时间；死区
dead water region 死水区
dead weight 静负载；固定负载
dead wind 逆风
dead wood 呆木；死木头
dead work 非直接性生产工作
dead zone 死角区；死区；盲区；静区
deadening 隔音材料
deadening effect 缓冲作用
dead-time 死区时间；空载时间
dead-time compensation 死区时间补偿
dead-time delay 空载时延
dead-time effect 死区效应

dead weight deflection 自重挠度
dead wind 逆风
dead zone effect 死区效应
debris 碎片；垃圾
debugging 排除故障
debugging phase 排除故障阶段
decay 衰减；衰变
decay constant 衰减常数
decay function 衰减函数
decay parameter 衰减参数
decay product 衰变产物
decay rate 衰减率
decaying oscillation 衰减振动
decaying wave 衰减流
deceleration 减速；制动
decelerator 减速器；缓冲装置；延时器
decelerometer 减速计
decentralized wind energy system 分散式风能系统
decibel 分贝
decode 译码
decontamination 净化（作用）
decontamination index 净化指数
dedendum 齿根高
dedendum angle 齿根角
dedendum circle 齿根圆
dedendum cone 圆锥齿轮啮合
dedendum line 齿根线
dedendum line of contact 齿根接触线
deep stall 严重失速
deep stall region 深度失速区
deepwater wave 深水波
defect 缺点；缺陷
definite 一定的；确切的
definite integral 定积分
definite proportion 定比
definite quality 定量
deflation erosion 风蚀
deflation opening 放气口
deflation vent 放气口
deflection 偏向；挠曲；偏差
deflection anemometer 偏转风速计
deflection angle 偏转角
deflection force of earth rotation 地球自转偏向力
deflection gauge 偏转计；挠度计
deflection magnitude 偏离幅度
deflection rod 偏转杆
deflection spectrum 偏转道；挠度谱
deflectional stiffness 抗弯刚度
deflector 导向装置；导流板
deformation 变形；畸变
deformation analysis 形变分析
deformation drag 变形阻力
deformation stress 变形应力
deformation theory of plasticity 塑性变形理论
deformation under load 载荷变形
degenerative 退化的；变质的
degenerative feedback 负反馈
degradation 退化；降落
degree 程度；等级
degree day 度日
degree of accuracy 精确度
degree of adaptability 配合度
degree of balance 平衡度
degree of confidence 置信度
degree of consistency 均匀度
degree of contamination 污染度
degree of curvature 曲率；弯度

degree of discomfort 不舒适度
degree of dispersion 弥散度
degree of eccentricity 偏心度
degree of freedom (DOF) 自由度
degree of hardness 硬度
degree of irregularity 不均衡度；不规则度
degree of moisture 湿度
degree of pollution 污染程度
degree of protection 防护等级
degree of regulation 调整精度；调准度
degree of reliability 可靠度
degree of safety 安全度
degree of saturation 饱和度
degree of swelling 膨胀量
degree of turn 转角
degree of twist 偏扭度
degree of unbalance 不平衡度
degree of wall porosity 壁开孔度
degree of ware 磨损度
degree of weathering 风化度
degree of wind sensibility 风敏感度
degree stability 稳定度
deicer 防冰器
delay 延迟；延缓；抑制
delay action 延迟作用
delay action push button 延时动作按钮
delay action relay 延时动作继电器
delay action switch 延时动作开关
delay constant 延时常数
delay control 延时控制
delay device 延时装置
delay element 延时元件
delay errors 延时误差
delay recorder 延时记录器
delay relay 延时继电器

delay system 延时系统
delay time 延时时间
delay time switch 延时时间开关
delay timer 延时计时器
delay unit 延时元件；延时装置
delay valve 延时阀
delaying separation 延迟分离
delicate adjusting 精调
delivered 递送
delivered power 输出功率
delivery 交付
delivery van 送货车
delivery work 输出功
delta 三角洲
delta connection 三角形联结
delta wing vortex 三角翼涡
demagnetization 去磁；退磁
demand 需求（量）
demodulation 解调
Den Hartog stability criterion 德哈托稳定判据
dense gas spill 浓气体溢出
dense plume 浓羽流
dense vapour spill 浓蒸汽溢出
densely built up city 建筑密集城市
densimeter 密度计
densimetric Froude number 密度弗劳德数
densitometer 光密度计；显像密度计；比重计
density 密度
density current 异重流
density diffusion 速度扩散率
density field 密度场
density gradient 密度梯度
density of air 空气密度
density of canopy 林冠郁闭度

density of charge 电荷密度
density of energy 能量密度
density of field 场强密度
density of filling gas 充气密度
density of load 载荷强度
density of momentum 动量密度
density of roughness element 粗糙元分布密度
density perturbation 密度扰动
density profile 密度廓线
density ratio 密度比
density resistance 密度阻力
density scale 密度刻度
density sensor 密度传感器
density slicing 密度分割
density spectra 密度谱
density spread 密度分布
density step tablet 密度分级片
density stratification 密度层结；密度分层
density testing 密度检验
density transducer 密度传感器
density tunnel 高压风洞；高密度风洞
density wave 密度波
density corrected strength 比强度；密度修正强度
density corrected ultimate tensile strength 比极限抗拉强度；密度修正极限抗拉强度
deoscillator 减震器
Department of Energy (DOE) （美国）能源部
departure 偏离；偏差
departure resistance 气流分离阻力
dependence 依赖；依靠
deposit gauge 沉积计；落灰计；降尘测定器
deposited drift area 沉积漂移区

deposition 沉积物
deposition coefficient 沉积系数；沉降系数
deposition flux 沉积通量
deposition region 沉积区
deposition surface 沉积表面
deposition velocity 沉积速度
depreciation 折旧；贬值
depreciation cost 折旧费
depreciation factor 折旧率；折旧因子
depreciation funds 折旧基金
depression 降压；低气压；洼地
deprivation 剥夺；损失
depth 深度
depth dose 深部剂量
depth of camber 弧线高度
depth of cutting 切削深度
depth of penetration 渗透深度
depth of tooth 齿高
derivation 引出；来历
derivative 衍生物；派生物
derivative coefficient 诱导系数
derivative discontinue 导数不连续性
derivative measurement 导数测量
derived air concentration 导出空气密度
derived limit 推定限值
desalination 盐水淡化
descending air 下沉空气
desert 沙漠
desert belt 沙漠带
desert climate 沙漠气候
desert deposit 沙漠沉积
desert devil 沙卷风
desert pavement 荒漠覆盖层；沙漠砾石表层
desert shelterbelt 沙漠防护林带
design 设计

design accuracy	设计精度
design activity	设计机关；设计组织
design aerodynamics	空气动力学设计
design assembly	设计装配图
design base accident	设计基准事故
design basis	设计基础
design concept	设计思想
design conditions	设计条件
design data	设计数据
design drawing	设计图
design feature	设计特点
design guide	设计指南
design instruction	设计任务书
design lift coefficient	设计升力系数
design limit	设计极限
design load	设计载荷
design load case	设计载荷工况
design manual (DM)	设计手册；制图手册
design of product	产品设计
design optimization	设计优化
design paper	设计文件
design parameter	设计参数
design patent	设计专利
design pay load	设计有效载荷
design performance	设计性能
design power	设计功率
design power efficiency	设计功率系数
design pressure coefficient	设计压力系数
design procedure	设计程序
design rating	设计额定量
design reference	设计参考
design requirement	设计要求
design rotor annual mean speed	设计风轮年平均转速
design rotor overspeed	设计风轮超速
design rule	设计规则
design scheme	设计方案
design service life	设计使用期限
design situation	设计工况；设计和安全参数
design specification	设计规范
design standard (DS)	设计标准
design step	设计步骤
design storm	设计风暴
design system	设计系统
design tip speed ratio	设计尖速比
design variable	设计变量
design wind load	设计风载
design wind speed	设计风速
designed waterline (DWL)	设计水线
desk fan	台(式风)扇
despatch	派遣
despatch centre	调度中心
destabilization	失稳，不稳定
destroy	破坏
destruction	破坏
destruction of insulation	绝缘损坏
destruction of soil	土壤破坏
destruction test	破坏实验
destructive	破坏的；破坏性的
destructive experiment	破坏实验
destructive malfunction	破坏性故障
destructive power	破坏力
destructive test	破坏性实验
destructive vibration	破坏性振动
detach	脱离；脱体
detached superstructure	独立上层建筑
detachment	分离；拆开
detachment of vortices	漩涡脱落
details	细部；细节
detection	侦查；探测

detection efficiency 探测效率
detection limit 探测极限
detection sensitivity 探测灵敏度
detection threshold 探测阈
detector 探测器
detector of defects 探伤仪
detector probe 探头
detent 止动装置
detent pin 定位销
detent plate 制动器板
detent plug 止动销
deterioration 恶化；退化；损坏；磨损
deterioration of weather 天气变坏
determinant 行列式
determinant calculation 行列式计算
determinant of matrix 矩阵行列式
determinant rank 行列式秩
determination of coordinate 坐标确定
determination of output 功率的测定
deterministic 确定性的
deterministic fluctuating wind speed 主脉动风速
deterministic gust 主阵风
deterministic variable 决定性变量
detuning 解调；去谐；解谐
developed 发达的；成熟的
developed blade area 叶片展开面积
developed power 发出功率
developed surface flaw 发展的表面缺陷
development 发展
development running 研制性运行
development stage 研制阶段
development test 研究实验
development testing 修整试验；开发测试；发展测试

development type 研制样品；试验样品
development work 试制工作
developmental length of snow drift 吹雪发育长度；雪漂移的发展长度
deviating 偏离
deviating force 偏转力；偏向力
deviation 偏差；偏离
deviation angle 偏角
deviatoric stress 偏应力
device 装置；策略
devil 尘暴；风暴
dew 露
dew point 露点
dew point hygrometer 露点湿度计
dew point spread 露点差
dew point temperature 露点温度
dewatering 排水
DF(dissipation factor) 损耗因素
DFIG (double fed induction generator) 双馈感应发电机；双馈异步发电机
DFSG (double fed synchronous generator) 双馈同步发电机
diagonal 对角线
diagonal cracking 对角裂缝
diagonal matrix 对角阵
diagonal triangle 对边三角形
diagram 图
diagram of curves 曲线图
diagram of gears 齿轮传动图
diagram of strains 应变图
diagram of work 示功图
diagrammatic drawing 草图
diagrammatic sketch 示意图
diagrammatic(al) 图解的
diagrammatic(al) chart 示意图

dial 转盘；刻度盘
DIAL (differential absorption lidar) 差异吸收光达
dial gauge 针盘量规；刻度盘；刻度计；指示器；测微仪；测微表
diameter 直径
diametral flow 径向流动
diamond penetrator hardness 维氏硬度
die 冲模；钢模
die cushion 模具缓冲器
die welding 模焊
dielectric test 介质试验
dielectric tip 绝缘材料翼尖
difference analogue 差别类比
difference equation 差分方程
difference method 差分法
differential 微分；差别；微分的；差别的
differential absorption 差异吸收
differential absorption cross section 差异吸收截面
differential absorption lidar (DIAL) 差异吸收光达
differential absorption technique 差异吸收技术
differential advection 差异平流
differential analysis 差值分析
differential and integral calculus 微积分
differential calculus 微分
differential chart 变差图
differential coefficient 微分系数
differential equation 微分方程
differential expression 微分式
differential heating 差异加热
differential kinematics 微分运动学
differential operator 微分算子

differential pressure 压差
differential pressure alarm 差压报警
differential quotient 微商
differential reflectivity factor 差异反射因子
differential thermal advection 差异热平流
differential thermal analysis 差热分析
differential thermometer 示差温度计
differentials of higher order 高阶微分
differentiation 微分；变异；分化
diffluence 分流
diffluent thermal ridge 分流温度脊
diffluent thermal trough 分流温度槽
diffluent trough 分流槽
diffracted ray 绕射线
diffracted wave 绕射波
diffraction 绕射
diffraction fringe 绕射条纹
diffraction pattern 绕射型
diffraction phenomenon 绕射现象
diffraction region 绕射区
diffraction spectrum 绕射谱
diffraction zone 绕射带
diffuse 扩射；漫射
diffuse boundary 扩散界面
diffuse field 漫射场
diffuse front 扩散锋
diffuse illumination 漫射照明
diffuse incident intensity 漫射入射强度
diffuse intensity 漫射强度
diffuse light 漫射光
diffuse radiation 漫辐射
diffuse reflection 漫反射
diffuse reflector 漫反射体
diffuse scattering 漫散射
diffuse sky radiation 天空漫辐射

diffuse skylight 漫射天光

diffuse solar radiation 太阳漫射

diffuser （风洞）扩散段；扩风器；扩散器

diffuser loss 扩散段损失

diffusion 扩散；漫射

diffusion bonding 扩散结合；扩散黏结

diffusion capacity 扩散能量

diffusion category 扩散类型

diffusion chamber 扩散云室

diffusion cloud 扩散云

diffusion coefficient 扩散系数

diffusion diagram 扩散图

diffusion equation 扩散方程

diffusion factor 扩散系数

diffusion field 扩散场

diffusion hygrometer 扩散湿度计

diffusion layer 扩散层

diffusion level 扩散高度

diffusion meanfree path 扩散平均自由程

diffusion model 扩散模式

diffusion of pollutants 污染物扩散

diffusion of smoke 烟扩散

diffusion parameter 扩散参数

diffusion theory 扩散理论

diffusion time 扩散时间

diffusion velocity 扩散速度

diffusiophoresis 扩散电泳

diffusiophoretic force 扩散迁移力；散泳力

diffusiophoretic velocity 散泳速度

diffusisphere 扩散层

diffusive equilibrium 扩散平衡

diffusive force 扩散力

diffusivity 扩散系数；扩散性

digital 数字

digital anemometer 数字式风速仪

digital clock 数字钟

digital control 数字控制

digital control system 数字式控制系统

digital controller 数字控制器

digital correlator 数字相关仪

digital data acquisition 数字数据采集

digital data recording system (DDRS) 数字数据记录系统

digital display 数字显示

digital filter 数字滤波器

digital image processing 数字图像处理

digital indicator 数字式指示器

digital input terminal 数字量输入端子

digital method 数字方法

digital output terminal 数字量输出端子

digital sensor 数字传感器

digital signal 数字信号

digital to analogue conversion 模数转换；D/A 转换

digitisation rate 数字转换速度

digitizing 数字化

digitizing tablet 数字面板

dihedral angle 二面角；双面角

dike 坝；堤；沟

dilatation 扩张

dilatation coefficient 膨胀系数

dilatation constant 膨胀常数

dilatation joint 膨胀缝

dilutability 稀释度

dilute concentration 稀释浓度

dilution 稀释（度）

dilution coefficient 稀释系数

dilution discharge 稀释排放

dilution factor 稀释率

dilution of pollution 污染稀释

dimension (DMN) 尺寸；维；量纲
dimension analysis 量纲分析
dimension chart 轮廓尺寸图
dimension equation 量纲方程
dimension figure 尺寸图
dimension relation 量纲关系
dimension scale 尺寸比例
dimensional 空间的；尺寸的
dimensional accuracy 尺寸精度
dimensional analysis 量纲分析
dimensional change 尺寸变化
dimensional characteristic 尺寸特性
dimensional equation 量纲方程
dimensional homogeneity 尺度均一性
dimensional interchangeability 尺寸互换性
dimensional method 量纲法
dimensional precision 尺寸精度
dimensional relation 量纲关系
dimensional theory 量纲理论
dimensional tolerance 尺寸公差
dimensionless 无量纲的
dimensionless coefficient 无量纲系数
dimensionless factor 无量纲系数
dimensionless parameter 无量纲参数
dimensionless quantity 无量纲值
dimensionless representation 无量纲表示法
dimensionless unit 无量纲单位
dint 凹痕；凹坑
diode 二极管
diode alternating current switch (DIAC) 双向触发二极管；二极管交流开关
diode bridge rectifier 二极管桥整流器
diode rectification 二极管整流
diode rectifier 二极管整流器
dip 下沉；下降

dip angle 倾角；俯角；磁倾角
dip moment 倾斜力矩
dipole 偶极子
dipping 浸渍；蘸
dipping and heaving 升沉运动
direct 直接的
direct acting 直接作用的
direct action (DA) 直接作用
direct action tunnel 直流式风洞
direct axis 直轴
direct axis component 直轴分量
direct axis reactance 主轴电抗
direct axis synchronous reactance 直轴同步电抗
direct axis transient time constant 直轴瞬变时间常数
direct bearing 导向轴承
direct clutch 直接离合器
direct component 直流部分
direct condenser 回流冷凝器
direct coupled motor drive 电动机直接传动
direct current (DC) 直流
direct current machine 直流电机
direct current motor 直流电动机
direct determination 直接测定法
direct digital control (DDC) 直接数字控制
direct drive 直接传动
direct drive generator 直驱发电机
direct drive permanent magnet generator 直接驱动永磁发电机
direct drive wind turbine 直接驱动风电机组
direct energy 定向能量
direct geared 齿轮直接传动
direct grid connection 直接并网
direct liquid cooling 直接液冷

direct load (DL) 直接载荷
direct quenching 直接淬火
direct reading instrument 直读式仪表
direct shock absorber 直接式减震器
direct short 短路
direct solar radiation 太阳直接辐射
direct solar radiation intensity 太阳直接辐射强度
direct stress 正应力
direct supply 直流电源
direct visual observation 直接目测
direct voltage 直流电压
directing 指导；导演
directing line 导线
directing property 定向性
direction 方向；指导
direction effect 方向效应
direction fluctuation 方向脉动
direction of feed 进给方向
directional air flow 定向气流
directional control 方向操纵性
directional radiation 定向辐射
directional response 方向反应
directional stability 方向稳定性
directive action 定向动作
directivity 指向性；方向性
dirt 污垢；尘土
dirt cloud 尘云
dirty environment 污秽环境
disabled 失去能力的
disabled time 不能工作时间
disaster 灾害
disc 桨盘；圆盘
disc area 桨盘面积
disc buoy （海上风车的）圆盘式浮台

disc clutch 盘形离合器
disc flow ratio 桨盘入流系数
disc load 桨盘载荷
disc plane 桨盘平面
disc ratio 叶盘面积比
discard 抛弃
discard electrode 焊条头
discharge 排放；卸货；放电
discharge condition 排放条件
discharge device 避电器
discharge fan 排风扇；抽风扇
discharge flue 排烟道；排气道
discharge point 排放点
discharge rate 排放率
discharge ratio 排放系数
discharge standard 排放标准
discharge velocity 排放速度
discharger 避电器
discharging rod 避雷针
discipline 学科
discomfort 不舒适（性）
discomfort index 不舒适指数
discomfort map 不舒适区图
discomfort parameter 不舒适参数
discomfort pattern 停止条件
discomfort threshold 不舒适阈
discomfort zone 不舒适区
disconnection 解列；断开；断路；切断
disconnection fault 断电故障
discontinue 停止
discontinue condition 不连续
discontinue of material 材料不均匀性
discontinue point 不连续点
discontinue stress 不连续应力
discontinue surface 不连续面

discontinuity	不连续性
discontinuous	不连续的；间断的
discontinuous load	突变载荷
discoupling	脱钩；切断
discrepancy	偏差；矛盾
discrete	离散的；分立元件；不连续
discrete excitation	离散激励
discrete gust	离散阵风
discrete solution	离散解
discrete source	离散源
discrete spectrum	离散谱
discrete value	不连续值
discrete vortex	散立旋涡；无联系的旋涡
disdrometer	雨滴谱仪
disengage	使脱离
disengage clutch	脱开式离合器
dish aerial	航空天线；碟形天线
disk	桨盘；圆盘
disk brake	盘式制动器；碟式刹车器
disk clutch	圆盘离合器
disorganized wake	不规则尾流
dispatcher's supervision board	控制板
dispersal of speed	扩散速度
dispersed generation system	分散式发电系统
dispersed phase	胶态；分散态；分散相；离散相
dispersed wind energy system	分散型风能转化系统
dispersible contaminant	弥散性污染物
dispersing medium	弥散介质
dispersion	弥散；分散
dispersion coefficient	弥散系数
dispersion degree of freedom	弥散自由度
dispersion parameter	弥散参数
dispersion rate	弥散速度
dispersion theory of smoke	排烟弥散理论
dispersity	弥散度；分散度
displaced fluid	被排开流体
displaced parcel theory	排开气体理论
displacement	取代；移位
displacement amplitude	位移幅值
displacement current	位移电流
displacement effect	位移效应
displacement mode	位移模态
displacement spectrum	位移谱
displacement thickness	位移厚度
displacement transducer	位移传感器
display	显示
display lamp	指示灯
display structure	显示结构
display unit	显示装置；显示器
disposal	处理；支配
disposal lift	有效升力
disposal load	活动载荷
disruptive	破坏的
disruptive field intensity	击穿电场强度
dissipated	消散的
dissipated power	散耗功率
dissipation	浪费；消散
dissipation coefficient	耗散系数
dissipation factor (DF)	耗散因子
dissipation function	耗散函数
dissipation loss	耗散损失
dissipation of energy	功率消散
dissipation rate	耗散率
dissipation term	耗散项
dissipation trail	耗散尾迹
distance	距离；远方
distance constant	距离常数
distance estimation	距离估计

distance factor 距离因子；里程因数
distortion 畸变；失真；变形
distortion of flow pattern 流谱畸变
distributed 分布式的；分散的
distributed drive train 分布式传动系统
distributed generation system 分布式发电系统
distributed load 分布荷载
distributed mass 分布质量
distributing 分配；散布
distributing apparatus 配电电器
distributing board 配电板
distribution 分布；分配
distribution block 接线板
distribution box 配电盒
distribution grid 配电网
distribution main 配电干线
distribution of wind directions 风向分布
distribution system 配电系统
distributor disk 配电盘
disturbance 干扰；扰动
disturbance pattern 扰动流谱；扰动模式
disturbance signal 扰动信号
disturbance vorticity 扰动涡量
disturbed region 扰动区
disturbing 干扰；烦扰的；令人不安的
disturbing acceleration 扰动加速度
disturbing force 干扰力
diurnal circle 日循环
diurnal variation 日际变化；昼夜变化
diurnal wind 日变风
divergence 发散；散度；辐散
divergence boundary 发散边界
divergence field 辐散场
divergence section （风洞的）扩散段
divergence wind speed 发散风速

divergent current 发散流；辐散流
divergent instability 发散不稳定性
divergent oscillation 增幅振动
divergent response 发散响应
diversity 多样性；差异
diversity effect 参差效应；分集效应
divided streamline 分路流线
divisor 公约数；分压变压器
divisor of integers 整数因子
DL (direct load) 直接载荷
D/L ratio(drag-lift ratio) 阻升比
DLC (dynamic load characteristic) 动载特性
DM (design manual) 设计手册，制图手册
DMN (dimension) 尺寸；维；量纲
DMPR (damper) 阻尼器
documentation 文件
DOE (Department of Energy) （美国）能源部
DOF (degree of freedom) 自由度
dog chuck 爪形夹盘
dog clutch 爪形离合器
doldrum 赤道无风带
domain 域；范围
dome 穹顶导流罩；圆屋顶
domestic building 居住建筑
domestic dwelling 民用住宅
dominant frequency 主频率
dominant mode 主模态
dominant wind 盛行风；主风
door 门
door frame 门框
door hinge pillar 门铰链立柱
door pillar 门立柱
door weather strip 门框防漏垫条
door window 门窗
door with lock system 带锁门

doorway 门道
doped sheet 涂漆层
doped fabric 涂油蒙布
doped fabric covering 涂漆布蒙皮
Doppler acoustic radar 多普勒声雷达
Doppler laser radar 多普勒激光雷达
Doppler lidar 多普勒激光雷达
Doppler radar 多普勒雷达
Doppler shift 多普勒频移
dose 剂量
dose build up factor 剂量累积因子
dose commitment 剂量负担
dose estimation model 剂量估算模式
dose limit 剂量限值
dose response curve 剂量响应曲线
dot product 标量积
double 两倍的
double acting brake 复式制动器
double acting damper 双动减震器
double amplitude 全幅值
double angular ball bearing 双列向心推力球轴承
double arc airfoil 双圆弧叶形
double-bladed windmill 双叶片风车
double clamp 双卡头
double coupling 双联轴节
double-direction angular contact thrust ball bearing 双向推力向心球轴承
double fed induction generator (DFIG) 双馈感应发电机；双馈异步发电机
double fed synchronous generator (DFSG) 双馈同步发电机
double flexible coupling 双弹性联轴器
double fluid theory 双流体学说
double-ground fault 双线接地故障
double helical gauge 人字齿轮
double helical gear 人字齿轮
double helix gearbox 双螺旋齿轮箱
double-humped curve 双峰曲线
double-humped distribution 双峰分布
double layer capacitor 双层电容器
double line 复线
double-pitch roof 双坡屋顶
double-reduction axle 复式减速轴
double-reduction bevel-spur gear 双极圆锥 - 圆柱齿轮减速器
double reduction gear (DRG) 双减速齿轮
double-reduction gear box 两级减速器
double-reduction unit 两级减速器
double-return wind tunnel 二次回流风洞
double-row angular contact spherical ball bearing 双列推力向心球面球轴承
double-row ball journal bearing 双列径向滚珠轴承
double row bearing 双列轴承
double row four point contact ball bearing 双列四点接触球轴承
double-row radial ball bearing 双列向心球轴承
double row tapered roller bearing 双列圆锥滚柱轴承
double-side spectrum 双边谱
double skinned construction 双层蒙皮结构
double spherical roller bearings 双球面滚柱轴承；双列滚子球面轴承
double tracer technique 双元示踪技术
double unit traction 双机牵引
double-water internal cooling 双水内冷
double-way rectifier 双相桥式线路整流器
double-wedge airfoil 双楔机翼

doublet 偶极子

doublet panel 偶极子鳞片

doublet sheet 偶极子面

doublet strength 偶极子强度

doubly 双重的；加倍的

Douglas-Neuman method 道格拉斯-纽曼法

down 沿着；往下

down conductor 引下线

down draft 下降气流

down hand welding 平焊

down time 维修时间；停机时间；故障时间；停工时间

down valley windflow 下坡风；出谷风

down wind 顺风；在下风；下风向

downburst 下猝发风；下击暴流；下暴流

downdraft 下沉气流；倒灌风；下吸式

downdraft plume 下吸式羽流

downfall （雨等的）大下特下

downflow 下沉气流

downslope 下坡的

downslope flow 下坡气流

downslope wind 下坡风

downslope windstorm 下坡风暴

downstream 顺流的；下游的

downstream fairing （风机的）下游整流器

downstream flow 顺流流动

downstream guide vanes 下游导向叶片

downtime 故障时间；停机时间

downtime percentage 休风率

downward lift 负升力

downwash 下洗

downwash correction 下洗修正

downwash effect 下洗效应

downwash field 下洗场

downwash flow 下沉洗流

downwash model 下洗模式

downwash parameter 下洗参数

downwash plume 下洗型羽流

downwash region 下洗区

downwash vortex 下洗涡

downwelling 沉降物

downwind 下风向

downwind rotor 下风式风轮

downwind sector 下风向扇形区

down wind type of WECS 下风式风力机（使风先通过塔架再通过风轮的风力机）

downwind wind turbine 顺风式风电机组

draft 通风；抽风；小股气流

draft core 抽吸核心

draft fan 排风扇

draft gear 牵引装置

draft meter 风压计

drag 阻力；曳力

drag acceleration 减速

drag anemometer 曳力风速计

drag area 风阻面积

drag balance 流阻平衡

drag brace （翼内）阻力张线

drag break 阻力刹车板

drag by lift 阻升比

drag center 阻力中心

drag coefficient 阻力系数

drag convergence 阻力减小

drag crisis 阻力危机

drag critical value 阻力临界值

drag cup anemometer 阻力型风杯风速计

drag due to lift 升致阻力

drag equation 阻力方程

drag force 阻力

drag friction 摩擦阻力

drag from pressure 压差阻力
drag increment 阻力增量
drag lift ratio 阻升比
drag loss 阻力损失
drag parameter 阻力系数
drag penalty 阻力增大
drag polar 阻力极曲线
drag principle 阻力原理
drag reducing device 减阻装置
drag reduction 减阻
drag reduction cowling 减阻整流罩
drag saddle 最小阻力点
drag saddle galloping 最小阻力驰振
drag shear 由迎面阻力产生的剪力
drag test 风阻试验
drag type rotor 阻力型风轮
drag type wind machine 阻力型风力机
drag variation with lift 阻力随升力的变化
drag vector 阻力矢量
drag wind load 风阻载荷
drain 泻油
drainage 排水
draught 气流
draught bead 窗挡风条
draught center 阻力中心
draught loss 压力损失
draw bar pull 拉杆牵引力；牵引杆拉力
draw gear 牵引装置；车钩
draw in bolt 拉紧螺栓
DRG (double reduction gear) 双减速齿轮
drift 吹扬；漂移；风沙流；风雪流
drift accumulation 吹雪堆积；吹沙堆积
drift angle 滑角；偏移角；偏差角
drift area 流沙区；吹雪区
drift control 堆雪控制；流沙治理

drift deposition 流沙沉积；堆雪沉积
drift drop plume 漂滴羽流
drift of zero 零点漂移
drift pattern 漂沙型式；堆雪型式
drift snow transport 吹雪输运
drift volume 吹扬体积
drifting dust 飘尘
drifting sand 流沙；风积沙
drifting snow 吹雪；堆雪
drill 钻孔；钻头；锥子；钻孔机；钻床；钻
drill gauge 钻规
driller 钻孔者；钻孔机
drilling 钻孔
drilling machines 钻床
drilling machines bench 钻床工作台
drilling machines high speed 高速钻床
drilling machines multi-spindle 多轴钻床
drilling machines radial 摇臂钻床
drilling machines vertical 立式钻床
drilling platform （海上）钻井平台
drive 驱动器；驾车
drive axle 主动轴
drive clutch 传动离合器
drive flange 传动法兰盘
drive gear 主动齿轮；驱动齿轮
drive gear box 传动齿轮箱
drive motor 驱动马达
drive pinion 传动小齿轮
drive ratio 传动比
drive shaft 主动轴
drive torque 驱动力矩
drive train 传动系统；驱动系统
drive train bearing 传动系统轴承
driven 驾驶；从动的
driven disk 从动轮

driven gear 从动齿轮
driver gear 传动齿轮
driver pinion 主动小齿轮
driving 驾驶；操纵
driving disk 主动轮
driving force 驱动力
driving gear 主动齿轮
driving rain 大风雨
driving rain index 大雨指数
driving shaft 驱动轴
driving snow 大风雪
driving test 道路试验
drizzle 细雨
drizzle fog 毛细雨
droop 下垂；弯曲
droop setting 调差
droop snoot 前缘襟翼
droop snoot blade 前缘下垂式叶片
drop 滴；落下
droplet 微滴
dry adiabatic equation 干绝热方程
dry adiabatic lapse rate 干绝热递减率
dry adiabatic process 干绝热过程
dry bulb temperature 干球温度
dry cooling tower 干式冷却塔
dry core cable 空气纸绝缘电缆；干芯电缆
dry deposition 干沉积
dry deposition rate 干沉积率
dry disk clutch 干式摩擦离合器
dry friction 干摩擦
dry paper insulated cable 空气纸隔绝缘电缆
dry powder fire extinguisher 干粉灭火器
dry snow zone 干雪地带
dry type transformer 干式变压器
dry wind 干风

DS (design standard) 设计标准
dual 双数的
dual Doppler radar 双多普勒雷达
dual film aspirating probe 双膜吸气式探头
dual glazing 双层玻璃
dual speed induction generator 双速异步发电机
dual speed squirrel cage induction generator 双速鼠笼式异步发电机
dual valve 复式阀
dual yaw servos 双摆伺服系统
ducted cooling 管道式冷却
ducted fan 涵道风扇
ducted oil cooler 管道式油冷却器
ducted wind turbine 涵道式风轮机
ductile 柔软的
ductile cast iron (DCI) 球墨铸铁
ductile cast iron casing 球墨铸铁缸套；延性铸铁缸套
ductile crack 延性破裂
ductile fracture 形变断裂
ductile metal 韧性金属
ductile rupture 韧性断裂
ductile to brittle transformation characteristics 韧脆转变特性
ductile to brittle transition behaviour 韧脆转变特性；韧性向脆性的转化特性
ductile yield 延性屈服
ductility 延展性
ductility limit 屈服点
ductility machine 延度仪
ductility test 延展试验
ductility transition 韧性转变
ductility value 延展值
dummy end stub 假尾支杆

dummy model 假模型
dummy conductor 无载导线
dummy strut 假支杆
dummy support 假支架
dune 沙丘
duplex 两倍的
duplex cable 双股电缆
duplex transmission 双工传输
duplex tunnel 双试验段风洞
duplicate tunnel 双试验段风洞
duplicatus 复云
durability 耐久性
durability test 耐久试验
dural alclad 包铝
duration 持续时间
duration limit wind wave 时间限制风浪
duration of braking 制动时间
duration of breaker contact 继电器接触时间；开关触头接触时间
duration of down time 停机时间
duration of pollution 污染持续时间
duration of release 释放持续时间
duration of sampling 取样持续时间
duration of strom 风暴持续时间
duration of sunshine 日照时间
duration running 持久试验
dust and fume 烟尘
dust and soot 煤尘
dust borne gas 含尘气体
dust bowl 尘暴
dust cloud 尘云
dust collecting fan 吸尘器
dust collection 集尘
dust collector 除尘器
dust concentration 粉尘浓度

dust content 含尘量
dust counter 尘埃计；测尘计；尘度计
dust devil 尘旋风；尘卷风；小尘暴
dust fall 尘降
dust fog 尘雾
dust horizon 尘埃层顶
dust laden air 含尘空气
dust loading 尘埃浓度
dust plume 尘埃流
dust pollution 尘土污染
dust protected 防尘
dust rain 尘雨
dust sampling 尘埃取样
dust scrubber 洗尘器
dust storm 尘暴
dust whirl 尘旋风
dustiness 尘雾度；尘雾浓度
dusty air 含尘空气
Dutch windmill 荷兰风车
duty 责任
duty cycle 占空系数；频宽比；工作比；负载循环；负载持续率；工作周期
duty period 工作周期
duty plate 机器铭牌
duty ratio 平均功率与最大功率之比；负载比；负荷比；负荷率；能效比
DVR (dynamic voltage restorer) 动态电压恢复器
dwelling district 居住区
DWL (designed waterline) 设计水线
DYAN (dynamic analysis) 动态分析
dye injection system 染色水注入装置
dye stream 染液流
dynaflow 流体动力传动
dynamic 动态；动力

dynamic accuracy 动态精度	dynamic equilibrium 动态平衡
dynamic action 动力作用	dynamic error 动态误差
dynamic adjustment 动态调整	dynamic excitation 动态激励
dynamic admittance 动力导纳	dynamic experiment 动态试验；动力学实验
dynamic aerodynamics 动态空气动力学	dynamic force 动力
dynamic aeroelasticity 动力气动弹性力学	dynamic friction 动摩擦
dynamic amplification 动力放大	dynamic hardness 冲击硬度
dynamic analog device 动态模拟装置	dynamic head 动压头
dynamic analysis (DYAN) 动态分析	dynamic heating 动力增温
dynamic balance 动态平衡	dynamic horsepower 动态马力；动态功率
dynamic balancer 动平衡器	dynamic ice pressure 动冰压
dynamic balancing test (balance running) 动平衡实验	dynamic imbalance 动态不平衡
dynamic ball indentation test 落球硬度试验	dynamic instability 动态不稳定性
dynamic behavior 动态特性	dynamic load 动载荷
dynamic boundary layer 动力边界层	dynamic load test 动力载荷试验
dynamic buckling 动力失稳；动力屈曲；动态屈曲	dynamic loading test 动态负荷试验
dynamic calibration 动态校准	dynamic loss 动态损失
dynamic load characteristic (DLC) 动载特性	dynamic matrix 动力矩阵
dynamic coefficient 动态系数	dynamic meteorology 动力气象学
dynamic control 动态控制	dynamic mockup 动力模型
dynamic control system 动态控制系统	dynamic model 动态模型
dynamic cooling 动力冷却	dynamic modelling 动态模拟
dynamic coupling 齿啮式联接；动态耦合	dynamic parameter 动态参数
dynamic cycling loading 动态周期载荷	dynamic performance 动态特性
dynamic deflection 动挠度	dynamic power consumption 动态功耗
dynamic derivative 动导数	dynamic pressure 动压
dynamic deviation 动态偏差	dynamic pressure anemometer 动压风速计
dynamic ecosystem 动态生态系	dynamic pressure recover 动压恢复
dynamic effect 动力效应；动态效应	dynamic pressure sensor 动压传感器
dynamic endurance test 动力耐久试验	dynamic property 动态特性
dynamic equation 动态方程	dynamic range 动态范围
dynamic equation of equilibrium 动态平衡方程式	dynamic ratio 动态比
	dynamic resistance 动态阻力
	dynamic response 动态响应
	dynamic response test 动力响应试验

dynamic rig 动力试验台
dynamic scale 动力缩尺
dynamic similarity 动力相似性
dynamic similarity principle 动力相似原理
dynamic similitude 动力相似
dynamic simulation 动态模拟
dynamic speed control 动态转速控制
dynamic stability 动态稳定度；动态稳定性；动力稳定性
dynamic stability derivative 动态稳定导数
dynamic stabilization 动态稳定
dynamic stall 动态失速
dynamic state 动态
dynamic state operation 动态运行
dynamic strain 动应变
dynamic stress 动应力
dynamic sublayer 动力副层；机械乱流层
dynamic temperature 动态温度（均匀流动的气体中，当动能等熵地转换为热能时，所增加的那部分温度）
dynamic temperature change 动态温度变化
dynamic temperature correction 动态温度修正
dynamic test 动力试验；动态测试
dynamic theory of gas 气体动力学理论
dynamic threshold 动力气动值

dynamic variation 动态变化
dynamic vibration reducer 动态减震器
dynamic viscosimeter 动黏度
dynamic viscosity 动黏性（系数）
dynamic voltage restorer (DVR) 动态电压恢复器
dynamic warming 动力增温
dynamic wind load 动态风载
dynamical 动力学的
dynamical friction 动摩擦
dynamical property 动态特性
dynamical similarity 动力相似性
dynamical stress concentration 动应力集中
dynamical system 动力系统
dynamically similar model 动力相似模型
dynamics 动力学
dynamo 发电机
dynamo brush 电刷
dynamo governor 发电机调节器
dynamo magneto 永磁发电机
dynamo output 发电机输出
dynamo sheet 电机硅钢片
dynamometer 功率计；测力器
dynamometer instrument 电力表
dynamometer test 测功试验

现代英汉风力发电工程

A Modern English-Chinese Dictionary of Wind Power Engineering

E

early effect 早期效应
earth 地球
earth conductor 接地线接地导线
earth connection 接地
earth electrode 接地体
earth fault 接地故障
earth free 不接地的
earth fixed axes system 地轴系
earth frame of balance 地球平衡架
earth ground 地；接地
earth radiation 地球辐射
earth resistance 接地电阻
earth system of axes 地轴系
earth termination system 接地装置；接地系统
earth's atmosphere 地球大气
earth's axis 地轴
earth's boundary layer 地球边界层
earth's crust 地壳
earth's surface 地表；地面
earthed 接地的
earthed circuit 接地电路
earthing 接地
earthing fault 接地故障
earthing reference points 接地基准点
earthing switch 接地开关

earthing system 接地系统
earthquake 地震
east wind 东风
easterlies 东风带
eave 屋檐
eave board 风檐板；屋檐板
EBF (externally blown flap) 外吹式襟翼
ECC (equipment configuration control) 设备外形检查
ECC (error checking and correction) 误差检验与校正
eccentric action 偏心作用
eccentric mass vibrator 偏心质量激振器
ecclesiastical building 宗教建筑
Eckert number 埃克特数
ecocycle 生态循环
ecological balance 生态平衡
ecological cycle 生态循环
ecological environment 生态环境
ecological equilibrium 生态平衡
ecological group 生态群
ecological indicator 植物风力指示
ecological process 生态过程
ecological system 生态系
ecology 生态学

economic analysis 经济性分析
economic assessment 经济性评价
economic evalution 经济性评价
economic feasibility 经济可行性
economic running 经济运行
economic viability 经济可行性；经济能力
ecosystem 生态系
ECPD (Engineers' Council for Professional Development) 工程师专业发展委员会
EDBS (engineering data base system) 工程数据库系统
EDC (engineering design change) 工程设计修改
EDC (engineering design collaboration) 工程设计协作
EDC (engineering drawing change) 工程图纸修改
EDCP (engineering design change proposal) 工程设计修改提议
EDCS (engineering design change schedule) 工程设计修改日程表
EDD (engineering design data) 工程设计数据
EDDP (engineering design data package) 工程设计资料包
eddy 涡流
eddy advection 涡动平流
eddy conductivity 涡动传导率
eddy correlation 涡旋相关
eddy current 涡流
eddy current damper 涡流阻尼器
eddy current flaw slave detector 涡流探伤仪
eddy current loss 涡电流损耗
eddy current method 涡流法
eddy current transducer 涡流传感器
eddy diffusion 涡流扩散

eddy diffusivity 涡动扩散率
eddy effect 涡流作用
eddy energy 涡动能量
eddy exchange coefficient 涡动交换系数
eddy flow 紊流；涡流
eddy flux 涡旋通量
eddy generation 发生涡流
eddy kinetic energy 涡旋动能
eddy length scale 涡旋长度尺度
eddy line 涡流线
eddy loss 涡流损失
eddy mill 涡流粉碎机；涡流式碾磨机；涡穴
eddy mixing 涡动混合
eddy momentum flux 涡动量通量
eddy motion 涡流运动
eddy Prandtl number 涡动普朗特数
eddy region 涡流区
eddy resistance 涡流阻力
eddy shear stress 涡动剪应力
eddy spectrum 涡谱
eddy stress 湍流应力
eddy transfer coefficient 涡动传递系数
eddy transfer theory 涡动传递理论
eddy velocity 涡流速度
eddy viscosimeter 涡流黏度；湍流黏度
eddy viscosity 涡动黏性(系数)
eddy water 涡流
eddy zone 涡流区
eddying 涡流的
eddying effect 紊流影响
eddying motion 紊流运动
eddying wake 尾涡流
edge 边缘；使锐利
edge condition 边界条件
edge effect 边界影响

edge fairing 边缘整流

edgewise 沿边

edgewise bending 缘向弯曲

edifice 大型建筑

EDM (electrical discharge machining) 电火花加工

EEA (Electrical Equipment Association) 电气设备协会

EEC (Electronic Equipment Committee) 电子设备委员会

EED (Electrical Engineering Department) 电机工程部

EEE (electrical & electronics engineering) 电气和电子工程

EEE (electronic engineering equipment) 电子工程设备

EER (energy efficiency ratio) 能效比；能源效率比值

effect 效应；效能

effect of current 流动影响

effect of streaming squeezing 流线密集效应

effect of surrounding 环境影响

effective 有效的；起作用的

effective aerodynamic downwash 有效气动下洗

effective angle of attack 有效迎角

effective area 有效面积

effective aspect ratio 有效展弦比

effective buoyancy 有效浮力

effective camber 有效弯度

effective case depth 有效硬化层深度

effective chimney height 有效烟囱高度

effective chord length 有效弦长

effective clearance 有效间隙

effective control 有效控制

effective cross section 有效截面

effective decay constant 有效衰减常数

effective decontamination 有效净化

effective deposition velocity 有效沉积速度

effective diameter 有效直径

effective diffusion coefficient 有效扩散系数

effective distance 有效距离

effective dose 有效剂量

effective downwash 有效下洗

effective elongation 有效伸长

effective entrainment coefficient 有效夹带系数

effective flow 有效流量

effective flow resistance 有效流动阻力

effective ground level 有效地面高度

effective height 有效高度

effective height of stack 有效烟囱高度

effective horsepower 有效功率

effective life 有效寿命

effective lifetime 有效寿命

effective load 有效负荷

effective mixing layer 有效混合层

effective output 有效输出；有效功率

effective passage throat 有效喉部截面；有效喉道截面

effective period 有效期

effective plume 有效羽流

effective plume height 有效羽流高度

effective plume rise 有效羽流抬升

effective power 有效功率

effective power factor 有效功率因数

effective power output 有效功率输出

effective pressure 有效压力

effective prevention distance 有效防护距离

effective range 有效（测量）范围；有效距离

effective section 有效截面
effective shearing rigidity 有效抗剪刚度
effective source area 有效源面积
effective source height 有效源高
effective source strength 有效源强
effective speed 有效速度
effective stack height 有效烟囱高度
effective strain 等效应变
effective summed horsepower 总有效功率；总有效马力
effective superstructure 有效上层建筑
effective surface 有效表面；有效表面积
effective terrain height 有效地形高度
effective throat 风洞工作段有效面积；焊缝有效厚度
effective value 有效值
effective wind speed 有效风速
effective work 有效功
effectiveness 效力
effects of saturation 饱和效应
efficiency 效率
efficiency characteristic 效率特性
efficiency estimation 有效估计
efficiency factor 效率因素
efficiency indication 效率指标
effluent 排放物；流出物
effluent emission 废气排放
effluent flow 废气流
effluent gas 废气
effluent plume 废气羽流
effluent standard 排放标准
efflux point 排放点
efflux velocity 排放速度
efflux-windspeed ratio 排放—风速比
effuser 扩散喷管；风洞收敛段

EFV (electric field vector) 电场矢量
eggbeater wind turbine 打蛋器型风轮机（即达里厄型风轮机）
EIA (environment impact assessment) 环境影响评估
Eiffel type wind tunnel 埃弗尔式开路风洞
eigen frequency 固有频率；特征频率；本征频率
eigen period 固有周期
eigen value 本征值
ejection type wind tunnel 引射式风洞
Ekman boundary layer 埃克曼边界层
Ekman layer 埃克曼螺线层
Ekman number 埃克曼数
Ekman spiral 埃克曼螺线
Ekman turning 埃克曼偏转
elastic 有弹性的；弹性
elastic axis 弹性轴
elastic body 弹性体
elastic buckling 弹性失稳；弹性屈曲
elastic buffer 弹性缓冲器
elastic center 弹性中心
elastic constant 弹性常数
elastic coupling 弹性连接
elastic displacement 弹性位移
elastic distortion 弹性畸变
elastic elongation 弹性伸长
elastic failure 弹性失效；弹性破坏
elastic force 弹力
elastic foundation 弹性基础
elastic instability 弹性不稳定性
elastic joint 弹性接头
elastic limit 弹性极限
elastic load 弹性载荷
elastic modulus 弹性模量

elastic packing 弹性密封

elastic plastic behaviour 弹塑性能

elastic plastic boundary in flexure 弯曲弹塑性边界

elastic plastic material 弹塑性材料

elastic range 弹性区

elastic recovery 弹性回复

elastic region 弹性区

elastic resistance 弹性阻力

elastic scaling 弹性缩尺

elastic scattering 弹性散射

elastic similarity 弹性相似

elastic stiffness 弹性刚度

elastic twist 弹性扭转

elastic vibration 弹性振动

elasticity 弹性力学

elasticity factor 弹性因数

elasticity modulus 弹性模量

elastomer 弹性体

elastomer element 弹性元件

elastomeric gasket 弹性垫圈；合成橡胶垫圈

electric 电的

electric actuator 电动执行器；电传动装置

electric arc welded tube 电弧焊缝管

electric block 触电；电击；电动葫芦

electric brake motor 电动制动马达

electric brazing 电热铜焊

electric chain hoist block 电动吊链

electric charge 电荷

electric circuit 电路

electric circuit complexity 电路复杂程度

electric conductance 电导

electric conductivity 导电性

electric conductor 导电体

electric control 电(气)控(制)

electric control valve 电气控制阀

electric coupling 耦合器

electric current 电流

electric dip 电偶极子

electric dynamometer 电动测力计；电动测功器

electric energy 电能

electric energy consumption 耗电量

electric energy transducer 电能转换器

electric engineer 电气工程师

electric engineering 电气工程

electric equipment 电气设备

electric fence 电篱笆；电丝网；电围栏

electric field vector (EFV) 电场矢量

electric field strength 电场强度

electric geared motor 齿轮驱动式马达；齿轮驱动式电动机

electric lift 电梯

electric lighting load 照明负荷

electric machine 电机

electric main 输电干线

electric monitor 电动监控器；电动监测器

electric pitching mechanism 电动调桨机构

electric power 电力；电功率

electric power control panel 电力控制面板

electric power grid 电力电网

electric power output 静电功率输出

electric power panel 电力控制面板

electric power quantity 发电量

electric power supply 供电

electric resistance 电阻

electric service lift 电动维护吊车

electric servo motor 电动伺服电机

electric shock 电击；电震

electric switchyard 电力开关站

electric test 电气试验
electric wire and cable 电线电缆
electrical 有关电的；电气科学的
electrical & electronics engineering (EEE) 电气和电子工程
electrical anemometer 电传风速计
electrical angular frequency 电角频率
electrical arc welding 电弧焊
electrical bridge 电桥
electrical cable 电缆
electrical conductivity 导电性
electrical contact 电触头
electrical contactor 电气接触器
electrical control and protection 电气控制和保护
electrical control room 电气控制室
electrical device 电气设备
electrical discharge 放电
electrical discharge machining (EDM) 电火花加工
electrical endurance 电气寿命；电气耐久性
Electrical Engineering Department (EED) 电机工程部
electrical equipment 电气设备
Electrical Equipment Association (EEA) 电气设备协会
electrical equipment hatch 电气设备舱门
electrical fuse 电力熔断器
electrical generating system 发电系统
electrical generator 发电机
electrical generator efficiency 发电机效率
electrical generator speed control 发电机转速控制
electrical horse power 电功率
electrical interface 电器接口

electrical locomotive 电力机车
electrical network 电网
electrical power cable 电缆线；输电线
electrical rotating machine 旋转电机
electrical stressing 电气应力
electrical substation 电力变电站
electricity 电
electro dynamometer 电功率计
electro gas dynamics 电气体动力学
electro gaseous dynamic wind driven generator 带电气体风力发电机
electro hydraulic servosystem 电动液压伺服系统
electrode 电极；电焊条
electrofluid 电流体
electrofluid dynamic augmented tunnel 带电流体动力加速装置的风洞
electrofluid dynamic wind driven generator 带电流体风力发电机
electrolyte 电解质
electrolyte battery 有机电解液蓄电池
electrolytic hydrogen energy storage 电解制氢贮能
electromagnetic flaw detector 电磁探伤仪
electromagnet 电磁铁；电磁体；电磁石
electromagnetic 电磁的
electromagnetic braking system 电磁制动系
electromagnetic compatibility filter (EMC-filter) 电磁兼容性滤波器；EMC滤波器
electromagnetic eddy current damper 电磁涡流阻尼器
electromagnetic energy 电磁能
electromagnetic induction 电磁感应
electromagnetic slip 电磁滑差
electromagnetic torque 电磁转矩

electromechanical 电动机械的；机电的
electromechanical brake 机电复合制动器；电子机械制动器
electromechanical disc brake 机电盘式制动器
electromechanical energy converter 机电能量转换器
electromechanical oscillation 机电振荡
electromechanical process 机电过程
electromotive force (EMF) 电动势
electron beam welding 电子束焊接
electronic 电子的
electronic anemometer 电子风速仪
electronic circuit 电子电路
electronic controller 电子控制器
electronic engineering equipment (EEE) 电子工程设备
Electronic Equipment Committee (EEC) 电子设备委员会
electronic surge arrestor 电子避雷器
electronics bench 电子设备操纵台
electroslag welding 电渣焊
electrostatic accumulator 电容器；静电累加器；静电存储器
electrostatic adherence 静电附着
electrostatic screen 静电屏蔽
electrostatics 静电学
element 元素
element buffer 缓冲元件
element failure 元件损坏
elementary derivation 初步推导
elementary flow pattern 基本流谱；基本流动模式
elevated house 高架房屋
elevated inversion 高空逆温
elevated line source 高架线源
elevated plume 抬升羽流
elevated release 高空释放
elevated source 高架源
elevation 高度
elevation limits 仰角变化范围
elevator 电梯
elliptic 椭圆形的
elliptic wind tunnel 椭圆试验段风洞
elliptical vortex generator 椭圆涡发生器
elongation 伸长；伸长率
elongation at rupture 断裂伸长
elongation factor 延伸系数
elongation in tension 受拉伸长
elongation test 拉伸试验
ELR (engineering laboratory report) 工程实验试报告
embedded 嵌入式的
embedded generation system 嵌入式发电系统
embrittlement 脆化；脆裂
embryonic dune 初期沙丘
EMC-filter (electromagnetic compatibility filter) 电磁兼容性滤波器；EMC滤波器
emergency 事故；紧急情况；突发事件；应急
emergency braking 紧急刹车
emergency braking system 紧急制动系统
emergency condition 事故工况；紧急状态
emergency control 应急控制
emergency exit 事故出口；紧急出口
emergency feather 应急顺桨
emergency generating unit 应急发电机装置
emergency generator 紧急发电机；备用发电机
emergency light 事故照明灯；事故信号灯
emergency lighting 事故照明

emergency lock 安全闸

emergency shutdown 紧急关机

emergency shutdown for wind turbine 紧急关机

emergency signal 事故信号

emergency stop 紧急停止

emergency stop push button 紧急停车按钮

emergency switch 紧急开关

emergency valve 安全阀；紧急阀

emery 金刚砂；刚玉砂

emery cloth 砂布；金刚砂布

emery wheel 金刚砂旋转磨石；砂轮

EMF (electromotive force) 电动势

eminently clean 良好流线型的

emission 排放；发射；排放物；散发物

emission concentration 排放物浓度

emission control 排放控制

emission control strategy 排放控制对策

emission factor 排放系数；排放因素

emission height 排放高度

emission inventory 排放源清单

emission point 排放点

emission rate 排放率

emission Reynolds number 排放雷诺数

emission source 排放源

emission standard 排放标准

emission temperature 排放温度

emissivity 发射率；辐射率

emittance 发射度；辐射度；辐射强度

emitter 发射管；放射器；发射极

empirical 经验主义的；完全根据经验的

empirical coefficient 经验系数

empirical correction 经验修正

empirical curve 经验曲线

empirical data 经验数据

empirical equation 经验方程式

empirical formula 经验公式

empirical law 经验定律；经验规律

empty 空的

empty running 空载运行

empty tunnel （没有装模型的）空风洞

empty working section 空风洞试验段；空水洞试验段

enameued leather 漆皮

enclosed wake 封闭尾流

enclosure 外罩；外壳

encode 编码

end 终止；结束

end effect 端部效应

end gas 尾气；废气

end outline 后视轮廓

end ring 端环

end winding 端部绕组

endplate 端板

endurance 耐久性

endurance bending strength 弯曲疲劳极限

endurance crack 疲劳裂纹

endurance failure 疲劳破坏

endurance life 疲劳寿命

endurance limit 疲劳极限

endurance ratio 耐久比

endurance running 持续运行

endurance strength 耐久极限

endurance test 耐久极限测试

endurance testing machine 持久试验机

energetic wind 强风

energize 励磁；通电

energize brake 通电制动器

energy 能量

energy absorbing capacity 能量吸收能力

energy absorbing material 吸能材料
energy absorption 吸能
energy advection 能量平流
energy breakdown 能量的品质降低
energy budget 能量收支
energy buffer 能量缓冲器
energy capacity 能量
energy conservation 能量守恒；能量节约
energy conservation equation 能量守恒方程式
energy conservation law 能量守恒定律
energy consumption 能耗
energy containing eddy 含能涡旋
energy content 内能；含能量
energy control system 能量控制系统
energy conversion 能量转换
energy converter 能量转换装置；电能转换器
energy crisis 能源危机
energy current 有效电流；有功电流
energy deficiency 能量不足
energy degradation 能量递降；能的递降；能量降级
energy demand 能量需求
energy density 能量密度；(蓄电池)贮能密度
energy dependence 能量依赖
energy deposition 能量吸收
energy development 能源开发
energy dissipation 能量损失
energy dissipation device 耗能装置
energy dissipation spacer （电缆线）耗能隔离子
energy distribution 能源分布
energy efficiency 能源效率
energy efficiency ratio (EER) 能效比；能源效率比值

energy equation 能量方程式
energy exchange 能量交换
energy extraction 获能
energy flow 能流
energy flow rate 能流速率
energy flux 能量通量
energy flux density 能量通量密度
energy frictional heat energy 摩擦热能
energy gain 能量收益
energy grade 能量等级
energy gradient 能量梯度
energy loss 能耗
energy management system 能量管理系统
energy meter 电度表；电能表；能量计；功率计；能量表
energy method 能量法
energy pattern factor (EPF) 能源结构因数
energy payback time 能量生产还本时间
Energy Research and Development Administration (ERDA) 美国能源研究开发署
energy resource 能量资源
energy return period 能量生产还本周期
energy rose 能量玫瑰图
energy saving 节能
energy source 能源
energy spectrum 能谱
energy storage 储能
energy storage device 储能设备
energy storage medium 储能介质
energy storage system of wind power 风力发电储能系统
energy tax credit 能量投资免税
energy thickness 能量厚度
energy thickness of boundary layer 边界层能量厚度

energy transfer 能量传递
energy transformation 能量转换
energy transmission 能量传送
energy transport 能量输送
energy utilization 能量利用
energy without pollution 无污染能源；清洁能源
engagement 约会
engagement mesh 啮合
engine 引擎；发动机
engine backplate 发动机隔板；发动机护板
engine compartment 发动机舱
engine cooling airflow 发动机冷却气流
engine cover 发动机罩
engine cowl 发动机罩；发动机整流罩
engine exhaust gas 发动机排气
engine exhaust trail 发动机排气尾迹
engine windmill 发动机风转
engine yaw 发动机偏航
engineering 工程；工程技术
engineering aerodynamic 工程空气动力学
engineering analysis 工程分析
engineering and technical services (ETS) 工程及技术服务
engineering council 工程协会
engineering cybernetics 工程控制论
Engineering data base system (EDBS) 工程数据库系统
engineering design 工程设计
engineering design change (EDC) 工程设计修改
engineering design change proposal (EDCP) 工程设计修改提议
engineering design change schedule (EDCS) 工程设计修改日程表

engineering design collaboration (EDC) 工程设计协作
engineering design data (EDD) 工程设计数据
engineering design data package (EDDP) 工程设计资料包
engineering drawing 工程制图
engineering drawing change (EDC) 工程图纸修改
engineering information 工程资料
engineering judgement 工程判断
engineering laboratory report (ELR) 工程实验试报告
engineering meteorology 工程气象学
engineering project 工程项目
engineering standard 工程标准；技术标准
engineering test (ET) 工程试验
engineering test facility (ETF) 工程试验设备
engineering test requirements (ETR) 工程试验要求
Engineers' Council for Professional Development (ECPD) 工程师专业发展委员会
enhancement 增强
enrichment factor 浓集因子
ensemble 集；系统
ensemble average concentration 系统平均浓度
ensemble mean field 系统平均场
entrained 夹带
entrained fluid 夹带流体
entrained mass 夹带质量
entrainment 夹带
entrainment coefficient 夹带系数
entrainment constant 夹带常数
entrainment effect 夹带效应
entrainment layer 夹带层

entrainment parameter 夹带参数
entrainment rate 夹带率
entrainment theory 夹带理论
enthalpy 焓
entropy 熵
envelope 包络线；外壳
enveloping box 外形包络箱
environ 近郊；围绕；环绕；包围
environment 环境
environment aerodynamics 环境空气动力学
environment condition 环境条件
environment factor 环境因素
environment impact assessment (EIA) 环境影响评估
environment science 环境科学
environment wind tunnel 环境风洞
environmental amenity 环境舒适性
environmental assessment 环境评价
environmental activity 环境放射性
environmental capacity 环境容量
environmental climate 室内小气候
environmental contamination 环境污染
environmental control 环境控制
environmental degradation 环境退化
environmental ecology 环境生态学
environmental engineering 环境工程
environmental field test 环境现场试验
environmental hazard 环境公害
environmental meteorology 环境气象学
environmental monitoring 环境现场监测
environmental noise 环境噪声
environmental pollution 环境污染
environmental project 环境规划
environmental protection 环境保护
environmental quality 环境质量

environmental quality index 环境质量指数
environmental quality pattern 环境质量模式
environmental quality standard 环境质量标准
environmental reform 环境改造
environmental test 环境试验
environmental testing facility 环境试验设备
environmental wind 环境风
EPF (energy pattern factor) 能源结构因数
epigene 外成的
epigene action 外力作用
epoch angle 初相角
epoxy 环氧树脂
epoxy glass 玻璃环氧树脂
epoxy resin 环氧树脂
epoxy resin pattern 环氧树脂模
equation 方程式
equation of continuity 连续方程
equation of continuous flow 流动连续方程
equation of diffusion 扩散方程
equation of dynamics 动力方程
equation of gaseous state 气体状态方程
equation of higher degree 高次方程
equation of ideal gas 理想气体方程
equation of mass conservation 质量守恒定律方程
equation of motion 运动方程
equation of small disturbance motion 小扰动运动方程
equation of state 状态方程
equation of state equilibrium 静平衡方程
equator 赤道；大圆
equatorial depression belt 赤道低压带
equatorial velocity 赤道速度；大圆速度
equilibrium 均衡；平衡

equilibrium boundary layer 平衡边界层
equilibrium condition 平衡条件
equilibrium constant 平衡常数
equilibrium convection 平衡对流
equilibrium diagram 平衡图
equilibrium distribution 平衡分布；均衡分布
equilibrium equation 平衡方程式
equilibrium factor 平衡因子
equilibrium layer 平衡层
equilibrium sand flow 均衡沙流
equilibrium valve 平衡阀
equipment 设备；装备
equipment capacity 设备容量；设备功率
equipment configuration control (ECC) 设备外形检查
equipment cost 设备费用
equipment design variable 设备设计变量
equipment failure 设备故障
equipment failure information 设备故障信息
equipment load 设备荷载
equipment maintenance 设备维修
equipment maintenance record (EMR) 设备维护记录
equipment maintenance record system (EMRS) 设备维护报告系统
equipment performance inspection 设备性能检查
equipment works 设备工程
equipotential 等势；等位
equipotential bonding 等电位连接
equipotential contour 等势线
equipotential surface 等势面；等位面
equivalent 等价的；相等的
equivalent circuit 等效电路
equivalent diameter 等效直径
equivalent diffusion velocity 等效扩散速度
equivalent mass 等效质量
equivalent roughness 当量粗糙度
equivalent stack height 等效烟囱高度
equivalent stress 等效应力
equivalent surface source 等效面源
equivalent T-circuit T型等值电路
equivalent time 等效时间
equivalent wind speed 等效风速
ERDA (Energy Research and Development Administration) 美国能源研究开发署
erection 吊装；耸立；装配；安装
erection bolt 装配螺栓
erection diagram 安装图；装配图
erection drawing 装配图；安装图
ergodicity 各态历经性；遍历性
ergogram 示功图
ergograph 示功器
ergometer 测功器
erosion 腐蚀；磨蚀；冲蚀（作用）
erosion pattern 风蚀图案
erosion picture 风蚀图案
erosion rate 侵蚀率
erosion resistance 耐腐蚀性
erosion technique 侵蚀技术
error 误差
error allowance 容许误差
error analysis 误差分析
error checking and correction (ECC) 误差检验与校正
error coefficient 误差系数
error control 误差控制
error correction 误差修正
error detector 误差检测器
error excepted 容许误差

error function table 误差函数表
error of caution 错误警告
error of measurement 测量误差
error of reading 读数误差
error of scale 刻度误差
error range 误差范围
error signal 误差信号
escape ladder 脱险梯
escape orifice 排泄口；逸出孔；逸出口
escaped plume 泄放羽流
escarpment 陡壁；悬崖
estimate 估计；估价
estimate data 估算数据
estimated 估计的
estimated design load 估算的设计负荷
estimated life 估算寿命
estimation 估计
estimation error 估算误差
estuary 河口；港湾
ET (engineering test) 工程试验
ETF (engineering test facility) 工程试验设备
ETMD (extended tuned mass damper) 扩展调谐质量阻尼器
ETR (engineering test requirements) 工程试验要求
ETS (engineering and technical services) 工程及技术服务
Euler number 欧拉数
Eulerian correlation coefficient 欧拉相关系数
Eulerian correlation function 欧拉相关函数
Eulerian covariance 欧拉协方差
Eulerian cross correlation 欧拉交叉系数
Eulerian system 欧拉系统
Eulerian time scale 欧拉时间尺度
Eulerian wind 欧拉风；测点风

European Wind Energy Association (EWEA) 欧洲风能协会
evaluate 评价
evaluation 评定；评价；估计
evaluation test 评价试验
evaporation 蒸发
evaporation heat 蒸发热
evaporation loss 蒸发损失
evaporation rate 蒸发率
evapotranspiration 蒸腾；蒸散
even fracture 平整断口；细粒状断面；平断口
event 事件
event information 事件信息
evently distributed load 均匀分布负载
everlube 耐寒性润滑油
exact similarity 精确相似
exaggerated test 超常试验
examination 考察
exceedance statistics 超过数统计
exceptional length 超长
excess load 过载
excess power peak 剩余功率的峰值
excessive 过多的
excessive grade 过陡坡度
excessive machining allowance 过大的加工余量
excessive peak electrical load 过度高峰电力负荷
excessive vibration 振动过大
exchange 交换；交流
exchange coefficient 交换系数
excitation 励磁；激磁
excitation current 激励电流；励磁电流
excitation energy 激励能
excitation field 励磁磁场

excitation force 激励力；激振力
excitation response 励磁响应
excitation source 励磁源
excitation system 励磁系统
excite 刺激
excite oscillation 励磁振动
excited mode 激励模态
exciter 励磁机；激振器
exciting current 励磁电流
exciting frequency 激励频率；激振频率
exciting oscillation 激振
exciting voltage 励磁电压
excitor 励磁器
exfiltration 渗漏；漏风
exhalation 呼气
exhalation coefficient 呼出系数
exhaust 排气；抽空
exhaust emission 废气排放
exhaust fan 排气扇；抽气扇；排风扇
exhaust flow 排烟；废气
exhaust fume 排烟；废气
exhaust gas 排气
exhaust jet 废气射流
exhaust opening 排气口
exhaust orifice 排气口
exhaust pipe 排气管
exhaust plume 废气羽流
exhaust port 排气口
exhaust pressure 排气压力
exhaust smoke 排烟
exhaust stack 排气烟囱
exhaust system 排气系统
exhaust trail 排气尾流
exhaust valve 排气阀
exhaust velocity 排气速度

exhibition structure 展示结构
exit 出口；排气管
exit flue 排烟道
exit region 出口区
exosphere 外大气层
expansion 膨胀；展开式
expansion angle 膨胀角
expansion bolt 伸缩栓；扩开螺栓；自攻螺丝
expansion joint 伸缩接头
expansional cooling 膨胀冷却
expected 预期
expected value 期望值
expected wind speed 期望风速
expenditure 费用；消耗；损耗
expenditure of energy 能量消费
experiment 试验；实验
experiment model 实验模型
experimental 实验的
experimental aerodynamic 实验空气动力学
experimental building 实验建筑物
experimental data 实验数据
experimental installation 实验装置
experimental medium 实验介质
experimental prototype 实验样机
experimental smoke plume 实验烟羽
experimental tank 实验池；实验水槽
exponential absorption law 指数吸收律
exponential decay 指数衰减
exponential distribution 指数分布
exponential equation 指数方程
exponential growth 指数增长
exponential law 指数律
exposed deck 露天甲板
exposed surface 暴露表面
exposure 暴露；照射

exposure factor 地貌开敞度
exposure period 暴露周期
expressional algebraic expression 表达的代数表达式
express way 高速公路；快车道
extended tuned mass damper (ETMD) 扩展调谐质量阻尼器
external 外部；外表
external aerodynamic 外流空气动力学
external armature circuit 电枢外电路
external calipers 外卡钳
external characteristic 外特性
external conditions (for wind turbines) (风力机) 外部条件
external crack 表面缝
external dimension 外形尺寸
external dose 外照射剂量
external exposure 体外照射
external field 外场；外部流场
external flow 外部绕流
external gear 外齿轮
external lighting protection system 外部防雷系统
external load 外部荷载
external perpendicular 外部垂直
external power supply 外部动力源
external pressure coefficient 外部压力系数
external pressure loss 外部压力损失
external riveting 外 (蒙皮) 铆接
external stator 外部定子；外定子
external suction 外部吸力
external wind load 外部风载
externally blown flap (EBF) 外吹式襟翼
extinction 光消散；熄火
extinction coefficient 消光系数

extinguisher 灭火器
extra 额外的
extractable energy 可获能量
extra high tension unit 超高压设备
extraneous loading 附加荷载
extraordinary wind 异常风
extrapolated 推测
extrapolated power curve 外推功率曲线
extrapolated value 外推值；外插值
extrapolation 外推法；外插
extratropical cyclone 温带气旋
extratropical storm 温带风暴
extreme 极端
extreme annual wind speed 年极端风速
extreme atmosphere events 极端大气现象
extreme climate 极端气候
extreme gradient wind 极端梯度风
extreme lifetime wind speed 寿命极端风速
extreme maximum 极端最高
extreme maximum temperature 极端最高温度
extreme mile wind speed 极端英里风速
extreme minimum temperature 极端最低温度
extreme return period (风速) 极端重现期；极端回报期
extreme surface wind 地面极端风
extreme wind 极端风
extreme wind speed 极端风速
extruded 压出的
extruded aluminium blade 挤压铝叶片
extruded load 挤压载荷
extrusion forming 挤压成型
eye estimation 目测
eye observation 目测
eye of storm 风暴眼

现代英汉风力发电工程

A Modern English-Chinese Dictionary of Wind Power Engineering

F

FAA (Federal Aviation Administration) 美国联邦航空管理局
fabric 织物
fabric roof 软屋顶
fabricating 制作；捏造
fabricating cost 制造费用
fabricating yard 施工现场
fabrication 加工；制造
fabrication drawing 制造图纸；制作图
fabrication procedure 加工工艺性
fabrication process 加工过程
fabrication technology 制造工艺
fabrication tolerance 制造容差
fabricator 装配面
façade wall 正面墙
face 面向；端面
face air velocity 工作面风流速度
face area 迎风面积
face bend test 表面弯曲试验
face coat 表面涂层
face crack 表面裂缝
face gear 平面齿轮
face line of teeth 齿顶线
face of tool 切削面
face of weld 焊缝表面
face-on-attack 迎风面
face tube 皮托管
face velocity 迎面风速
face ventilation 工作面通风
face width 齿宽
faceplate 面板；花盘
facility 设备；设施；工具；装置
facility charge 设备费
facility cost 设备成本
facility failure 设备故障
facility of repair 维修设备
facility management system 设备管理系统
facility modernization 设备现代化；设施现代化
facility power control 设备功率控制
facility power panel 设备电源板
facility reliability 设备可靠性
facility request 设备要求
facing 饰面
factor 因素；要素
factor analysis 因子分析；因素分析
factor capacity 功率；容量因子
factor of adhesion 黏着系数
factor of air resistance 空气阻力系数
factor of fatigue 疲劳系数

factor of safety 安全系数
factored resistance 设计风阻
factored wind load 设计风载
FACTS (flexible AC transmission system) 柔性交流输电系统
fail 失败
fail close 出故障时自动关闭
fail open 出故障时自动打开
fail point 破坏点；失效点
fail safe 故障安全防护装置；安全装置；失效安全
fail safe brake pressure 故障保护制动压力
fail safe control 防止控制装置
fail safe device 故障自动防护装置
fail safe disc brake 故障自动防护盘式制动器
fail safe facility 安全装置；保险装置
fail test 可靠性试验；故障测试
failing load 破坏荷载
failure 失效；故障
failure absorbent actuator 故障防护装置
failure analysis 故障分析
failure by shear 剪切破坏
failure condition 故障条件
failure costs 损失费用
failure crack 断裂纹
failure criterion 破坏判据
failure free operation 无故障运行；正常运行
failure free period 无故障工作期
failure in service 使用中的故障
failure load 破坏载荷
failure mode 失效模式；故障模式；故障种类；故障形式
failure monitor 故障检测仪
failure rate 事故频率
failure record 故障记录

failure recovery 故障排除
failure stress 破坏原应力
failure test 故障试验；可靠性试验；破坏性试验
failure warning 故障警告
failure warning indicator 故障警告器
failure warning relay 故障警报继电器
FAIR (fairing) 整流罩；导流帽
fair wind 顺风
fairing (FAIR) 整流罩；(风洞支架) 风挡
fairing cap 整流罩；导流帽
fall off 开始顺风；转向下风处
fall off on one wing 横侧失速
fall out (微粒) 回降；(放射性) 沉降
fall wind 下吹风；下降风；下坡风
fallback 低效运行
fall-back state 低效运行状态
falling current of air 气流下降
falling protection 下跌保护；坠落保护
fallout 沉降；放射性尘埃
fallout front 沉降前锋
fallout wind 沉降风
fallwind 下降风
false floor (风洞) 假地板
false jaw 虎钳口
false ogive 整流罩
false spar 假梁
false tripping 误动；误启动
false wall (风洞) 假洞壁
family of curves 曲线族
fan 风扇；鼓风机
fan belt 风扇皮带
fan blade 风扇叶片
fan heater 风扇加热器
fan hub 风扇轮毂
fan inlet 通风机进气口

fan noise 风扇噪声

fan power factor 风扇功率系数

fanning plume 扇形羽流

fantail 尾风轮；扇状尾

fantail roof 扇形屋顶

fan-type stay cable 扇形拉索

far back maximum thickness （翼剖面）靠后的最大厚度（位置）

far field 远场

far field concentration 远场浓度

far field condition 远场条件

far field flow 远场流动

far field plume dispersion 远场羽流弥散

far wake 远尾流

Faraday's law 法拉第定律

faradic current 感应电流

fast 快速的

fast coupling 硬性联轴节

fast Fourier transformer (FFT) 快速傅里叶变换

fast motion gear 变速齿轮

fast response instrument 快速响应仪表

fastest 最快速的

fastest mile 最大英里风速

fastest mile of wind 最大英里风速

fastest mile wind speed 最大英里风速

fatigue 疲劳

fatigue behaviour 疲劳特性

fatigue bending machine 弯曲疲劳试验机

fatigue break down 疲劳破坏

fatigue criterion 疲劳判据

fatigue data 疲劳实验数据

fatigue durability 耐疲劳性

fatigue failure 疲劳破坏

fatigue fracture 疲劳破裂

fatigue life 疲劳寿命

fatigue limit (FL) 疲劳极限；疲劳寿命；疲劳限界

fatigue loading 疲劳加载

fatigue rating (of material) （材料）疲劳额定值

fatigue resistance 疲劳抗力

fatigue strength 疲劳强度

fatigue strength at n cycle 在反转 n 周次下的疲劳强度

fatigue strength under oscillation stresses 振动疲劳强度

fatigue strength under reversed stresses 交变疲劳强度

fatigue range 疲劳限度

fault 故障；缺陷；损伤

fault current 故障电流；事故电流

fault detect 故障检测

fault detection 探伤

fault diagnosis 故障诊断

fault earthing 故障接地

fault finder 探伤仪

fault finding 故障查找

fault fissure 断层裂纹

fault ground bus 故障接地母线

fault indicator 故障指示器

fault line 裂纹线

fault localization 探伤；故障定位；故障测距技术；障碍勘测

fault location 故障位置测定；故障定位

fault protection device 故障防护装置

fault relay 故障继电器

fault ride through 故障穿越

fault signal 故障信号

fault switch 故障模拟开关

fault time 故障时间；停机维修时间
fault transient 暂态故障
faulty 有错误的
faulty component 故障件
faulty unit 次品
favourable 有利的
favourable interference 有利干扰
favourable pressure gradient 顺压梯度
favourable pressure interference 顺压差
FCM (fiber composite materials) 纤维复合材料
feasibility 可行性；可以实现
feasibility analysis 可行性分析
feasibility evaluation 可行性评价
feasibility reliability check 可行可靠性检验
feasibility report 可行性报告
feasibility study 可行性研究；技术经济论证
feather 羽毛
feather direction 顺桨方向
feather key 滑键
feather position 顺桨位置
feathered pitch 顺桨桨距
feathering 顺桨
feature 特点；特征；零件；外貌
feature size 形体尺寸
Federal Aviation Administration (FAA) 美国联邦航空管理局
feedback 反馈
feedback compensation 反馈补偿
feedback component 反馈元件
feedback control 反馈控制
feedback control system 反馈控制系统
feedback loop 反馈回路
feedback signal 反馈信号
feedback system 反馈系统

feeder 馈电线；供电户
feeder line 馈电线；馈线；支线
FEM (finite element method) 有限元法
fence 挡板；栅栏
fence rail 栏杆；栅栏
fender 翼子板；挡泥板
fender board 翼子板；挡泥板
fender skirt 翼子板裙板
fender support 翼子板支架
Ferrari ridge 法拉利脊；鸭尾脊
ferroconcrete tower 钢筋混凝土塔架
ferromagnetic 铁磁的
ferromagnetic flux return path 铁磁通量回路
ferromagnetic material 铁磁性材料
FET (field effect transistor) 场效应晶体管
fetch 风区；吹程；风区长度
fetch limited wind wave 有限风区风浪
FFT (fast Fourier transformer) 快速傅里叶变换
FGS-FP (fixed generator speed fixed-pitch) 恒速定桨距
FGS-VP (fixed generator speed variable-pitch) 恒速变桨距
fiber 纤维
fiber composite 纤维加强的复合材料
fiber composite blade 纤维复合材料叶片
fiber composite materials (FCM) 纤维复合材料
fiber glass 玻璃纤维
fibre glass epoxy blade 玻璃钢叶片；纤维玻璃环氧叶片
fiber glass filament 玻璃纤维丝
fibre glass material 玻璃纤维材料
fiber glass reinforcement 玻璃纤维加强
fibre insulation 纤维绝缘

fibre material 纤维材料
fibre metal 纤维状金属；金属丝
fibre optic sensor 光纤传感器
fibre placement 纤维铺放
fiber reinforced composite 玻璃纤维增强复合材料
fiber reinforced composite materials (FRC) 纤维加强复合材料
fiber reinforced plastics (FRP) 纤维增强塑料；玻璃纤维加强聚酯
fiber reinforced rubber (FRR) 纤维加强橡胶
fibreglass 玻璃丝
fibreglass blade 玻璃钢叶片
fibreglass braided wire 玻璃丝编织线
fibreglass mat 玻璃纤维薄毡
fibreglass reinforced plastics 玻璃纤维增强塑料
fibrous 纤维的
fibrous glass 玻璃纤维
fibrous material 纤维材料
fibrous weld 纤维状焊缝
Fickian diffusion 斐克扩散
fictitious load 假荷载；模拟荷载
fictitious power 虚功率
fidelity 保真度
field 场；野外
field apparatus 现场设备
field assembled 现场组装
field assembly 现场装配
field bolt 现场安装螺栓
field checking 现场检验
field coil 励磁线圈；场线圈
field conditions 现场条件
field connection 现场联结
field copper loss 磁场铜耗

field core 磁场铁心
field current 场电流；励磁电流
field data 现场数据
field demonstration model 现场演示模型
field density 场密度；磁通量密度；场强
field effect 电场效应
field effect transistor (FET) 场效应晶体管
field fabricated 工地制造的；现场装配的
field installation 现场安装
field instrument 携带式仪表；野外仪器
field maintenance 现场维修
field maintenance equipment 现场维修设备
field maintenance personal 现场维修人员
field measurement 野外测量
field modulation generator 现场调制发电机
field monitoring 野外监测
field monitoring instrument 现场监测仪表
field mounted 现场安装
field note 现场记录
field observation 野外观测
field of force 力场
field painting 现场喷涂
field pattern 场图
field performance 现场工作性能
field reliability test 现场可靠性试验
field report 现场报告
field service 现场服务
field splice (塔架)现场接合面；现场拼接
field strength 场强
field survey 现场调查
field test 野外试验
field test with turbine 外联机试验
field testing 现场试验
field voltage 励磁电压
field welding 现场焊接

field winding 磁场绕组；励磁绕组
field-effect 电场效应
figure 数字
figure of loss 能量损耗系数
figure of noise 噪声指数
figure of performance 性能指数
filament 细线；单纤维
filament line 流线
filament reinforced metal 纤维加强金属
filament winding 纤维缠绕
filamentary structural composite 纤维结构复合材料
fill 装满
fill away 顺风行驶
fill factor 填充率，填充系数；装填系数；占空因子
filled composite 填充复合材料
filler 充填剂；填充物
filler metal 焊料；焊丝
filler ring 垫圈
filler rod 焊条
fillet weld 角焊
fillet welding 角焊；填角焊
filling 填充；填料
filling in ring 垫圈
filling material 充填材料
film cooled blade 膜冷却叶片；油膜冷却叶片
film lighting arrester 膜片避雷器
film plotting 摄影测绘
filter 滤波器；过滤器
filter circuit 滤波器电路
filter clog 过滤器堵塞
filter function 滤波函数
filter stop band 滤波器阻滞
filtered model 滤波模式

filtering frequency 滤波频率
filth 垃圾
filtration 滤除性；过滤；筛选
filtration rate 滤除率
fin 散热片
fin area 尾翼面
final conditions 边界条件
final plume rise height 羽流最终抬升高度
finder 探测器
fine balance 精调
fine coal 粉煤
fine granular 细团粒
fine gravel 细砾
fine scale turbulence 微尺度湍流
fine streaming body 细长流线体
fineness 精度；细长比；径长比
finish 完成；结束
finish work 精加工
finished bright 抛光
finished product 成品
finishing 最后的
finishing cut 精加工
finishing strip 补强胶条
finishing temperature 最终温度；终轧温度
finit element method (FEM) 有限元法
finit life 有限寿命
finite 有限的
finite difference method 有限差分法
finite span effect 有限展长效应
finny tail 鳍状尾巴
fire 火
fire alarm sounder 火灾报警器
fire barriers 防火间隔
fire extinguisher 灭火器
fire fighting system 消防系统

fire hazard 火灾
fire hydrant 防火栓
fire material 防火材料
fire pressure 火压
fire resistance 耐火性
fire resistance test 防火性试验
fire safety 防火安全措施
fire safety equipment 防火安全设备
fire safety switch 防火安全开关
fire sign 火灾信号
fire spread rate 火灾蔓延速度
firebrick 耐火砖
fireman 消防员
fireman's axe 消防斧
firing 开火；烧制
firing angle 触发角；点火角；点弧角
firing range 射程；靶场；着火范围
firm 坚定的；牢固的
firm load 固定负荷
firm output 恒定输出
firm peak discharge 恒定最大输出
firm power 恒定功率
firm power energy 恒定电能
firmer 凿子
firmer chisel 凿子；直边凿
first aid repair 紧急修理
first derivative 一阶导数
first gust 阵风前阵
first integral 初积分
first law of thermodynamics 热力学第一定律
first mode 一阶模态；基本振型
first order closure method 一阶封闭法
first order approximation 一阶近似
fish bolt 鱼尾螺栓
fishtail wind 不定向风

fission product contaminant 裂变产物污染物
fissuration structure 裂缝结构
fit 配合；装配
fit clearance 配合间隙
fit in 使适合
fit key 配合键
fit quality 配合等级
fit tolerance 配合公差
fits and tolerances 配合与公差
fitted bolt 配合螺栓；定位螺栓
fitted capacity 装配容量；设备容量
fitted curve 拟合曲线
fitting 拟合
fitting allowance 装配余量；配合公差
fitting assembly 装配总成
fitting constant 拟合常数
fitting fixture 装配工具
fitting metal 配件；附件
fitting parameter 拟合参数
fitting part 配件；零件
fitting piece 配件
fitting surface 安装面
fitting tolerance 配合公差
fitting work 装配工作
five digit series airfoil NACA 五位数字系列翼型
fixed constant 不变常数
fixed coordinate system 固定坐标系
fixed coupling 固定联轴结
fixed flange 固定法兰盘
fixed generator speed fixed-pitch (FGS-FP) 恒速定桨距
fixed generator speed fixed-pitch wind turbine 恒速定桨距风电机组；FGS-FP风电机组

fixed generator speed variable-pitch (FGS-VP) 恒速变桨距

fixed generator speed variable-pitch wind turbine 恒速变桨距风电机组；FGS-VP风电机组

fixed ground board （风洞）固定地板

fixed hub 固定桨毂

fixed joint 刚性连接

fixed load 固定负荷

fixed pitch blade 定桨距叶片

fixed pulley 定滑轮

fixed separation point 固定分离点

fixed tab 固定翼片

fixed time test 定时寿命试验

fixed wing (FW) 固定翼

fixed(-incidence) vane 固定叶片

fixed yaw rotor 定向风轮

fixing dimension 装配尺寸

fixity 稳定性；硬度

fixting bolt 固定螺栓

fixture 固定架

FL (fatigue limit) 疲劳极限；疲劳寿命；疲劳限界

flag bridge 信号桥楼

flagging 旗状（植物风速指示）；旗形羽流

flagging plume 旗形羽流

flame 火焰

flame ability 可燃性

flame arrester (damper) 灭火器

flame cutting 气矩切割

flame hardening 火焰硬化

flame resistance 耐燃性

flame welding 气焊

flammability 可燃性

flammable gas 可燃气体

flammable substance 可燃物质

flammable vapor 可燃蒸汽

flange 法兰

flange area 法兰面积

flange beam 工字钢

flange bolt 凸缘螺栓

flange bushing 法兰衬套

flange clamping plate 法兰夹固板

flange connection 凸缘连接

flange coupling 法兰盘联轴结

flange detector 轮缘检测器

flange gasket 法兰垫片

flange joint 法兰接合

flange of coupling 联轴器凸边

flange tee 法兰三通

flanged 装有法兰的

flanged beam 工字梁

flanged nut 凸缘螺母

flanged union 凸缘连接；法兰联管节；折缘管节

flank 侧面；侧墙

flank of tooth 齿面

flap 襟翼

flap chord 襟翼弦

flap drag 襟翼阻力

flap drive shafting 襟翼传动轴系

flap electric motor 襟翼操纵电动机

flap gauge 襟翼操纵机构

flap hinge moment 襟翼铰矩

flap leading edge 襟翼前缘

flap reference plane (FRP) 襟翼基准平面

flap retraction 收襟翼

flap-lag 挥舞—摆振

flapping 抖动；挥舞

flapping angle 挥舞角

flapping force 拍打力
flash butt weld 电弧对接焊；闪光焊；闪光对接焊
flashover 闪络；飞弧；击穿
flat 平坦的
flat key 平键
flat plate drag 平板阻力
flat plate flow 平板绕流
flat plate flutter 平板颤振
flat position welding （顶面）平卧焊
flat response 平坦响应
flat roof 平顶屋
flat roofed building 平顶建筑
flat shaped building 板状建筑物
flat steel 平钢；扁钢
flat terrain 平坦地形
flatwise 平放地
flatwise bending 板状弯曲；平面弯曲
flaw 生裂缝
flaw detecting hook 探伤钩
flaw detection 探伤
flaw detector 探伤仪；裂纹探测器
flaw in casting 铸件裂痕
Fletter rotor 富勒特转子
flexibility 柔韧性；适应性
flexibility factor 挠曲系数
flexibility test 挠度试验
flexible 软接头
flexible AC transmission equipment 柔性交流输电设备
flexible AC transmission system (FACTS) 柔性交流输电系统
flexible ball joint 活动球状接头
flexible bearing 柔性轴承
flexible blade 柔性叶片
flexible building 柔性建筑物
flexible cable 柔性电缆；软性电缆；软电线
flexible conduit 软管
flexible connector 软性连接管；挠性连接器；弹性接头
flexible coupling 弹性联轴器；弹性连接；挠性联结器；挠性管接头
flexible electrode 软焊条
flexible fastening 弹性固定
flexible foam blanket 柔性泡沫材料衬垫；柔性泡沫毯
flexible gear 柔性齿轮
flexible joint 软活接；柔性接头挠性接头；柔性接缝
flexible life 弯曲疲劳期限
flexible pliers 万向套筒扳手
flexible rigidity 抗弯刚度；弯曲刚性
flexible rolling bearing 柔性滚动轴承
flexible roof 柔性屋顶
flexible structure 柔性结构物
flexible suspension bridge 柔性悬索桥
flexible walled wind tunnel 柔壁风洞
flexural 弯曲的
flexural axis 弯曲轴
flexural central 弯曲中心
flexural damping 弯曲阻尼
flexural fatigue 弯曲疲劳
flexural loading test 弯曲载荷试验
flexural measurement 弯曲测量
flexural mode vibration 弯曲振动；弯曲型振动
flexural oscillations 弯曲振动
flexural rigidity 弯曲刚度；抗挠刚度；抗弯刚度
flexural stiffness 抗弯刚度
flexural strength 抗弯强度

flexural stress 挠曲应力
flexure 屈曲
flexure strength 抗弯强度
flexure stress 弯曲应力
flexure test 弯曲试验
flexure test machine 挠曲试验机
flexure torsion flutter 弯扭颤振
flicker 闪变
flicker coefficient for continuous operation 持续运行的闪变系数
flicker severity factor 闪烁劣度系数；闪烁强度系数；闪烁严重程度因子
flicker step factor 闪变阶跃系数
floater 浮球；漂珠；浮子
floating action 无定向动作；不稳作用
floating dock 浮坞
floating dust 飘尘
floating platform 浮动平台
floating wind turbine 漂浮式风电机组
flow 流动；流量
flow amount 流量
flow angularity 气流偏角；流角
flow area 流动截面积；流通面积
flow around regime 绕流状态
flow back 逆流；回流
flow blockage 流动阻塞
flow boundary layer thickness 流动边界层厚度
flow calibration 流场校测
flow calibration in working section 风洞试验段的流场校准
flow capacity 流量
flow characteristic 流动特性
flow chart 流程图
flow choking 流动壅塞

flow continuity 流动连续性
flow control circuit 流量控制回路
flow control device 流量控制装置
flow control regular 流量控制
flow control system 流量控制系统
flow control unit 流量控制装置
flow control valve 流量调节阀；带单向阀的流量控制阀
flow controlling gate 节流阀
flow curvature 流动弯曲
flow direction detector 流向检测装置
flow discontinue 流动不连续性
flow discontinuity 流动不连续性
flow distortion 气流畸变
flow divergence 流动发散
flow drag 流动阻力
flow driven oscillation 流动激振
flow duration 风洞工作时间；流动连续时间
flow dynamics 流体动力学；流体力学
flow equation 流量方程
flow expansion 气流膨胀
flow field 流场
flow field calibration 流场校测
flow field quality 流场品质
flow field representation 流场表示法
flow fluctuation 气流脉动
flow form 流动形态
flow friction 流动摩擦
flow friction characteristics 流动摩擦特性
flow governor 流量调节器
flow graph 流线图
flow in 流进（入）
flow in continuum 连续流
flow in convection 对流气流
flow in momentum 动量变化

flow in three dimension 三维（空间）流动
flow in two dimension 二维流动
flow in vortex 涡流
flow inclination angle 气流偏角；气流倾角
flow induced vibration 流致振动；流动激振
flow irregularity 流动的不均匀性
flow layer 流层
flow limit 塑性流动（极限）
flow loss 流动损失
flow map 流谱
flow mass 质量流量
flow measurement 流量测量
flow mechanics 流体动力学
flow mechanism 流动机理
flow model 流动模型
flow modelling technique 流动模拟技术
flow monitor 流动监测器
flow noise 流动噪声
flow oscillation 流动振荡
flow passage 流道
flow path 流迹
flow pattern 流谱
flow perturbation 气流扰动
flow processing section （风洞）整流段
flow quality 流动特性
flow quantity 流量
flow rate 流量
flow reattachment （分离的）气流再附着
flow recirculation zone 回流区
flow regime 流动状态
flow resistance 流阻
flow reversal 回流
flow Reynolds number 流动雷诺数
flow separation 流动分离
flow separation phenomenon 气流分离现象

flow simulation 流动模拟
flow soldering 射流焊接
flow speed 流速
flow stability 流动稳定性
flow streamline 流线
flow stress 塑流应力
flow stress ratio 塑流应力比
flow structure 流动结构
flow superposition 流动叠加法
flow survey 流动测量
flow switching 流向转换
flow test 流动试验；流量试验
flow tube 流管
flow turbulence 流动紊流
flow uniformity 流动均匀性
flow value 流值
flow velocity 流速
flow visualization 流动显示
flow welding 铸（浇）焊
flowed 流动
flowed energy 气流能量
flowed equation 流动方程
flowed fluctuation 流量波动
flowed friction 流动阻力
flowing 流动的
flowing property (power) 流动性
flowing stream 流；气（液）流
flown line 流纹；流线
fluctuating 波动的
fluctuating aerodynamic force 脉动气动力
fluctuating concentration 脉动浓度
fluctuating deflection 脉动挠度
fluctuating external wind loading 外部脉动风载
fluctuating flow pattern 脉动流谱

fluctuating force 脉动力

fluctuating frequency 脉动频率

fluctuating internal wind loading 内部脉动风载

fluctuating load 变动载荷

fluctuating moment 脉动力矩

fluctuating plume model 脉动羽流模式

fluctuating pressure 脉动压力

fluctuating reattachment 脉动再附

fluctuating response 脉动响应

fluctuating separation 脉动分离

fluctuating wind 脉动风

fluctuation 脉动；波动

fluctuation of service 运行不稳定性

fluctuation period 脉动周期

fluctuation rate 波动率；变动率

fluctuation statistics 脉动统计（量）

fluctuation stream 脉动流；不稳定流

flue 烟道

flue dust 烟尘

flue gas 烟气；废气

fluid 流体

fluid behavior 流动特性

fluid body 流体

fluid boundary 流体边界

fluid boundary layer 流体边界层

fluid condition 流体状态

fluid coolant 流体冷却剂

fluid coupling 液力耦合器；液力联轴器

fluid cycle 液力循环

fluid damper 流体减震器

fluid damping 流体阻尼

fluid dynamic damping 流体动力阻尼

fluid dynamic parameter 流体力学参数

fluid dynamics 流体动力学

fluid equation 流体方程

fluid feedback 流体反馈

fluid flow 流体流量

fluid friction 流体摩擦

fluid horsepower 流体功率

fluid induced vibration 流体致振

fluid jet 流体射流

fluid kinetics 流体动力学

fluid level indicator 液位指示器

fluid mechanics 流体力学

fluid mechanics principle 流体力学原理

fluid medium 流体；流体介质

fluid memory effect 流体记忆效应

fluid parcel 流体团；流体块

fluid particle 流体质点

fluid power 流体动力

fluid power motor 液力马达

Fluid Power Society (FPS) 流体动力学会

fluid pressure 流体静压力

fluid property 流体性能

fluid state 流态

fluid static pressure 流体静压力；静水压力

fluid velocity 流体流速

fluid viscosity 流体黏性（系数）

fluid whirl 流体旋涡

fluidity 流动性

fluidized bed 流化床

fluorescence analysis 荧光分析

fluorescent crack (flaw) detection 荧光探伤

fluorescent fault detector 荧光探伤仪

fluorescent filament 荧光微丝

fluorescent oil flow method 荧光油流法

fluorescent particle tracer 荧光示踪粒子

fluorescent tracer technique 荧光示踪技术

fluorescent pigment tracer 荧光示踪颜料

flush hole 蒙皮破孔
fluted covering 波纹蒙皮
flutter 颤振
flutter coefficient 颤振系数
flutter derivative 颤振导数
flutter frequency 颤振频率
flutter mode 颤振模态；颤振模型
flutter moment 颤振力矩
flutter stability 颤振稳定性
flutter tendency 颤振趋势
flutter wind speed 颤振风速
flux 磁通
flux density 磁通密度
flux linkage 磁链；磁通匝连数
flux of exhaust buoyancy 排气的浮力通量
flux profile 通量廊线
flux Richardson number 通量理查森数
flux wave 磁通波
flux weakening 弱磁；磁通降低
flux weakening region 磁场弱化区；弱磁区
fly ash 扬灰；飞尘
flying boat 飞艇
flying bridge 驾驶台
flying gangway 天桥
flying spot scanner 飞点扫描器
flywheel 飞轮
flywheel action 飞轮效应
flywheel energy storage 飞轮储能法
flywheel inertial storage 飞轮惯性储能
FMS (force measuring system) 测力系统
foam 泡沫；起泡沫
foam article 泡沫制品
foam fire extinguisher 泡沫灭火器
foam glass 泡沫玻璃
foam insulation 泡沫绝缘材料；泡沫隔热

foam insulation material 泡沫隔热材料
foam plastics 泡沫塑料
foam sandwich 泡沫夹芯结构
foamed 泡沫状的
foamed materials 泡沫材料；多孔材料
foamed plastic 泡沫塑料
foaming substance 发泡物质
foehn wind 焚风
fog 雾
fog frequency 雾频
fog lamp 雾灯
foil 反射板；箔片
foilcraft 水翼艇
following distance 尾随距离
following edge 后缘
following train 后行列车
following wind 顺风
food dye 食用染料
foot of a perpendicular 垂足
footbridge 人行街
force 力量
force balance 力平衡；测力天平
force cell 测力传感器
force coefficient 力系数
force component 分力
force derivative 力导数
force diagram 作用力示意图
force due to friction 摩擦阻力
force equilibrium 力平衡
force excited oscillation 力激振动
force fan 压风机；增压风扇；强压通风机；强力风扇
force friction 摩擦力
force line 力线
force measuring system (FMS) 测力系统

force moment	力矩
force polygon	力多边形
force resolution	力的分解
force spectrum	力谱
force test	测力试验
force triangle	力三角形
force unbalance	力失衡
force vector	力矢量
forced	被迫的
forced air cooler	强制空气冷却器
forced air refrigeration system	强制空气循环制冷系统
forced circulating	强制循环
forced circulation air-cooling	强制循环空气冷却
forced commutated inverter	强迫换流逆变器；强制转换逆变器
forced cooling	强制冷却
forced flow	强迫流动
forced lubrication	强制润滑；压力润滑法
forced oscillation	受迫振荡
forced rolling	强制横摇
forced stop	强制停机
forced surface air cooling	强迫表面空气冷却
forced ventilation	强制通风
forced vibration	受迫振动
forced vortex	强迫涡
forced yaw system	强迫偏航系统
forcing frequency	强迫频率
forcing function	强迫函数
fore	前部
fore and aft rigged vessel	纵帆船；纵帆帆船
fore and aft sail	纵帆；前后帆
fore axle	前桥；前轴
fore bearing	前轴承
forebody	前体
forebody drag	前体阻力
foresail	前帆
foreshore	海岸
forest	森林
forest canopy	林冠
forest community	森林群落
forest fire smoke	森林火灾烟
forest meteorology	森林气象学
forge	锻造
forged	锻造的
forged high carbon steel	锻造高碳钢
forged hollow shaft	锻造空心轴
forged hollow steel shaft	锻造空心钢轴
forging	锻造
forging aluminium	锻铝
forging cold	冷锻
forging copper	铜锻
forging dies	锻模
forging steel	钢锻
forging welding	锻焊；锻接
form	形状；构成
form brace	模板支撑；模板拉条
form control image	格式控制图像
form drag	形阻
form factor	型数；体型系数
form freeboard	型体干舷
form resistance	型阻
former	从前的；模型
forming	形成
forming resilient joint	弹性联结
forward	向前的
forward curved blade	前弯式叶片
forward facing front surface	向前正面锋面
forward facing surface	向前锋面

forward flow 顺流；正向流
forward flow zone 顺流区
forward mounted flap 前缘襟翼
forward power loss 正向功率损耗
forward stagnation point 前驻点
forward stagnation streamline 前驻点流线
forward transfer function 正向传递函数
fossil energy resource 矿物能源
fossil fuel 矿物燃料
foul 犯规的；犯规
foul air 污浊空气
foul gas 臭气
foul wind 恶风
foundation 基础
foundation bolt 地脚螺栓
foundation connection 基础连接
foundation earth electrode 基础接地体
four point ball bearing 四点球轴承
four digit series airfoil NACA 四位数字系列翼型
Fourier 傅里叶
Fourier amplitude spectrum 傅里叶幅值谱
Fourier analysis 傅里叶级数展开；傅里叶分析
Fourier coefficient 傅里叶系数
Fourier component 傅里叶分量
Fourier phase spectrum 傅里叶相位谱
Fourier transformation 傅里叶变换
Fourier's series (FS) 傅里叶级数
Fourier's integral 傅里叶积分
Fourier's law 傅里叶定律
Fourier's number 傅里叶数
four paws 四爪
four quadrant back-to-back PEC 四象限背靠背 PEC
four-jaw chuck 四爪卡盘

four-jaw concentric chuck 四爪同心卡盘
four-jaw independent chuck 四爪单动卡盘
four-jaw plate 四爪卡盘
FPS (Fluid Power Society) 流体动力学会
fraction 分数
fraction gauge 组合齿轮
fractional 部分的
fractional heat 摩擦热
fractional load 轻载；部分负荷
fractional pressure 分压
fractional tailed test 局部破坏试验
fracture 断裂
fracture behaviour 断裂特性
fracture strength 断裂强度
fracture stress 断裂应力
fracture test 断裂试验
fracturing 使破裂
fracturing load 致断负载
fractus nimbus 碎雨云
fragility 脆性
fragment 碎片；片段
frame 框架
frame shielding coefficient 框架遮蔽系统；框架屏蔽系数
frame structure 框架结构
frame shear wall structure 框架剪力墙结构
framework 骨架；框架
FRC (fiber reinforced composite materials) 纤维加强聚酯
freak 畸变；变异
free air 大气；自由大气；自由空间；自由空气
free air data 大气数据
free air facility 空气动力试验设备
free air tunnel 大气湍流；自由空气风洞
free area 有效截面

free atmosphere 自由大气
free atmosphere wind 自由大气风
free beam 简支梁
free bearing 球形支座
free boundary 自由边界
free convection 自由对流
free convection cooling 自由对流冷却
free cooling 自由冷却
free decay 自由衰减
free energy 自由能
free expansion 自由膨胀
free flight wind tunnel 自由飞行风洞
free flow 自由流
free flow check valve 自由流动单向阀
free frequency 固有频率
free from vorticity 无涡的；自由涡
free jet 自由射流
free jet wind tunnel 自由喷射风洞
free lift 自由升力
free oscillation 自由振荡
free play 齿隙；空隙；空转
free pulley 惰轮；动滑轮
free rolling 自由横摇
free running 空转
free running frequency 固有频率
free running operation 自由震荡；自由振荡
free settling 自由沉降
free shear layer 自由剪切层
free slack between couplers 耦合器之间游隙
free stagnation point 自由驻点
free state 游离状态；单体状态
free standing stack 独立烟囱
free stream 自由来流；未受扰动流
free stream boundary 自由流边界
free stream surface 自由流面

free stream turbulence 自由来流湍流；自由来流湍流度
free stream value 自由来流值
free stream wind 非扰动气流
free stream wind speed 自由流风速
free streamline 自由流流线；势流流线；无旋流线；位势流线
free thermal convection 自由对流
free turbulence boundary 自由湍流边界
free vector 自由矢量
free vibration 自由振动
free volume 净容积
free vortex 自由涡
free vortex blading 自由涡流叶片组
free vortex sheet 自由涡流面
free vortex type 自由涡流
free vorticity 自由涡量
free way 快速道
free wheeling 单向离合器；自由旋转；自由转轮机构
free wind 顺风
free yaw rotor 定向风轮；自由偏航转子
free yawing 自由偏航
free(simple) beam 简支梁
freedom of motion 运动自由度
freewheeling 惯性滑行
freezing 冰冻的；严寒的；冷冻用的
freezing fog 冻雾；雾凇
freezing rain 冻雨
frequency 频率
frequency analyser 频率分析仪
frequency analysis 频率分析
frequency band 频带
frequency bandwidth 频率频宽；带宽
frequency changer 变频器

frequency characteristics 频率特性
frequency controller 频率控制器
frequency convertor 变频器；频率变换器
frequency dependent damper 依赖于频率的阻尼器
frequency distribution 频率分布
frequency distribution of the wind speed 风速频率分布
frequency domain 频域
frequency drift 频移
frequency effect 频率效应
frequency meter anemometer 频率表式风速计
frequency modulation 调频
frequency of gust 阵风频率
frequency of occurrence 出现频率
frequency of sampling 取样频率
frequency of turbulence 湍流频率
frequency of vortex shedding 旋涡脱落频率
frequency of wind direction 风向频率
frequency of windspeed 风速频率
frequency response 频率响应
frequency shift keying (FSK) 移频键控
frequency spectrum 频谱
frequency variation 频率变动
frequent wind speed 常现风速
fresh air 新鲜空气
fresh breeze 五级风 (29~38千米/时)
fresh gale 大风；八级风 (62~74千米/时)
freshen 增强；变强
fret 回纹饰；腐蚀处；磨损处
fretting 微振磨损；侵蚀
friction 摩擦
friction clutch 摩擦离合器
friction coat 耐磨涂层

friction coefficient 摩擦系数
friction disk 摩擦片
friction drag 摩擦阻力
friction energy loss 摩擦能量损失
friction factor 摩擦力；摩擦因子；摩擦率
friction index 摩擦指数
friction layer 摩擦层
friction loss 摩擦损失
friction of motion 滑动摩擦
friction of rest 静摩擦
friction of rolling 滚动摩擦
friction pad 摩擦垫；摩擦块
friction resistance 摩擦阻力
friction sheave 摩擦盘
friction type torque 摩擦式扭矩
friction value 摩擦值；摩擦系数
friction velocity 摩擦速度
friction wake 摩擦尾流
friction wear 摩擦损耗
friction work 摩擦功
frictional boundary layer 摩擦边界层
frictional drag 摩擦阻力
frictional heat 摩擦热
frictional moment 摩擦力矩
frictional power 摩擦功率
frictional pressure 摩擦压力
frictional pressure loss 摩擦压力损失
frictional resistance 摩擦阻力
frictional stress 摩擦应力
frictional torque 摩擦扭矩
frictional work 摩擦功
frictionless liquid 无摩擦流体；理想流体
frictionless wind 无摩擦风；理想风
frigostabile 耐低温的
fringe region of atmosphere 大气边缘区

front 前锋；锋面；前端面
front axle 前轴；前桥
front end processor 前端通信处理机；前端处理机
front end style 头部式样
front fender 前翼子板；前叶子板
front gear box 前齿轮箱
front inversion 锋面逆温
front lift 前轮升力
front pillar A支柱；前风窗支柱
front side force 前轮侧力
front span 前节距
front spoiler 前扰流板
front standing pillar A支柱；前风窗支柱
frontage 屋前空地；正面空地
frontal 正面的；前面的
frontal appearance 前部外形
frontal area 正面面积；迎风面积
frontal cyclone 锋面气旋
frontal drag 迎面阻力
frontal inversion 锋面逆温
frontal line 锋线
frontal passage 锋面过境
frontal projected area 迎面投影面积
frontal resistance 迎面阻力
frontal surface 锋面
frontal version 锋面逆温
frontal zone 锋带
frost 霜
frost point 霜点
frost smoke 冻烟
Froude number 弗劳德数
Froude scaling 弗劳德缩尺
Froude similarity 弗劳德相似(性)
frozen 冻结的
frozen battery 不充电电池
frozen gust 冻结阵风
frozen rain 冻雨
frozen turbulence 冻结湍流
FRP (fuselage reference plane) 机身水平基准面
FRR (fiber reinforced rubber) 纤维加强橡胶
FS (Fourier's series) 傅里叶级数
fuel cell battery 燃料组元电池
fuel consumption 燃耗；燃料消耗
fuel storage 燃料储存
full 完全的
full aeroelastic approach 全气动弹性法
full and by 满帆；扯满帆地；满帆顺风地
full automation and cybernation 完全自动控制
full body exposure 全身照射
full bridge converter 全桥变流器
full bridge model 全桥模型
full depth 全高
full depth gear 标准齿高齿轮
full depth involute system 全高齿渐开线制
full depth simulation 全厚度模拟
full depth tooth 全齿高齿
full developed boundary layer 充分发展边界层
full developed flow 充分发展流动
full feather 完全顺桨
full feathered position 完全顺风位置
full feathering 顺桨
full flap 全襟翼
full gale 强风
full ground reflection 地面全反射
full hardening 淬透
full load 满载

full load characteristic	满负荷特性
full load current	满载电流
full load efficiency	满负荷效率
full load loss	满负荷损失
full load operation	满载运行
full load power peak	最大负荷；功率峰值
full load rating	额定全负荷
full load run	满载运行
full load running	满载运转
full load speed	满负荷速度
full load test	满负载试验
full load torque	满载转矩
full running speed	全速运行
full scale data	全尺寸数据
full scale experiment	全尺寸实验
full scale measurement	全尺寸测量
full scale model	全尺寸模型
full scale prototype	全尺寸原型物
full scale Reynolds number	全尺寸雷诺数
full scale time	全尺寸时间
full scale wind tunnel	全尺寸风洞
full scale wind tunnel test	全尺寸风洞试验
full sized model	全尺寸模型
full span	整个叶片
full span blade pitch	变桨距；全翼展叶片间距
full span bridge model	全桥模型
full span flap	全翼展襟翼
full span flutter	全跨颤振
full span galloping	全跨驰振
full span ground board	全跨地板
full span pitch control	全跨度桨距控制
full stall	全失速
full wave rectification	全波整流
full wave rectifier (FWR)	全波整流
fully attached flow	完全附着流(动)
fully controlled device	全控型器件
fully developed boundary layer	充分发展边界层
fully developed profile	完全发展速度廓线
fully developed wake	完整的尾流
fully laminar	全部层流的
fully rigid model	完全刚性模型
fume	烟雾
fume abatement	烟雾消除
fume emission	烟气排放
fume height	烟柱高度
fume shape	烟雾形
fumigating plume	下熏型羽流
fumigation	熏烟；熏沉
functional dependence	函数关联
functional determinant	函数式行列式
fundamental	基本的
fundamental active power	基波有功功率
fundamental assumption	基本假设
fundamental current	基波电流
fundamental frequency	基频；固有频率
fundamental harmonic	基波
fundamental law	基本定律
fundamental mode	主模态；基谐模
fundamental oscillation	固有振动；基本振荡
fundamental resonance	基频谐振
fundamental wave	基波
fundamental wave length	基波长
funnel	烟囱；漏斗
funnel cloud	漏斗云
funnelling effect	漏斗效应
funnelling plume	漏斗型羽流
furacana	飓风
furling	收拢
furling device	（风力机）停车装置；收帆装置

furling handle （风力机）停车手柄

furling wind speed 收帆风速；（风力机）停车风速

fuse 保险丝；熔断器

fuse block 保险丝盒

fuse point 熔点

fuse rated voltage (FRV) 保险丝额定电压

fuse rating 保险丝额定值

fuse wire 保险丝

futurology 未来学

FW (fixed wing) 固定翼

FWR (full wave rectifier) 全波整流器

现代英汉风力发电工程

A Modern English-Chinese Dictionary of Wind Power Engineering

G

gage 计量器
gain 增益
gain changer 传动比变换装置
gale 大风（暴）(8级风，风速 17.2~20.7米/秒)
gale damage 大风损失；风灾
gale pollution 大风污染
galling 表面机械损伤
galloping 驰振
galloping criterion 驰振判据
galloping excitation 驰振激励
galloping flutter 驰振型颤振
galloping force coefficient 驰振力系数
galloping instability 驰振不稳定性
galloping response 驰振响应
galloping torsion 驰振扭转
galloping vibration 驰振型振动
galvanic battery 原电池
galvanic cell 原电池
galvanic current 稳定的直流电；伽伐尼电流；动电电流
galvanic potential 电位；伽伐尼电位
galvanic series 电势序；电位序；电压序列
galvanizing 通电流于…；电镀给…；镀锌
galvanometer 电流计
gamma distribution γ分布

gamma function γ函数
gamnitude 倒幅度
gang saw 直锯
gantry 起重机架
gantry crane 龙门起重机
gap 间隙；峡口；山口
gap adjustment 间隙调整
gap clearance 间隙；对缝间隙
gap frame C型框架
gap seal 间隔封罩；间隙密封
garboard 龙骨翼板
garden planning 园林规划
gas 气体
gas absorption 气体吸附
gas black 气黑；气烟末；天然气炭黑
gas cavity 气孔
gas chromatographic column 气相色谱柱
gas cleaning 气体净化
gas constant 气体常数
gas contamination 气体污染
gas current 电流
gas cutting 气割
gas diffusion 气体扩散
gas dispersion 气体弥散；气体扩散
gas dynamic behavior 气体动力特性

gas dynamic equation 气体动力方程
gas dynamic facility (GDF) 气体动力研究设备
gas dynamic function 气体动力函数
gas dynamic theory 气体动力学理论
gas eddy 气涡
gas efflux speed 气体排放速度
gas equation 气体方程式
gas equilibrium 气体平衡
gas flame welding 气火焰焊接
gas flow expansion 气流膨胀
gas insulated switchgear (GIS) 气体绝缘开关设备
gas kinetics 气体动力学
gas law 气体定律；气体状态方程
gas layer produced surface shear 气流附面层引起的表面剪切力
gas phase 气相
gas plume 气态羽流
gas poisoning 煤气中毒
gas rise 气体抬升
gas sampler 气体取样器
gas shield welding 气体保护弧焊
gas sphere 气界；气圈
gas transfer 气体传递；气体输运
gas turbine locomotive 燃气轮机车
gas welding 气焊接
gas dynamics 气体动力学
gasdynamics tunnel 气体动力研究风洞
gaseous 气态的
gaseous diffusion 气体扩散
gaseous effluent 气体排放物
gaseous emission 气体排放；气体排放物
gaseous fuel 气体燃料
gaseous impurity 气态杂质
gaseous plume 气态羽流

gaseous pollutant 气体污染物
gaseous state 气态
gaseous waste 废气
gasification 气化
gasket 垫片；垫圈；接合垫；衬垫
gate 门
gate charge 选通电极充电；门控充电；门电荷
gate current 门电流
gate pole 门极
gate signal 门信号；选通信号
gate terminal 门极端子
gate turn off thyristor (GTO) 门极可关断晶闸管
gate valve 门阀
gauge 标准尺；规格；量规；量表；测量
gauge board 样板；模板规准尺
Gaussian diffusion 高斯扩散
Gaussian distribution 高斯分布；正态分布
Gaussian plume 高斯羽流
Gaussian plume equation 高斯羽流方程
Gaussian plume model 高斯羽流模式
Gaussian plume parameter 高斯羽流参数
Gaussian white noise 高斯白噪声
gauze screen 金属丝网；(风洞)阻尼网
gavel 槌；锤子
GB 国家标准
GBS 国家标准局
GBX (gear box) 齿轮箱
gear 齿轮；传动装置
gear backlash 齿轮啮合背隙
gear bearing not driving end 齿轮轴承非驱动端
gear box (GBX) 齿轮箱
gear box ratio 齿轮箱变比
gear case 齿轮箱

gear chamfering 齿轮倒角
gear chatter 轴承颤动
gear clutch 齿轮离合器
gear composite 齿轮组合
gear cutting machines 齿轮切削机
gear cutting operation 切齿操作
gear down 齿轮减速；换低挡；起落架放下
gear efficiency 传动装置效率
gear fan 齿轮箱冷却风扇
gear grease 齿轮润滑脂
gear housing 齿轮箱体
gear hub 齿轮毂
gear lever 变速杆
gear lubricant 齿轮润滑油
gear motor 齿轮马达
gear oil pump high-speed 齿轮油泵高速
gear oil pump low-speed 齿轮油泵低速
gear pair 齿轮副
gear pair with parallel axes 平行轴齿轮副
gear pump 齿轮泵
gear ratio 齿数比；齿轮传动比
gear set 齿轮组
gear train 齿轮系
gear transmission 齿轮传动
gear type motor 齿轮式马达
gear up 加速传动装置；齿轮增速
gear water pump 齿轮水泵
gear wheel 齿轮
gearbox 齿轮箱
gearbox bearing 齿轮箱轴承
gearbox casing 齿轮箱体；变速箱体
gearbox lubricant oil 齿轮箱润滑油
gearbox manufacturer 齿轮箱制造厂商
gearbox nominal power 齿轮箱额定功率
gearbox oil 齿轮箱油

gearbox oil volume 齿轮箱润滑油体积
gearbox ratio 传动比
gearbox shock absorber 齿轮箱减震器
gearcase (GRC) 齿轮箱
geared 用齿轮传动的
geared ratio 速比；传动比
geared wind turbine 齿轮驱动风电机组
gearhousing 齿轮箱壳
gearing 传动；传动装置；齿轮传动
gearing down 减速传动
gearing in 啮合
gearing up 增速传动；齿轮增速
gears with addendum modification 变位齿轮
gearshaft 齿轮轴
Geiger counter 盖格计数器
gel coat 表面涂漆
gel coated 胶衣
general 一般的
General Assembly 联合国大会
general assembly 总装配
general circulation 环流；大气环流
general dimension 主要尺寸
general environment 总环境；一般环境
general machining centers 通用加工中心
general overhaul 总翻修
general planning 总体规划
general proportions 总体比例
general rolling country 一般丘陵地区；缓坡丘陵
general stream 主流；自由流
generalized 广义的；普遍的
generalized aerodynamic force 广义气动力
generalized aerodynamic moment 广义气动力矩
generalized coordinate 广义坐标

generalized load 广义荷载
generating 发电
generating capacity 发电量
generating unit 发电机组
generator 发电机
generator circuit breaker 发电机电路断路器
generator current limiter 发电机限流器
generator fan external 发电机外部风扇
generator fan internal 发电机内部风扇
generator power angle 发电机功角
generator rotor shaft 发电机转子轴
generator speed range 发电机转速范围
generator terminal 发电机端子
generator voltage 发电机电压
generator winding thermistor protection 发电机绕组热敏电阻保护装置
gentle 温和的
gentle breeze 轻风；三级风
gentle wind 和风
geographic distribution 地理分布
geographic location 地理位置
geographic region 地理区
geography 地理学
geometric 几何学的；几何体
geometric accuracy 几何准确度
geometric airfoil 几何翼型
geometric angle of attack 几何迎角
geometric boundary condition 几何边界条件
geometric center 几何中心
geometric chord of airfoil 几何弦长
geometric dimension 几何尺寸
geometric height 几何高度
geometric incidence 几何迎角
geometric leading edge 几何前缘
geometric power diagram 矢量功率图

geometric progression 几何级数；等比级数
geometric proportion 等比
geometric scale factor 几何缩尺因子
geometric scale model 几何缩尺模型
geometric series 几何级数；等比级数
geometric similarity 几何相似性
geometric stiffness 几何刚度
geometric twist 几何扭转
geometrical error 几何误差
geometrical position 几何位置
geometry 几何学；几何形状
geomorphic blow 地貌渐变
geomorphic feature 地貌；地理特征
geomorphic occurrence 地貌突变
geomorphic profile 地貌剖面图
geomorphy 地貌
geophysical vortex 地球物理旋涡
geopotential height 位势高度
geostrophic acceleration 地转加速度
geostrophic advection 地转平流
geostrophic balance 地转平衡
geostrophic current 地转风气流
geostrophic departure 地转偏差
geostrophic drag coefficient 地转阻力系数
geostrophic equilibrium 地转平衡
geostrophic flow 地转风气流
geostrophic force 地转力
geostrophic Richardson number 地转理查森数
geostrophic shear 地转切变
geostrophic transport 地转运输
geostrophic wind 地转风
geostrophic wind field 地转风场
geostrophic wind height 地转风高度
geostrophic wind vector 地转风矢量

geostrophic wind velocity 地转风速
geosynchronous meteorological satellite 地球同步气象卫星
GF (glass fiber) 玻璃纤维
GFR (glass fiber reinforced) 用玻璃纤维加强的
GFRP (glass fiber reinforced plastic) 玻璃纤维增强塑料
giant nuclei 巨核
giant particle 巨大粒子
gigawatt hour 十亿瓦特时
gill 加强筋；散热片；腮
gimbals 平衡环；平衡架
gimlet 手钻；螺丝锥
girder 梁；桁架
girder bridge 桥梁桥
girder frame 横梁；桁架梁
girder structure 大梁式结构；梁式结构
girder truss 桁架梁
girth 围绕
girth welding 环缝焊接
glaciation 冰蚀
gland bonnet 密封盖；轴端密封盖
gland retainer plate 轴密封盖
glass break 玻璃状断口
glass curtain 玻璃幕墙
glass cutter 玻璃刀
glass epoxy 玻璃钢板；环氧树脂玻璃
glass fabric 玻璃布
glass fiber (GF) 玻璃纤维
glass fiber coat 玻璃纤维敷层
glass fiber epoxy laminate 玻璃纤维环氧层压板
glass fiber mat wool 玻璃纤维棉垫
glass fiber reinforced (GFR) 用玻璃纤维加强的
glass fiber reinforced composite materials (GRC) 玻璃丝加强复合材料
glass fiber reinforced epoxy resin (GRE) 玻璃纤维增强环氧树脂
glass fiber reinforced plastic (GFRP) 玻璃纤维增强塑料
glass fiber reinforced thermoplastic materials (GRTP) 玻璃纤维加强的热塑材料
glass fibre reinforced polyester resin (GRP) 玻璃纤维增强聚酯树脂
glass reinforced plastics 玻璃增强塑料
glasshouse effect 暖房效应
Glauert-Den Hartog criterion 葛劳渥-德哈托判据
glaze 雨凇
glaze ice 雨冰
glazed frost 雨凇；冻雨
glazing panel 玻璃幕墙
glider 滑翔机；滑行艇；滑翔运动员
gliding 滑翔滑行的；流畅的；滑顺的
gliding ratio 滑翔比
glime 半透明冰；雨雾凇
global air pollution 全球大气污染
global circulation 全球环流
global circulation pattern 全球环流型
global climate 全球气候
global contaminant 全球性污染物
global dispersion 全球性弥散
global dispersion model 全球弥散模式
global network of research station 全球研究观测站网
global radiation 总辐射
global roughness 宏观粗糙度
global rule 全球尺度；普遍原则

global scale 全球尺度
global solar radiation 总辐射
global tide 全球潮汐
global warming 全球增温
globe bearing 球面轴承
glow corona 电晕
glycol water 乙二醇溶液
goodness factor 品质因数
gooseneck ventilator 鹅颈通风筒
gorge 峡谷;河谷
Görtler instability 戈特勒不稳定性
Göttingen type wind tunnel 哥丁根型(单回路)风洞;开口回流风洞
governing 调节
governing equation 控制方程
governing parameter 控制参数
governor 调节器
grab sample 定时取样样品
grade 坡度;等级;粒径
grade change 粒径变化
grade estimation 质量评定
graded blockage 分层变阻塞;用于风速廓线风洞模拟的分层变阻塞
gradient 梯度;坡度
gradient of gravity 重力梯度
gradient of temperature 温度梯度
gradient Richardson number 梯度理查德森数
gradient theory 梯度理论
gradient transfer theory 梯度运输理论;梯度转移理论
gradient wind 梯度风
gradient wind height 梯度风高度
gradient wind speed 梯度风速
grading diagram 级配图;分级图
grading ring 均压环

grain size 粒径
granular snow 粒雪
granule 颗粒料;颗粒;团粒
graph 图
graphite fiber composite blade 石墨纤维复合材料叶片
grass barrier 草障
grass shelter belt 防风草带
grassland 草地
gravel 砂砾;砾石
gravitation 重力;引力
gravitational 重力的
gravitational acceleration 重力加速度
gravitational field 重力场
gravitational potential energy 重力势能
gravitational settling 重力沉降
gravitational water 重力水
gravity 重力
gravity correction 重力修正
gravity current 重力流
gravity effect 重力效应
gravity sag 重力下垂
gravity settling 重力沉降
gravity spread 重力散布
gravity wave 重力波
gravity wind 重力风
grazing angle 掠射角
GRC (gearcase) 齿轮箱
GRC (glass fiber reinforced composite materials) 玻璃丝加强复合材料
GRE (glass fiber reinforced epoxy resin) 玻璃纤维增强环氧树脂
grease 油脂
greasing 涂油脂;润滑
great inversion 对流层顶层

Green's strain tenser 格林应变张量

greenhouse effect 温室效应

Greenwich mean time 格林尼治标准时间

grey body 灰体

greyhound 快速船；远洋快船

grid 电网

grid code 电网导则；电网规程

grid compatibility 电网兼容性

grid connected 并网运行

grid connected wind turbine 并网风电机组

grid coupling 电网连接器；栅极耦合

grid dropout 掉网

grid failure 电网故障

grid frequency 电网频率

grid mesh 网眼

grid point 网格点

grid side converter 电网侧变流器

grid status 电网状态

grid voltage 电网电压

Griggs-Putnam index 格里戈-普特南极数（植物风力指示）

grill(e) 格栅

grind 磨碎

grind off 磨掉

grinder 磨床

grinder bench 磨床工作台

grinder cylinder 碎木机压力缸

grinders thread 螺纹磨床

grinders tools and cutters 工具磨床

grinders ultrasonic 超声波打磨机

grinding disk 磨擦盘

grinding machines 磨床

grinding machines centerless 无心磨床

grinding machines cylindrical 外圆磨床

grinding machines universal 万能磨床

grinding tools 磨削工具

grinding wheels 磨轮

grip gauge 夹紧装置

grit 砂砾

grittiness 砂砾性

grommet 垫圈；密封垫；衬垫；索环；绝缘环

groove 凹槽；最佳状态

groove weld 坡口焊缝

gross Richardson number 总体理查森数

ground 地面；(风洞)地板

ground adhesion 地面附着力

ground axes 地轴

ground blizzard 地面雪暴

ground board (风洞)地板

ground clearance 离地净高；离地距离

ground concentration 地面浓度

ground drag 地面阻力

ground effect 面效应

ground experiment 地面实践

ground floor passageway 底层车道；底层通道

ground interference 地面干扰

ground inversion 地面逆温

ground layer 近地层

ground level 地面

ground level concentration 地面浓度

ground level inversion 地面逆温

ground level source 地面源

ground level wind environment 地面风环境

ground of building 建筑群

ground of cooling tower 冷却塔群

ground plane 地平面；(风洞)地板

ground radiation 地面辐射

ground reflection factor 地面反射因子

ground resonance 地面共振

ground roughness height 地面粗糙元高度

ground support equipment 地面支撑设备
grounding conductor 接地导体
groundwork 路基；基础
group of boundary layer 边界层组
group of building 建筑群
grout 薄泥浆；水泥浆
grouting 灌浆
grower washer 弹簧垫圈
growth of boundary layer 边界层增长
GRP (glass fiber reinforced polyester resin) 玻璃纤维增强聚酯树脂
GRTP (glass fiber reinforced thermoplastic materials) 玻璃纤维加强热塑材料
GTO (gate turn off thyristor) 门极可关断晶闸管
guide bars 导向棍；导杆；导向杆
guide blade 导流片；导流叶片
guide block 导向块
guide line 导线
guide ring 导向绳
guide wire 尺度定距索；准绳
guiding shaft 导向轴
gull wing sail 海鸥翼式帆
Gurney flap 格尼襟翼
gusset plate 角撑板；加固板
gust 阵风
gust (wind) tunnel 阵风风洞
gust alleviation factor 阵风衰减因子
gust amplitude 阵风变幅
gust and lull 阵风阵息；风阵风歇
gust anemometer 阵风风速计
gust averaging time 阵风平均时间
gust component 阵风分量
gust decay time 阵风衰减时间
gust downwash 阵风下洗

gust duration 阵风延时；阵风持续时间
gust effect 阵风效应
gust effect factor 阵风响应因子
gust energy factor 阵风能量因子
gust environment 阵风环境
gust excitation 阵风激励
gust factor 阵风因子
gust factor approach 阵风因子法
gust formation time 阵风形成时间
gust frequency 阵风频数
gust front 阵风锋面；阵风锋
gust generator 阵风发生器
gust influence 阵风影响
gust intensity 阵风强度
gust lapse rate 阵风递减率
gust lapse time 阵风递减时间
gust load 突风荷载
gust loading 阵风荷载
gust measuring anemometer 阵风风速计；阵风测量风速计
gust of rain 阵雨
gust peak speed 阵风最大风速
gust recorder 阵风记录仪
gust response factor 阵风响应因子
gust scale 阵风尺度
gust size 阵风尺寸
gust spectrum 阵风谱
gust velocity variation 阵风速度变化
gust volume 阵风容积
gustiness 阵风性
gustiness effect 阵风效应
gustiness factor 阵风因子
gustiness wind 阵风
gutter 雨水槽；排水沟；槽；栏距
guy 拉索

guy cable 拉索；缆风
guy cable anchor 拉索地锚
guy clip 线卡子
guy rope 拉线
guy tension 拉索张力
guy wire 拉线
guyed building 拉索建筑
guyed cantilever 拉索悬臂
guyed structure 拉索结构
guyed tower 拉索塔架
gymnasium 体育馆
gymnasium structure 体育馆结构

gyration 回转；环动
gyration radius 回转半径
gyre force 环动力
gyroscope 陀螺仪；回旋装置；回转仪；纵舵调整器
gyroscopic effect 陀螺效应
gyroscopic force 陀螺力
gyroscopic load 回转负荷
gyroscopic moment 回转力矩
gyroscopic precession 陀螺仪进动
gyroscopic torque 回转力矩

现代英汉风力发电工程

A Modern English-Chinese Dictionary of Wind Power Engineering

H

H steel 宽缘工字钢
habitation 住宅
hacksaw 可锯金属的弓形锯；钢锯
hail 雹
hail shooting 降雹
hailstorm 雹暴；雹暴般的降临；风雹
hair crack 细裂纹
hair felt 毛毡
hair line crack 毛细裂缝
hairpin 夹发针；发夹
hairpin bend 急弯
hairpin vortex 发卡涡
hairspring 细弹簧；游丝
half angle 半角
half coupling 联轴节
half duplex transmission 半双工传输
half frequency 半频
half model test 半模试验
half peak width 半峰宽度
half value layer 半值层
half value width 半值层宽度
half wave mode 半波模态
half wave rectification 半波整流
hammerhead crane 塔式起重机；锤头式起重机

Hamming data window 数据海明窗
hand tally 计数器
hand held wind anemometer 手持式风速计
handhold 扶手；栏杆握住；线索；把柄
handiness 操纵性轻便；灵巧；敏捷
handleability 操纵性
handling quality 操纵性能
handrail 扶手
handy rule of thumb 经验定律
hanger 吊架
hanging cable 吊索
hanging railway 悬索铁道
hanging roof 悬挂屋盖
hanging structure 悬挂结构
hard alloy 硬质合金
hard and free expansion sheet making plant 硬板（片）材及自由发泡板机组
hard brazing 硬钎焊
hard carbide 硬质合金
hard cushioning 硬式减震
hard damping 强阻尼
hard facing 表面硬化
hard facing alloy 表面硬化用合金
hard flutter 强颤振
hard hat 安全帽

hard metal 硬性金属

hard metal product 硬质合金制品

hard quench 淬硬

hard service 超负荷工作状态；不良使用；困难工作条件

hard site 硬质场地；坚固发射场

hard solder 硬焊料

hard surfacing 表面淬火

hardenability band 淬透性带

hardenability characteristic 淬透性

hardenability limits 可淬透性极限

hardenability test 淬透性试验

hardenability value 硬化指数

hardened and tempered steel 调制钢

hardened case 硬化表面

hardened plate 硬钢板；淬硬钢板

hardened resin 硬树脂

hardened right out 使完全硬化

hardener 固化剂；硬化剂

hardening and tempering 调质

hardening at subcritical temperature 低温淬火

hardening bath 淬火浴

hardening break 淬火开裂

hardening by isothermal heat treatment 等温处理硬化

hardening capacity 硬化度

hardening distortion 硬化畸变

hardening flaw 淬火裂纹

hardening strain 淬火应变

harding behaviour 淬硬特性

hardness level 硬度等级

hardness meter 硬度计

hardness penetration 淬透性

hardness standard block 硬度标准块

hardness test 硬度试验

hardware 硬件

hardware platform 硬件平台

harmful gas 有害气体

harmful impurity 有害杂质

harmful waste 有害废物

harmless odour 无害气体

harmonic 和谐的；谐波的

harmonic cancellation 谐波消除

harmonic correction 谐波校正

harmonic current 谐波电流

harmonic distortion 谐波失真；谐波畸变

harmonic number 谐波数

harmonic oscillation 谐振

harmonic voltage 谐波电压

harnessable power 可用风能；有效风能

harnessing wind （美国俗语）风能利用

harp type cable 竖琴形拉索

harvesting energy 获能

hatchet 短柄斧

hauling 牵引

HAWT (horizontal axis wind turbine) 水平轴风力机

hazard 危害；公害；危险

hazard assessment 危害评价；灾害评估

hazard beacon lamp 危险警告信号灯；危害航标灯

hazard evalution 危险性评价

hazard probability 受害概率

hazard rate 故障率

hazard rating 危害等级

haze 轻雾；霾

haze horizon 霾层顶

haze layer 霾层

HB 布氏硬度

head 头
head down 顺风航行；朝向；向下
head light 大灯
head on rotor 上风型风轮
head on wind 迎面风；逆风迎面风
head on wind machine 上风型风力机；迎风风力机
head pickup 热敏元件
head resistance 迎面阻力
head screw 主轴螺杆
head sea 顶浪
head sensor 温度传感器；传感器
head sink 散热片；压头沉落
head up 迎风航行；艏向上；船首线向上
head wind 逆风；顶风
headboard 斜帆首板
headed sphere anemometer 球状头部风速计
headlamp 大灯
headstock 主轴箱
headwind 顶头风；逆风
health hazard 对健康危害
heat absorber 吸热器
heat alarm 过热报警信号
heat balance 热平衡；热收支
heat budge 热收支
heat conductivity 导热性；热导系数
heat content 热含量；焓；含热量
heat convection 热对流
heat dissipation circuit 散热回路
heat efficiency 热效率
heat exchange 热交换器
heat flux 热通量；热流
heat island 热岛
heat island center 热岛中心
heat island circulation 热岛环流

heat island effect 热岛效应
heat loss 热损失
heat of friction 摩擦热
heat of radition 辐射热
heat of vaporization 汽化热
heat pollution 热污染
heat preserving furnaces 保温炉
heat pump 热泵
heat resistance paint 耐热漆
heat resistant material 耐热材料
heat resistant metal 耐热金属
heat resistant paint 耐热涂料；耐热漆
heat resistant quality 耐热特性；隔热性能
heat resisting glass 耐热玻璃
heat resisting material 耐热材料；耐高温材料
heat resisting property 耐热性能
heat resisting quality 耐热性；耐高温性
heat resisting test 耐热试验
heat sensing device 热灵敏装置；感温器；热敏器
heat sensitive component 热敏成分；热敏元件
heat sensitive element 热敏元件
heat sensitive paint 热变涂料；示温漆
heat sensitive sensor 热敏传感器
heat sensitivity 热敏性
heat sensitization 热敏化
heat sensor 热传感器；热敏元件
heat sink 热汇；散热片；吸热部件；冷源；热沉
heat source 热源
heat transfer 热传递
heat transmission 传热；热传导；热传递；热传送
heat transport 热量输运；传热；热交换
heat treated forged steel 热处理锻压钢
heated thermometer anemometer 热风速仪

heated wire's air speed anemometer 热线风速仪
heated wire's type anemometer 热线风速仪
heater 加热器
heating appliance 电热器
heating boxes 加热室
heating fuse 热熔丝
heating treatment furnaces 熔热处理炉
heating wire 电热丝
heavenly tunnel 高空大气湍流
heaving motion 起伏运动
heaving oscillation 起伏振荡；升沉振荡
heaving pitching motion 升沉纵摇运动
heavy crane 重型吊车
heavy current 强电流
heavy duty 高功率；重型；重载
heavy duty power electronic 高功率的电力电子器件；重型电力电子
heavy gust 强阵风
heavy oscillation 剧烈振荡
heavy repair 大修
heavy separation 严重分离
hedge 篱笆；障碍物
hedgerow 树篱；绿篱
heel of tooth 齿根面
heeling 横倾
height of release 排放高度
height of smoke outlet 排烟口高度
height of source 源高
Hele-Shaw cell 霍尔-肖盒
helical 螺旋形的
helical bevel gear 斜齿锥齿轮
helical gear 斜齿轮；柱齿轮；螺旋齿轮
helical gear shaft 螺旋齿轮轴
helical protuberance 螺旋突条

helical spur gear 斜齿正齿轮
helical strake 螺旋箍条
helical vortex sheet 螺旋涡面
helical wheel 斜齿轮
helicoidal anemometer 螺旋桨式风速计
helicoidal vortex sheet 螺旋面形涡面
helicopter rotor 直升机旋翼
heliostat cluster 定日镜簇；定日镜组
heliotropic wind 日成风；日转风
heliport 直升机场；直升机停机坪
helix 螺线；螺旋箍条
helix volute 螺旋蜗壳
helm 舵
helm angle 舵角
helm wind 舵轮风
helmet 盔；帽；安全帽
helmet shield 焊工面罩；护目头罩；盔式护罩
Helmholtz resonance frequency 亥姆霍兹共振频率
herringbone 人字形条纹
herringbone gauge 人字齿轮
herringbone gear 人字齿轮
herringbone tooth 人字齿
herringbone wheel 人字齿轮
heterogeneous atmosphere reaction 非均相大气反应；多相大气反应
heterogeneity 多相性；不均匀性；异构性
heteropolar dynamo 异极电机
hexagon 六角形
hexagon screw die 六角板牙；六角螺丝钢板
hexagon spanner 六方
hexagonal 六边的
hexagonal nut 六角螺母
high aspect ratio building 细长建筑
high cloud 高云

high current 高强度电流
high dose corridor 高剂量走廊
high drag turbulent-flow regime 大阻力紊流状态
high efficiency filter 高效过滤器
high efficiency flap 高效襟翼
high energy magnet 高能磁体
high frequency end 高频端
high frequency generator 高频发电机
high frequency response 高频响应
high gain 高增益
high grade energy 高级能
high level anticyclone 高空反气旋
high level inversion 高空逆温
high lift airfoil 高升力翼型
high lift flap 高升力襟翼
high performance 高性能的
high pitch cone roof 大坡度圆锥形顶盖
high pitch roof 陡坡屋顶
high precision 高精度
high pressure 高压；高气压
high pressure flange 高压管法兰
high pressure draft 高压抽吸；高压气流
high pressure wind tunnel 增压风洞；高压风洞
high rate battery 高速放电蓄电池；高速电池
high Reynolds number water tunnel 高雷诺数水洞
high Reynolds number wind tunnel 高雷诺数风洞
high rise block 高层大楼
high rise building 高层建筑
high rise structure 高层结构
high shoulder 超高路肩
high slip induction generator 高滑差异步发电机
high solidity rotor 高实度风轮
high speed aerodynamic 高速空气动力学
high speed airfoil 高速翼型
high speed brake 高速制动器
high speed low torque shaft 高速低扭矩轴
high speed photographic technique 高速摄影技术
high speed rotor 高速风轮
high speed shaft 高速轴
high speed shaft system 高速轴系统
high speed switching 高速切换
high speed WECS 高速风力机；额定叶尖速率比不小于3的风力机
high speed wind tunnel 高速风洞
high speed wind tunnel test 高速风洞试验
high speed winding 高速绕组
high strength bolt 高强度螺栓
high strength bolting 螺栓高强度固定
high strength tension bolt 高抗拉强度螺栓；高强度拉紧螺栓
high strength toughened heat-treated steel 高强度韧化热处理钢
high temperature brazing 高温钎焊
high temperature cemented carbide 耐高温硬质合金；高温硬质合金
high tension cable 高压电缆
high voltage 高压
high wind 大风；疾风
high wind fumigation 强风下熏
higher mode 高阶模态；高阶振型；高次模
higher order closure 高阶封闭
highest common divisor 最高公约式
highly cambered blade 大曲度叶片
hightensile 高强度

hill 丘陵

hill brow 山顶；山眉

hill cluster 群山；丘陵群

hill crest 山峰

hillock plain 低丘平原

hillside 山坡

hillslope 山坡

hilly cross country 丘陵原野

hindcasting 追算；倒推法（借鉴往事预测未来）

hinged 有铰链的

hinged blade 铰接叶片

hinged moment 铰接力矩

hinged moment balance 铰链力矩天平

hip 坡屋顶屋脊；斜脊

histogram 直方图

HMI (human machine interface) 人机界面

hoist 起重机

hoist tower 起重塔；塔式卷扬机；提升塔

holder 扶手；支架

holding current 吸持电流；保持电流

hole 孔

hollow 空的

hollow blade 空心叶片

homoentropic flow 均熵流

homogeneity 均匀性；同一性

homogeneity fluid 均质流体

homogeneity rule 同一性相似率

homogeneity substance 等质体

homogeneous 均匀的

homogeneous distribution 均匀分布

homogeneous field 均匀场

homogeneous flow model 均质流模式

homogeneous fluid 均匀流体

homogeneous isotropic turbulence 均匀各向同性湍流

homogeneous terrain 均布地形

homogeneous tunnel 均匀紊流度

homologous turbulence 均匀湍流

honeycomb board 蜂窝夹心胶合板

honeycomb cracks 网状裂缝

honeycomb laminate 蜂窝夹层板；蜂窝状叠层布

honeycomb sandwich construction 蜂窝夹芯结构

honeycomb sandwich structure 蜂窝夹层结构

honing machines 搪磨机

hood 发动机罩；烟囱风帽

hook 挂钩

hook bolt 吊耳；钩头螺栓；钩形螺栓；钩螺栓

hook spanner 钩，弯脚扳手

Hooke's law 胡克定律

hook up wire 架空电线

hoop 集电弓；箍条箍；铁环

horizontal 水平的

horizontal & vertical machining centers 卧式及立式加工中心

horizontal axis 水平轴

horizontal axis wind turbine (HAWT) 水平轴风力机

horizontal buoyancy correction 水平浮力修正

horizontal coherence 水平相干性

horizontal input to vertical axis output wind turbine (HVAWT) 水平输入垂直轴输出式风电机组

horizontal machining centers 卧式加工中心

horizontal resultant 水平合力

horizontal turbulent diffusion 水平湍流扩散

horizontal wind 水平风

horizontal wind field 水平风场

horizontal wind shear 水平风切变
horizontal axis rotor 水平轴风轮
horizontal axis—rotor WECS 水平轴风力机（风轮轴线的安装位置与水平面夹角不大于15°的风力机）
horizontal axis wind machine 水平轴风力机
horse latitude high pressure 副热带高压；马纬高压
horse latitudes 副热带无风带；回归线无风带；马纬度
horsepower (HP) 马力；功率
horsepower characteristic 功率特性
horsepower curve 功率曲线
horsepower hour 马力小时
horsepower input 马力输入；功率输入
horsepower loading 动力负荷；马力负荷；功率负荷
horsepower of equipment 设备马力；设备功率
horsepower of transmission 传动马力
horsepower output 输出马力；功率输出
horsepower rating 额定功率；计算功率
horseshoe magnet 马蹄形磁铁
horseshoe vortex 马蹄涡
horseshoe vortex system 马蹄涡系
hose 软管；胶皮管；蛇管
hose clip 管夹
hot dipped galvanized 热浸镀锌
hot film anemometer 热膜风速计
hot quenching 热淬火
hot wind tunnel 热风洞
hot wire anemometer 热线风速计
hot wire direction meter 热线风向计
hot wire probe 热线探针
hot wire technique 热线技术
hourly mean wind speed 每小时平均风速
hourly wind speed 每小时风速
hours of wind 刮风小时数
house microclimate 室内小气候
household wind turbine 家用风电机组
housing 壳体；机舱；住房
housing density 房屋密度
housing estate 住宅区；居民点
housing project 住房建造规划
HP (horsepower) 马力；功率
HRV (hydraulic relief valve) 液压安全阀
HS 硬度；肖氏硬度
hub 毂
hub assembly 毂组件
hub controller 轮毂控制器
hub height 轮毂高度
hub material 轮毂材料
hub plate 毂衬
hub precone 桨毂预锥角
hub ratio 轮毂比
hub rigidity 轮毂刚度
hub spider 辐射形桨毂
hub spider assembly 桨毂辐状装配
hub type 轮毂类型
hubcap 轮毂罩
hull 船体；外壳
human comfort 人体舒适性
human discomfort 人体不舒适性
human engineering 人体工程学；运行工程学
human error 人为误差
human machine interface (HMI) 人机界面
human perceptible 人可觉察的
human responsiveness 人体反应
human sensitivity 人体敏感性
human tolerance to wind 人体耐风性
humid air 湿空气
humidity 湿度

humidity sensitive element 湿敏元件
hump 驼峰；峰值
hurricane 飓风（十二级以上，风速≥32.7米/秒）
hurricane boundary layer 飓风边界层
hurricane core 飓风核心
hurricane eye 飓风眼
hurricane wind 飓风（12级以上）
Hütter wind turbine 赫特风轮机
HV 维氏硬度
HVAWT (horizontal input to vertical axis output wind turbine) 水平输入垂直轴输出式风电机组
hybrid 杂种；混合；混合动力；混合型
hybrid reinforced plastics 混合纤维增强塑料
hybrid tower 混合式塔架
hydraulic 液力的
hydraulic accumulator 液压蓄能器
hydraulic actuator 液压制动器；液压传动
hydraulic balance device 液压平衡装置
hydraulic block 液压块
hydraulic brake 液压制动器
hydraulic braking system 液压制动系统
hydraulic buffer 液压缓冲器
hydraulic clutch 液压离合器
hydraulic components 液压元件
hydraulic control 液压控制
hydraulic cylinder 液压缸
hydraulic damper 液压阻尼器
hydraulic diameter 水力直径
hydraulic engineering 水利工程
hydraulic filter 液压过滤器
hydraulic flap 液压操纵襟翼
hydraulic flow 湍流
hydraulic fluid 液压油
hydraulic governing 液压调节
hydraulic jump 水跃
hydraulic model 水力模型
hydraulic motor 液压马达
hydraulic oil 液压油
hydraulic operated valve 液压阀
hydraulic piston 液压活塞
hydraulic pitch lever mechanism 液压调桨杠杆机构
hydraulic pitching lever 液压调桨杠杆
hydraulic pitching mechanism 液压调桨机构
hydraulic pitching rod 液压调桨杆
hydraulic pitching system 液压调桨系统
hydraulic power system 液压动力系统
hydraulic power tools 液压工具
hydraulic power units 液压动力元件
hydraulic pressure 液压
hydraulic pressure control system 液压控制系统
hydraulic pump 液压泵
hydraulic pumping circuit 液压泵送回路
hydraulic ram 液压油缸
hydraulic relief valve (HRV) 液压安全阀；液压减压阀
hydraulic resistance 流体阻力；水阻力；水力阻力
hydraulic resistance balance 流动阻力平衡
hydraulic rotary cylinders 液压回转缸
hydraulic servo 液压伺服机构
hydraulic servomotor 液压伺服电动机
hydraulic set 液压装置
hydraulic system 液压系统
hydraulic trouble 液压系统故障
hydraulic tube 液压管
hydraulic unit 液压装置
hydrodynamic characteristics 流体动力特性

hydrodynamic constant torque converter 液力恒转矩变矩器
hydrodynamic coupling 液力联轴器
hydrodynamic form 流线型
hydrodynamic governor 液压调节器
hydrodynamic head 流体动压头
hydrodynamic power drive 动液压传动
hydrodynamic shock 流体动力冲击
hydrodynamic torque converter 液力变矩器
hydrodynamics 流体动力学
hydroelectric system 水电系统
hydrofoil 水翼
hydrogen energy storage 制氢贮能
hydrokinematics 流体运动学
hydromechanics 流体力学
hydrometeor 水汽凝结体
hydromotor 液压马达
hydropower station 水电站
hydrosphere 水圈；水界
hydrostatic 水静压；流体静力学的；流体静力的
hydrostatic equation 流体静力学方程
hydrostatic equilibrium 流体静力平衡
hydrostatic pressure 流体静压力
hydrostatic pressure ratio 流体静压率；静水压力系数
hydrostatic pressure test 流体静压力试验；水压试验
hydrostatic stability 流体静稳定性
hydrostatic test 流体静力学试验；水压试验
hydrostatics 流体静力学
hydrostatics test 流体静力试验
hydrostorage 水能储蓄
hygrograph 自动湿度记录计；自动湿度计
hygrometer 湿度计
hygroscopicity 吸湿性
hyperbolic cooling tower 双曲冷却塔
hypercritical flow regime 过临界流态；高超临速流态
hypersonic tube 高超音速风洞
hypothetical event 假想事件
hysteresis 滞后作用；磁滞现象
hysteresis band 磁滞带；滞环
hysteresis damping 迟滞阻尼
hysteresis loop 迟滞回线
hysteresis loss 磁滞损耗

现代英汉风力发电工程

A Modern English-Chinese Dictionary of Wind Power Engineering

I

IAM (International Association of Machinists) 国际机械师协会
IAMAP (International Association of Meteorology and Atmospheric Physics) 国际气象与大气物理协会
IAWE (International Association for Wind Engineering) 国际风工程学会
IBL (internal boundary layer) 内边界层
IC (integrated chip) 集成芯片
ice 冰
ice accretion 积冰
ice adiabat 冰绝热线
ice boating 冰帆运动
ice cap 冰盖
ice coated power cable 裹冰电缆
ice coated transmission line 裹冰输电线
ice coating 裹冰
ice crystal 冰晶
ice deposit 积冰
ice evaporation level 冰汽转相高度；高空冰汽转相高度
ice fog 冰雾
ice formation 结冰
ice nucleus 冰核
ice prisms 冰针；飘降冰晶
ice saturation 冰饱和
ice sensor 结冰传感器
ice sheet 冰盖
ice storm 冰暴
ice wind tunnel 结冰风洞
ice yachting 冰帆运动
icing index 积冰指数
icing intensity 积冰强度
icing level 积冰高度
ICWE (International Conference on Wind Engineering) 国际风工程会议
ideal 理想的；完美的
ideal air 理想空气
ideal black body 理想黑体；绝对黑体
ideal body 理想体
ideal boundary 理想边界
ideal conditions 理想条件；标准条件
ideal efficiency 理想效率；理论效率
ideal flow 理想流动
ideal fluid 理想流体
ideal fluid theory 理想流体理论
ideal formula 理想公式；标准公式
ideal function 理想函数
ideal gas 理想气体
ideal gas constant 理想气体常数

ideal gas equation of state 理想气体状态方程
ideal gas law 理想气体定律
ideal gas state equation 理想气体状态方程
ideal liquid 理想液体
ideal load curve 理想负荷曲线
ideal source 理想电源
ideal viscous fluid 理想黏性流体
identification 鉴定；识别
identification mark 识别标志
identity matrix 单位矩阵
idle 闲置的
idle capacity 空载功率；备用功率
idle current 无功电流
idle equipment 闲置设备
idle hour 停机时间
idle load 空载
idle motion 空转
idle running 空转
idle wheel(gear) 惰轮
idling 空转
idling running 空载运行
idling torque 空转扭矩
IEA (International Energy Agency) 国际能源机构
IEC (International Electrotechnical Commission) 国际电工委员会
IEC 61400-1 Ed2 Wind Turbine Safety and Design Revision IEC 61400—1风电机组安全和设计（修订版）
IEC 61400-1 Wind Turbine Safety and Design IEC 61400—1 风电机组安全和设计
IEC 61400-11 Acoustic Noise Measurement Techniques IEC 61400—11 噪音测量技术
IEC 61400-12 Wind Turbine Power Performance Testing IEC 61400—12风电机组功率特性试验
IEC 61400-13 Mechanical Load Measurements IEC 61400—13 机械载荷测量
IEC 61400-2 Small Wind Turbine Safety IEC 61400—2 小型风电机组安全性
IEC 61400-21 Measurement and Assessment of Power Quality Characteristics of Grid Connected Wind Turbines IEC 61400—21 并网风电机组功率质量特性测试与评价
IEC 61400-22 Wind Turbine Certification IEC 61400—22 风电机组认证
IEC 61400-23 Blade Structural Testing IEC 61400—23 叶片结构试验
IEC system for conformity testing and certification of wind turbines 风电机组的合格试验和认证的IEC系统
IEE (Institute of Electrical Engineers) 电气工程师协会
IEEE (Institute of Electrical and Electronics Engineers) 电气与电子工程师协会
IGBT (insulated gate bipolar transistor) 绝缘栅双极型晶体管
IGCT (integrated gate commutated thyristor) 集成门极换流晶闸管；集成门极换向晶闸管
image 镜像；映像
image method 镜像法
image source 镜像源
image system 镜像系
imaginary emission point 镜像排放点
imaginary part 虚部
impact damper 冲击阻尼器；缓冲器；减震器
impact elasticity 冲击韧性；冲化性；冲击弹性
impact induced vibration 冲击至振
impact resistant battery 抗震电池
impact stress 撞击力；冲击应力

impact torque 冲击扭矩
impact tube 皮托管
impact velocity 冲击速度
impact wrench 套管扳手
impaction loss 碰撞损失
impactometer 碰撞式空气取样器
impactor 冲击器
impedance 阻抗
impedance angle 阻抗角
impedance voltage 阻抗电压
impeller 叶轮
impeller passage 叶轮通道
imperfect earth 接地不良
imperfect gas 非理想气体
impingement 拍撞；冲击；影响；侵犯
impingement height （羽流）拍撞高度
impinger 尘埃测定器；冲击式采样器
impinging plume 拍撞羽流
impulse 推动；冲动；脉冲
impulse load tests 冲击动载荷试验
impulsive excitation 脉动冲
impurity 杂质
impurity level 杂质度
in line 成一直线；一致；协调；有秩序
in line array 行阵
in line displacement 顺风向位移
in line response 顺风向响应
in plane bending 面内弯曲
in situ 在原位置；在原处
in situ concrete tower 现浇混凝土塔架
in wind response 顺风（向）响应
inactive area 非放射性区
inception of oscillation 起振；起始振荡
incidence 迎角；安装角
incidence angle 入射角；迎角；安装角

incident 入射的
incident beam 入射流
incident flow 来流
incident flow turbulence 来流湍流（度）
incident flux 入射通量
incident gust 来流阵风
incident turbulence 来流湍流（度）
incident wind 来流风
incipient galloping 初始驰振
incipient instability 初始不稳定性
incipient motion 初始运动
inclination 倾斜；倾角
inclined multitube manometer 倾斜多管压力计
inclined thermal 倾斜热泡；倾斜热
inclined axis rotor WECS 斜轴风力机（风轮轴线的安装位置与水平面夹角在15°～90°的风力机）（不包括90°）
inclosed body 箱式车身
incoming 引入的
incoming beam 入射束
incoming flow 来流；迎面流
incoming solar radiation 射入太阳辐射
incoming turbulence 来流湍流（度）
incoming wind 来流风；迎面风
incompressibility 不可压缩性
incompressible flow 不可压流动
incompressible fluid 不可压流体
incompressible wake 非压缩流尾流
incompressible wind 来流风；迎面风；不可压缩风
independent event 独立事件
independent power producer(IPP) 独立电力生产者
index of pollution 污染指数

index of resistance 阻力指数；风阻指数
indicated wind speed 指示风速
indicial motion 示性运动
indirect 间接的
indirect carrier 间接载体
indirect circulation 逆环流
indisturbed flow 未扰流动
individual 个人的
individual axis drive 单轴驱动；独立驱动
individual control 独立控制
individual control system 分级控制系统
individual dose 个人剂量
individual drive 单独驱动；单独传动
individual gust 单阵风
individual measurement 单独测量
individual system 单独系统
individual test 单件试验；个别试验
individual weight 自重；净重
indoor air pollution 室内空气污染
indoor environment 室内环境
indoor gust 室内阵风
indoor temperature 室温
indoor thermal comfort 室内热舒适性
indraft 吸气；引入
induced 感应的；诱发的
induced air 吸气；引入空气
induced angle of attack 诱导迎角
induced current 感应电流
induced downwash 诱导下洗
induced drag 诱导阻力
induced draft cooling tower 吸风冷却塔
induced drag coefficient 诱导阻力系数
induced drag of lift 升力诱导阻力
induced electric current 感生电流；感应电流
induced flow 诱导流动

induced flow effect 诱导气流影响
induced power 诱导功率
induced pressure 诱导压力
induced pressure gradient 诱导压力梯度
induced velocity 诱导速度
induced vibration 诱导振动
inductance figure 电感系数
induction coefficient 诱导系数；电感应系数
induction coil 电感线圈
induction disc relay 感应圆盘式继电器
induction effect 感应作用；诱导作用
induction generator 感应发电机
induction machine 感应电机
induction meter 感应式电表
induction motor 感应电动机
induction phenomenon 感应现象
induction velocity 诱导速度
inductive 诱导的；感应的
inductive acceleration 感应加速度计
inductive action 感应作用
inductive component 电感分量；感应分量；电感性分量；感性（无功）分量
inductive effect 感应效应；诱导效应
inductive effect index 诱导效应指数
inductive reluctance 感应磁阻
inductor 电感器
industrial aerodynamics 工业空气动力学
industrial aerodynamics wind tunnel 工业空气动力学风洞
industrial air pollution 工业空气污染
industrial building 工业建筑物
industrial chimney 工厂烟囱
industrial city 工业城市
industrial climatology 工业气候学
industrial complex 工厂群

industrial density 工厂密度
industrial dust 工业粉尘
industrial haze 工业薄雾
industrial meteorology 工业气象学
industrial PC 工业用PC机
industrial plume 工业烟羽
industrial pollution 工业污染
industrial smoke 工业烟雾
industrial stack 工厂烟囱
industrialization 工业化
inelastic deformation 非弹性变形
inelasticity coefficient 非弹性系数
inert carrier 惰性载体
inert gas 惰性气体
inert gas metal-arc welding 惰性气体保护金属（熔化）极弧焊
inertia 惯性
inertia balance 动平衡
inertia moment 惯性矩；转动惯量
inertia shear 惯性剪力
inertial axes system 惯性轴系
inertial coordinate system 惯性坐标系
inertial effect 惯性效应
inertial impaction 惯性碰撞
inertial instability 惯性不稳定性
inertial mass 惯性质量
inertial settling 惯性沉降
inertial subrange 惯性子区；惯性次区
inertial subrange law 惯性子区定律
inertial switch 惯性开关
inertial transfer of energy 惯性能量输送
inertial wave 惯性波
inestimable 极贵重的；（大得）无法估计的
inferior arc 劣弧
inferior limit 下限

infiltration 渗透；漏风
infiltration rate 渗透速率
infinite cloud model 无限烟云模式
infinite line source 无限线源
infinite plane source 无限平面源
infinite slab source 无限平面源
infinite span 无限翼展；无限展长
infinite voltage gain 无穷大电压增益
infinite vortex street 无限涡街
infinite wake 无限尾流
infinitesimal 无穷小的
infinitesimal analysis 微积分
infinitesimal calculus 微积分学
infinitesimal deformation 无穷小形变
infinitesimal displacement 无穷小位移
infinitesimal geometry 微分几何
infinitude 无穷
inflow 流入
inflow angle 入流角
influence 影响
influence area 影响区；影响面积
influence by the tower shadow 塔影响效应（塔架造成的气流涡区对风力机产生的影响）
influence by the wind shear 风切变影响
influence coefficient 影响系数
influence zone 影响区
influx 流入量
information 信息
infrared tracking 红外线跟踪
infrequent wind 异常风
ingestion dose 摄入剂量
inhalation hazard 吸入危害
inherent 固有的
inherent regulation 自平衡调节
inherent regulation of controlled plant 调节

对象自平衡
inherent stability 固有稳定性
inherent stress 内在应力
inherent vice 内部缺陷
inherently safe 自身安全的
inhibiting effect 抑制效果
inhibiting factor 抑制因素
inhibition 抑制作用
inhomogeneous 不均匀性
inhomogeneous flow 非均匀流动
inhomogeneous turbulence 不均匀乱流
initial concentration 起始浓度
initial condition 初始条件
initial data 原始数据
initial disturbance 初始扰动
initial performance 最初性能
initial plan 初步设计
initial plume dimension 初始羽流尺度
initial plume rise 初始羽流抬升
initial position 初始位置
initial pressure 初始压力
initial release turbulence 初始释放湍流度
initial stage 初始阶段
initial tunnel turbulence 风洞初始湍流度
initial velocity 初速度
initial wind speed 初始风速
injection molding 喷射造型法；喷射模塑法；注压法；注模
inland 内陆
inland desert 内陆荒漠
inland lake 内陆湖
inland plain 内陆平原
inleakage 渗入
inleakage of air 漏风
inleakage of wind 漏风

inlet opening 入口；进气孔
inlet port 入口；进气口
inmost 最深的
inmost layer 最内层
inner race （滚动轴承）内（座）圈；内环
innovative 革新的；创新的
innovative wind energy conversion system 创新型风能转换系统
innovative wind system 革新性风能转换系统
inorganic toxic material 无机毒物
input 输入
input nominal speed 额定输入转速
input power 输入功率
input shaft 输入轴
inrush 涌入；浸入
inrush current 涌入电流；冲击电流；起动电流；突入电流；瞬间起峰电流
inshore wind 向岸风
inside pitch line length 齿根高度
insolation parameter 日照参数
insolation 日照；太阳辐射
inspecting standard 检查标准
inspection 检验；监督
inspection certificate of quality 品质检验证书
inspection clause 检验条款
inspection code 故障等检验规程
inspection door 检修门
inspection earthing 检修接地
inspection hole 检查孔；观察孔
inspection instrument 检查仪器
instability 不稳定性；不稳定度
installation 安装；装配；装置
installation capacity 安装能力
installation cost 安装费
installation diagram 安装图

installation dimension 安装尺寸
installation engineer 安装工程师
installation error 安装误差
installation expense 安装费用
installation instruction 安装说明
installation material 安装材料
installation procedure 安装程序
installation shop 装配车间
installation site 安装现场；安装位置
installation size 安装尺寸
installation specification 安装检修规程
installation standard 安装标准
installation test 装配试验
installation tool 安装工具
installation weight 装置重量；设备重量
installation work 安装工作
installations and equipments 设施和设备
installed capacity 装机容量
installed gross capacity 总装机容量
installed load 安装荷重
installed power 装机功率
instantaneous area source 瞬时面源
instantaneous capacity 瞬时功率；瞬时容量
instantaneous electric power 瞬时电功率
instantaneous incidence 瞬时迎角
instantaneous measured 瞬时测值
instantaneous mechanical power 瞬时机械功率
instantaneous output 瞬时输出功率
instantaneous over current 瞬时过电流
instantaneous overturning moment 瞬时倾覆力矩
instantaneous peak demand 瞬时高峰需求量
instantaneous peak load 瞬时高峰负荷
instantaneous point source 瞬时电源；瞬时点源
instantaneous power 瞬时功率
instantaneous power output 瞬时功率输出
instantaneous pressure 瞬时压力
instantaneous turbulence energy 瞬时湍流能量
instantaneous value 瞬时值
instantaneous velocity 瞬时速度
instantaneous volume source 瞬时体源
instantaneous wind speed 瞬时风速
Institute of Electrical and Electronics Engineers (IEEE) 电气与电子工程师协会
Institute of Electrical Engineers (IEE) 电气工程师协会
institutional constraint 制度约束；制度限制
instrument 仪器；工具
instrument air 仪表气源
instrument inertia 仪表惯量
instrument rack 计测器支架
instrumentation 使用仪器
instrumentation averaging time 仪表平均时间
instrumentation engineering 仪器仪表工程
instrumentation noise 仪表噪声
insulant 绝缘物
insulated 使绝缘；绝缘的；隔热的
insulated aluminium bar 绝缘铝棒
insulated copper bar 绝缘铜棒
insulated gate bipolar transistor (IGBT) 绝缘栅双极型晶体管
insulated materials 绝缘材料
insulated panel 隔热板
insulating ability 绝热能力；绝缘本领；绝缘能力
insulating barrier 绝缘障；绝缘挡板；绝缘层

insulating blanket 绝缘镀层；绝缘垫层
insulating board 隔热板；绝缘板
insulating boots 绝缘靴
insulating bushing 绝缘套管
insulating glove 绝缘手套
insulating material 隔热材料；绝缘物质
insulating oxide 绝缘氧化物
insulating panel 绝热板；绝缘板
insulating resistance 绝缘电阻
insulation 绝缘
insulation classes (stator/rotor) 绝缘等级（定子/转子）
insulation level 绝缘水平；绝缘等级
insulation ratio 绝缘比
insulation resistance 绝缘电阻
insulation wedge 绝缘楔
insulator 绝缘子
insulator string 绝缘子串
intake 进气口；摄入量；进风量
intake air 进气；进风流
integral 积分
integral body 整体式车身
integral calculus 积分
integral equation 积分方程
integral length scale of turbulence 湍流积分长度尺度
integral sensitivity 积分灵敏度
integral time 积分时间
integral transport equation 积分运输方程
integrated 综合的
integrated chip (IC) 集成芯片
integrated coupling 固定连接
integrated deposition density 综合沉积密度
integrated dose 积分剂量
integrated exposure 积分照射量

integrated gate commutated thyristor (IGCT) 集成门极换流晶闸管；集成门极换向晶闸管
integrated puff model 积分喷团模式；集成烟团模式
integrated value 累计值
integration 整合；集成；积分
intensity 强度；强烈；亮度；紧张
intensity of turbulence 湍流（强）度
interaction 相互作用；干扰
interaction effect 干扰效应；交互作用；互作用效应
interactive 交互式的
interactive system 相互作用系统
inter-annual variation 年际变化
interconnection(for WTGS) 互连（风力发电机组）
interdependent flow 横连流动
interdisciplinary 各学科间的
interdisciplinary study 跨学科研究；多学科研究
interface 接口
interfacial oscillation （气流）分界面振荡
interfacial sublayer 交界面副层
interference 干扰
interference correction 干扰修正
interference drag 干扰阻力
interference galloping 干扰驰振
interference free model 无干扰模型
interfering flow field 干扰流场
interharmonic 间谐波
interior pressure 内压
interior zone 内区
interlocker 联锁装置
interlocking 连锁；咬合作用
interlocking control 连锁控制

interlocking device 连锁装置
interlocking system 连锁系统
intermediate 中间的；中间物
intermediate capacitor 中间电容器
intermediate gear box 中间减速器；中间齿轮箱
intermediate repair 中间修理；架修
intermediate scale turbulence 中尺度湍流
intermediate shelterbelt 过渡防护林带
intermediate tower platform 中间塔架平台
intermediate wake 中间尾流
intermediate wind machine 中型风力机
intermit 暂停；间歇
intermittency 间歇性
intermittency factor （湍流）间歇因子
intermittent 间歇的
intermittent duty 间歇负载；间歇工作方式
intermittent load 间歇负荷
intermittent nature of wind 风的间歇性
intermittent operation 间歇运行
intermittent pollution 间歇污染
intermittent power source 中继动力源；间歇性电源
intermittent process 间歇过程
intermittent wind tunnel 间歇式风洞
intermittently turbulent region 间歇紊流区
internal 内部的
internal aerodynamic 内流空气动力学
internal ascent 内部爬梯
internal balance 内平衡；内部均衡
internal ballistics 内弹道学；内弹道
internal boundary layer (IBL) 内边界层
internal calipers 内卡钳
internal combustion engine 内燃机
internal coupling 内部耦合
internal damping 内阻尼
internal degree of freedom 内自由度
internal drag 内阻
internal drag bracing 翼内阻力张线
internal energy 内能
internal environment 内部环境
internal exposure 体内照射
internal flow 内流
internal fluid friction 流体内摩擦
internal fluid mechanics 内流流体力学
internal friction 内摩擦
internal friction coefficient 内摩擦系数
internal friction heat 内摩擦热
internal gear 内齿轮
internal gear pair 内齿轮副
internal impedance 内阻抗
internal lightning protection facility 内部防雷设施
internal lightning protection system 内部防雷系统
internal load 内部负荷
internal losses 内部损失
internal pressure 内部压力
internal pressure coefficient 内压系数
internal pressure loss 内压损失
internal pressure test 内部压力试验
internal resistance 内阻
internal stack flow 烟囱内流
internal stator 内部转子；内转子；内定子
internal strain gauge balance 内式应变式天平
internal turbulence 内紊流
internal volute 内蜗壳
International Association for Wind Engineering (IAWE) 国际风工程学会
International Association of Machinists

(IAM) 国际机械师协会
International Association of Meteorology and Atmospheric Physics (IAMAP) 国际气象与大气物理协会
International Conference on Wind Engineering (ICWE) 国际风工程会议
International Electrotechnical Commission (IEC) 国际电工技术委员会
International Energy Agency (IEA) 国际能源机构
international standard atmosphere (ISA) 国际标准大气
International Standardization Organization (ISO) 国际标准化组织
interoffice 局间的
interoffice communication 办公室间通信；局间通信
interpolation 内插；插值
interpolation method 插值法
interrelationship 相互关系
intersected 分割的
intersected country 起伏地形
intersection 相交；交叉；十字路口；交集；交叉点
intervening spacing 交错分布干预间距
into the wind 逆风；迎风；顶风
intolerance 无法忍受的
intrinsic viscosity 特性黏度；固有黏度；本征黏度
inverse 倒数；倒置，反相的；逆的
inverse back coupling 负反馈
inverse current 反相电流
inverse fast Fourier transform 快速傅里叶逆变化
inverse feedback 负反馈
inverse flow 逆流
inverse function 反函数

inverse ratio 反比
inverse time relay 反时限继电器
inversion 逆变；反转；倒置；反向；倒转
inversion base 逆温层底
inversion base height 逆温层底高度
inversion break up fumigation 逆温破坏型下熏
inversion dissipation 逆温耗散
inversion fumigation plume 逆温下熏型羽流
inversion heat flux 逆温热通量
inversion height 逆温层高度
inversion layer 逆温层
inversion lid 逆温层顶
inversion of energy 能量转化
inversion penetration 逆温渗透；逆渗透
inverted draft 反向气流
inverted roof 凹屋面；倒置屋顶
inverter 逆变器；变换器
inviscid flow 无黏流
inviscid fluid 无黏流体
inviscid shear flow 无黏剪切流
involute gauge 渐开线齿轮
involute profile 渐开线齿形
ionosphere 电离层
iron loss 铁损
irradiance 辐照度
irradiation damage 辐照损伤
irradiation hazard 辐照危害
irregular 不规则物
irregular fluctuation 不规则脉动
irregular grading 不规则的分级
irregular terrain 不平地形
irregular wind 风多变
irreversibility 不可逆性
irreversible motion 不可逆运动
irrigation pumping 灌溉提水

irrotational flow 无旋流
irving type balance 欧文式补偿；翼内空气动力补偿
ISA (international standard atmosphere) 国际标准大气
isallobar 等气压图
isallobaric wind 等变压风
isallohypse 等变高线
isallotherm 等变温线
isanemone 等风速线
isanomal 等距常线；等地平
isentropic 等熵线的
isentropic flow 等熵流
isentropic process 等熵过程
island 岛屿
ISO (International Standardization Organization) 国际标准化组织
isobar 等压线
isobaric 等压线
isobaric line 等压线
isobaric spin 同位旋
isobaric surface 等压面
isobront 等雷(日)线
isoceraunic 等频雷暴的；等雷雨的
isoceraunic line 等雷频线；等雷雨强度线
isocheim 等冬温线
isochoric 等层厚的；等体积的；等容的
isoconcentration 等浓度线
isogradient 等梯度线
isohel 等日照线
isohume 等湿线；等水分线
isohyet 等雨量线
isolate 隔离
isolated aerofoil 孤立翼型
isolated building 孤立建筑
isolated chimney 孤立烟囱
isolated construction 孤立建筑
isolated cooling tower 孤立冷却塔
isolated hill 孤山
isolated source 孤立源
isolated thermal 孤立热泡
isolation 隔离；分离；绝缘；隔振
isolator 隔离开关；隔振器；绝缘体；隔离器；隔离物
isoline 等值线
isopiestics 等压线
isopleth 等值线；等浓度线；等成分面；等值线图
isopycnic 等密度线
isostatic equilibrium 各向等压平衡
isosurface 等值面
isotach 等风速线
isotherm 等温线
isothermal atmosphere 等温大气
isothermal energy storage 等温贮能
isothermal layer 等温层
isothermal line 等温线
isothermal process 等温过程
isotope 同位素
isotope abundance 同位素丰度；同位素分布量
isotropic 各向同性的
isotropic dispersion 各向同性弥散
isotropic point source 各向同性点源
isotropic turbulence 各向同性湍流
isotropic turbulence scale 各向同性素流度
isovel 等速线；等速度曲线
isovelocity 等速线
iteration 迭代
iterative 迭代的；重复的
iterative analysis 迭代分析
iterative earth 重复接地

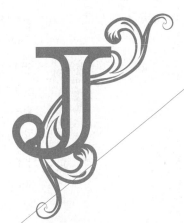

现代英汉风力发电工程

A Modern English-Chinese Dictionary of Wind Power Engineering

J

jacket cooling 水套式冷却
jaw coupling 爪盘联轴节
jet 射流；喷流；喷射；喷嘴；喷气式飞机
jet boundary 射流边界
jet effect wind 急流效应风；喷射效应风；喷流效应风
jet molding 喷（射）模（塑）法
jet pair 射流对
jet pump 喷射泵；射流泵
jet rise 射流抬升
jet spread 喷射分散
jet stream 急流（风速 50~200海里/时，又称 jet air stream）；喷流；射流
jet switching 射流转换；射流开关现象
join 连接；接合点
joining bolt 连接螺栓
joining pipe 连接管
joining rivet 连接铆钉；接合铆钉
joint 连接；接头
joint acceptance function 联合接纳函数；结合受纳函数

joint coupling 连接器；活节连接器；电缆接头套管；管接头；接头套管
joint current 总电流
joint flange 连接法兰
joint frequency 联合频率
joint part 连接部分；连接件
joint probability distribution 联合概率分布
joint ring 连接垫圈；连接环
joint slack 联轴节
joint thread 连接螺纹
Joukowski airfoil 儒科夫斯基翼型
Joule's law 焦耳定律
jumbo windmill 巨型风车
jump rope shaped blade 跳绳形叶片
jumper clamp 跳线线夹
junction 连接；接点；接合体
junction battery 结型电池
junction box 分线箱；接线盒
junction pipe 连接管
junction point 连接点；接触点
juncture 连接；接头

现代英汉风力发电工程

A Modern English-Chinese Dictionary of Wind Power Engineering

K

Kamm back 卡姆背
Kamm tail 卡姆尾
Karman constant 卡门常数
Karman spectrum function 卡门谱函数
Karman turbulence spectrum 卡门湍流谱
Karman('s) vortex street 卡曼涡街
katabatic 下降的
katabatic flow 下降流动
katabatic wind 下降风；下吹风
katafront 下滑锋
keel beam 龙骨梁
keel bracing 龙骨撑杆
keen draft 穿堂风；强空气流
Kelvin scale 开式温标；绝对温标
kernel function 核函数
kerosene vapor visualization 煤油烟显示法
keyhole saw 铨孔锯；键孔锯；栓孔锤
key direct access 键直接存取
key operate 键盘操作
keyway 键槽
khamsin 喀新风；喀新热浪（每年从撒哈拉沙漠吹向埃及的一种干热南风）
kilowatt hour 千瓦时
kinematic 运动学上的
kinematic coefficient 运动系数

kinematic equation 运动方程
kinematic similarity 运动相似性
kinematic viscosity 运动黏性系数（流体的动力黏度与密度的比值）
kinematic(al) viscosimeter 运动黏度
kinematics 运动学
kinetic 运动的
kinetic characteristic 动力学特性；动态特性
kinetic characteristic curve 运动特性曲线；动态特性曲线
kinetic control 动态控制
kinetic control system 动态控制系统
kinetic effect 动态效应
kinetic energy 动能
kinetic energy absorption 动能吸收
kinetic energy flow 动能流
kinetic energy storage system 动能贮能系统
kinetic equation 动力学方程
kinetic equilibrium 动态平衡
kinetic friction 动摩擦
kinetic friction coefficient 动摩擦系数
kinetic friction factor 动摩擦因数
kinetic friction torque 动摩擦力矩
kinetic parameter 动力学参数；动态参数
kinetic potential 运动势

kinetic power 动态功率

kinetic principles 动力学原理

kinetic property 动力学性质

kinetic system 动力学系统；动态系统

kinetic theory of fluid 流体动力学理论；流体分子运动论

kinetic viscosity 动力学黏度

kinetics 动力学

king pin bush 主销衬套

kink 结点，拐点

Kirchhoff flow 基尔霍夫流

kit 成套工具；用具包；工具箱；成套用具

kite 风筝

kite anemometer 风筝风速计

kite ascents 风筝探测

kite balloon(=kitoon) 风筝气球

K-model K模式

kneading action 揉搓作用

knee point 拐点

knitted fabric 针织布

knob 按钮；调节器

knuckle 关节

knuckle radius 转角半径；过渡半径

knuckle joint 万向接头；指骨关节；铰链接合

knurled nut 凸螺母[机]；滚花螺母

knurling 滚花；压花纹；滚花刀；压花刀

knurling cylinder 滚花圆柱

Kolmogorov's hypothesis 科尔莫格罗夫假说

Kolmogorov's similarity theory 科尔莫格罗夫相似理论

konimeter 计尘器；尘度计

K-profile K廓线

Kronecke delta 克罗内克 δ

krypton tracer 氪示踪剂

K-theory K扩散理论

kurtosis 峰态；峭度

Kussner function 库斯纳函数

Kutta condition 库塔条件

Kutta-Joukowski condition 库塔-儒科夫斯基（后缘）条件

Kutta-Joukowski theorem 库塔-儒科夫斯基定理

现代英汉风力发电工程

A Modern English-Chinese Dictionary of Wind Power Engineering

L

LA 避雷器
labile equilibrium 不稳定平衡
labile flow 不稳定流
labile region 不稳定区
labile state 不稳定状态
lability 不稳定性；易变性
laboratory 试验室；试实验室
laboratory apparatus 实验室仪器
laboratory bench 试验室工作台
laboratory data 实验室数据
laboratory examination 实验室检验
laboratory instrument 实验室仪表
laboratory report 实验室报告
laboratory result 实验室结果
laboratory simulation 试验室模拟
laboratory type tube 实验型风洞
labyrinth 迷宫
labyrinth seal 迷宫密封
lack voltage alarm 欠压报警
lacquer 漆
lacquer coat 漆涂层
ladder 梯子
lag time 滞后时间
lagged 延迟
lagged incidence 滞后迎角
lagging 绝缘层材料；落后的
lagging feedback 延迟反馈
lagging motion 迟滞运动
Lagrangian autocorrelation tensor 拉格朗日自相关张量
Lagrangian autocovariance 拉格朗日自协方差
Lagrangian covariance 拉格朗日协方差
Lagrangian equation 拉格朗日方程
Lagrangian integral time scale 拉格朗日积分时间长度
Lagrangian similarity theory 拉格朗日相似理论
Lagrangian spectral function 拉格朗日谱函数
Lagrangian strain 拉格朗日应变
lake breeze 湖风
laminar 层流的；层式的；层流式
laminar air flow 空气层流；空气片流
laminar area 层流区
laminar boundary layer 层流边界层
laminar cellular convection 层流环型对流
laminar composite 层状复合材料
laminar condition 层流状态；层流条件
laminar convection 层流对流
laminar current 层流
laminar drag 层流阻力

laminar Ekman boundary layer 埃克曼层流边界层
laminar film 层流膜
laminar flow 层流
laminar flow aerofoil 层流翼面
laminar flow airfoil 层流翼型
laminar flow model 层流模型
laminar flow motion 层流；层流运动
laminar flow regime 层流状态
laminar flow stability theory 层流稳定性理论
laminar flow state 层流流态
laminar free convection 层流自由对流
laminar mass transfer 层流传质
laminar motion 层流运动
laminar plastic flow 片型塑性流变
laminar plume 层流羽流
laminar region 层流区
laminar separation 层流分离
laminar sublayer 层流底层
laminar turbulent transition 层流向湍流转捩
laminar viscosity 层流黏性
laminar vortex shedding 层流旋涡脱落
laminar vortex street 层流涡街
laminar wake 层流尾流
laminate 薄片制品
laminate quality 层压质量
laminated 层压的
laminated construction 叠层结构
laminated core 叠片铁芯
laminated iron core 叠片铁芯
laminated molding 层压模塑（法）
laminated soft iron 叠片软铁
laminated steel 层压钢；复膜铁；覆膜铁；镜面钢板
lamination 层压；叠片

land 国土；陆地
land and sea breeze 海陆风
land availability 可利用土地
land Beaufort scale 陆地蒲福风级
land breeze 陆风
land spill 地面溢漫
land wind 陆地风
landform 地貌；地形
landscape 景观
landscape modification 景观改造；地貌改造
land and sea breeze 海陆风
landspout 陆龙卷
landward 近陆的
landward wind 陆地风
lap weld 搭焊
Laplace equation 拉普拉斯方程
Laplace transformer 拉普拉斯变换
lapping 研磨
lapping compound 研磨剂；抛光剂
lapping machines 精研机
lapping machines centerless 无心精研机
lapse 推移；递减；垂直梯度
lapse limit 对流层顶
lapse rate 递减率；直减率；温度垂直梯度
large angle scattering 大角度散射
large diameter multi-pole stator 大直径多极定子
large diameter variable speed separately excited multi-pole WRSG 大直径变速它励多极式 WRSG
large disturbance method 大扰动法
large nuclei 大核
large particle 大粒子
large power drill 大型钻
large scale 大尺度；大缩尺比

large scale circulation 大尺度环流
large scale eddy 大尺寸涡流
large scale model 大比例模型
large scale pollution 大范围污染
large scale structure （湍流）大尺度结构
large scale turbulence 大尺度湍流
large scale wind energy conversion system (LWECS) 大型风能转换系统
large scale wind turbine 大型风电机组
large sized wind machine 大型风力机
laser 激光
laser alignment measuring system 激光准直测量系统
laser altimeter 激光测高计
laser anemometer 激光风速计
laser cutting 激光切割
laser cutting for steel plate 激光钢板切割机
laser detector 激光探测器
laser scintillometer 激光闪烁计数器
laser Doppler anemometer 激光多普勒测速计
laser Doppler velometer 激光多普勒测速计
lasting quality 耐久性
latch 锁存器；门闩线路；凸轮；闩锁
latent energy 潜能
latent fault dormant failure 潜伏故障
latent heat 潜热
lateral attitude 倾斜姿态；坡度
lateral axis 横轴
lateral buckling 侧向屈曲；弯扭屈曲；横向压屈
lateral correlation function 横向相关函数
lateral deflection 侧向挠度；横向偏转；旁向偏转
lateral diffusion 横向扩散
lateral dispersion 横向扩散
lateral displacement 侧向位移

lateral flapping 侧向挥舞
lateral force 侧力
lateral galloping 侧向驰振
lateral growth of plume 羽流横向增长
lateral gust 横向阵风
lateral length scale 横向长度尺度
lateral load 横向荷载
lateral oscillation 横向振动
lateral play 横向游隙
lateral plume spread 羽流横向展宽
lateral resistance 横向阻力
lateral scale 横向尺度
lateral shear 横向剪切
lateral stability 横向稳定性
lateral vibration 横向振动
lathe 车床
lathe bench 车床工作台
lathe tool 车床工具
lathes automatic 自动车床
lathes heavy-duty 重型车床
lathes high-speed 高速车床
lathes turret 六角车床
lathes vertical 立式车床
latitude 纬度
lattice 网格；格构；晶格；格子；格架
lattice aerial 网状天线
lattice plate 格板
lattice point 格网点；点阵
lattice structure 点阵结构；网格结构
lattice tower 桁架式塔架
lattice tower foundation 桁架式塔架基础
lattice truss 格构桁架
lattice work 格构结构；栅格结构
Laval nozzle 拉瓦尔喷管
law 定律

law of conservation 守恒定律
law of conservation of energy 能量守恒定律
law of conservation of mass energy 质能守恒定律
law of electromagnetic induction 电磁感应定律
law of energy conservation 能量守恒定律
law of mass action 质量作用定律
law of mass conservation 质量守恒定律
law of momentum conservation 动量守恒定律
law of motion 运动定律
law of partial pressure 分压定律
law of perdurability of matter 物质守恒定律；物质不灭定律
law of similarity 相似律
law of wake 尾流定律
law of wall 壁面定律；界壁定律
laws of flux conservation 磁通守恒定律
Lawson criterion 劳森判据
lay down yard 放置场
lay up procedure （增强塑料）敷层方法
layer 层
layer molding 分层压制模制
layered construction 叠层结构
L/D 升阻比
lead accumulator 铅蓄电池
lead acid battery 铅酸蓄电池
lead acid battery energy storage 铅酸蓄电池贮能
lead and lag motion 超前滞后运动
lead spring 前导弹簧
lead time 超前时间；提前期；订货至交货的时间；研制周期；交付周期
leading dimension 主要尺寸；轮廓尺寸

leading edge 前缘
leading edge discontinue 前缘转折点
leading edge droop 前缘下垂
leading edge flap 前缘襟翼
leading edge radius 前缘半径
leading edge separation 前缘分离
leading edge stagnation 前缘驻点
leading edge suction 前缘吸力
leading edge vortex 前缘涡
leading edge vortices 前缘（旋）涡系
leading features 主要特征
leading power factor 超前功率因数
leading wind 顺风
leaf actuator 刀形断路器
leaf spring 板弹簧；叶片弹簧；片簧
leak test 漏泄试验
leakage 泄漏
leakage current 漏电流
leakage flux 漏磁通
leakage gap 泄漏缝；泄漏间隙
leakage rate 泄漏率
leakage reactance 漏磁电抗
lean to mansard roof 单面折线屋顶
lean to roof 单坡屋顶
least action 最小作用量
least cost operation 最低费用运转
least drag body 最小阻力体
least error 最小误差
least limit 最小极限
least resistance body 最小阻力体
least square 最小二乘方
least work 最少功率
leather machine belting （机用）皮带
lee 背风面；下风面
lee depression 背风坡低压

lee eddy 背风涡
lee face 背风面
lee side 背风面
lee side vortex 背风面旋涡
lee slope 背风坡
lee wave 下风波；背风波
leeward 背风面；下风
leeward face 背风面；下风面
leeward side 背风侧；背风面
leeward slope 背风坡
leeward wall 背风壁
leeward yacht 下风船
leeway(=sideslip) 风压差
left hand rule 左手定则
legal constraint 法律制约
legislation 法制；法规
Lego baseboard 莱戈块粗糙元板；鱼鳞板
Lego block 莱戈方块
Lego block roughness 莱戈方块粗糙元
length of blade 叶片长度
length of working cycle 工作周期；工作循环期
Lenz's law 楞次定律
leptokurtic 尖峰的
let fly 突然展开；发射；攻击
LETS (linear energy transfer system) 能量线性传递系统
level 阶段；水平；水准仪；液位
level cross country 平原交切地区
level gauge 水平规；水准仪
level instrument 位面计；水平仪
level of air quality 空气品质级
level of pollution 污染水平
level of stability 稳定水平
level of zero wind （有效）零风面
level switch （信号）电平开关；液位开关
level terrain 平坦地形
leveling 校平
leveling base 基准面
leverage 扭转力矩；杠杆作用
LG (low gear) 低速齿轮
lidar 激光雷达
LIDAR (light detection and ranging) 光波探测与测距
life 寿命
life belt 安全带
life cycle 寿命周期；产品寿命
life cycling test （循环加载的）疲劳寿命试验；耐久性试验
life cycle time 生命周期时间
life duration 耐用寿命
life of equipment 设备寿命
life of thermal 热泡寿命
life test 寿命试验
life time 使用期限；试验时间
life time dilation 寿命延长
life zone 生物带
lifetime 寿命
lift 升力；电梯；升降机
lift associated admittance 升力导纳；升力相关导纳
lift bolt 提升螺栓
lift(ing) capacity 升力
lift center 升力作用点；升力中心
lift coefficient 升力系数
lift component 升力分量
lift convergence 升力减小
lift curve 升力曲线
lift curve slope 升力曲线斜率
lift dependent drag 诱导阻力
lift diameter ratio 升力直径比

lift direction 升力方向
lift dissymmetry 升力不对称
lift distribution 升力分布
lift drag ratio 升阻比
lift effect 升力效应
lift efficiency 升力特性
lift equation 升力方程
lift force 升力
lift increment 升力增量
lift induced drag 升力诱导阻力
lift loading 升力分布
lift oscillator model 升力振子模型
lift point 升力点
lift principle 升力原理
lift range 升力变化范围
lift ratio 升力系数
lift ring 升力圈
lift type device 升力型装置
lift type rotor 升力型风轮；升力型转子
lift type wind machine 升力型风力机
lift vector 升力矢量
lift(ing) vortex 升力（旋）涡
lifter 升降器；升降机
lifting body 升力体
lifting force moment 升力矩
lifting line 升力线
lifting line theory 升力线理论
lifting moment 升力矩
lifting movement 上升运动
lifting plane theory 升力面理论
lifting reentry 升力重返
lifting surface 升力面
lifting surface theory 升力面理论
light 光；轻的
light air 高空大气；软风（一级风）
light beacon 灯塔
light breeze 轻风（二级风）
light condition 空载状态
light detection and ranging (LIDAR) 光波探测与测距
light duty 小功率工作状态；小功率的
light duty electrical cable 轻载电缆
light emitting diode 发光二极管
light flicker 闪变
light house 灯塔
light metal spirit level 光金属水平仪
light overhaul 小翻修
light wind 轻风（二级风）
lightening rod 避雷针；电极棒
lighting 照明设备
lighting conductor 避雷器；避雷针
lighting fixture 照明器材
lighting gap 避雷器放电间隙
lighting protection 雷击防护
lighting storm 雷暴
lightning 闪电
lightning arrestor 避雷针；避雷装置
lightning conductor 避雷装置；避雷针
lightning current 雷电流
lightning overvoltage 闪电过电压
lightning protection 防雷接地；防雷保护；避雷保护；避雷装置
lightning protection rod 避雷针；防雷杆
lightning protection system (LPS) 防雷系统
lightning protection zone (LPZ) 防雷区
lightning protector 避雷器
lightning rod 避雷针
lightning shielding 避雷
lightning storm 雷暴
lightning strike 雷击；闪电攻击

lightweight cladding 轻型维护结构；轻量级包层

lightweight structure 轻型结构

limit 极限

limit analysis 极限分析

limit case 极限情况

limit condition 极限状态

limit control 极限控制

limit cycle 极限周期

limit design 极限设计

limit distribution 极限分布

limit equilibrium 极限平衡

limit load 临界载荷；限制载荷

limit load factor 限制负荷因数；最大使用负荷因数

limit of accuracy 精度极限

limit of elastic 弹性极限

limit of error 误差极限

limit of integration 积分范围

limit of pressure 压力极限

limit point 极限点

limit speed switch 限速开关

limit state 极限状态

limit switch 限位开关

limit temperature 极限温度

limit value 极限值

limitation 限制；限度；极限

limitation capacity 极限容量

limited amplitude response 限幅响应

limited availability 有限利用度

limited current circuit 限流电路

limited mixing fumigation 有限混合型下熏

limiter 限幅器

limiting dimension 极限尺寸

limiting intensity 极限强度

limiting orbit 约束轨道

limiting point 极限点

limiting pressure 极限压力

limiting stress 极限应力

limiting value 极限值

limiting velocity 极限速度

limiting viscosimeter number 特性黏度；极限黏度计值

limiting viscosity 特性黏度

limousine 大型高级轿车

line 路线

line blow 强风；直线性强风

line check 小检修

line commutated inverter 线换流逆变器；线换向逆变器

line detuner 行解调器

line frequency 行频；行频率；线路频率；工频

line gap 线路避雷器

line generator 直线发生器；矢量发生器

line like building 细长建筑

line like structure 细长结构

line number 行数

line of cloud 云线

line of flow 流量线

line of pressure 压力线

line regulation 电源电压调整率；线路调节

line source 线光源；线发射源；线污染源

line source equation 线源方程

line switching voltage 线路交换电压

line trap 线路陷波器；限波器；线列陷波电路；导线插头

line vortex 线涡

linear 线型的

linear correlation 线性相关

linear dilatation 线性膨胀

linear dynamic response 线性动态响应
linear elasticity 线性弹性
linear energy transfer 线性能量传递
linear energy transfer system (LETS) 能量线性传递系统
linear equation 线性方程
linear expansion 线性膨胀
linear function 线性函数
linear oscillator 线性振子
linear perturbation 线(性)扰(动)理论
linear regression 线性回归
linear relation 线性关系
linear scale 线性比例
linear stratification 线性层结；线性层理
linear vector equation (LVE) 线性矢量方程
linear vector function (LVF) 线性矢量函数
linear velocity (LV) 线速度
linear viscous fluid damper 线性黏滞阻尼器；线性黏性减振器
linear zone 线性区
lineariser 线化器
linearity error 线性误差
linearized aerodynamic 线(性)化空气动力学
linearized equation 线(性)化方程
linearized hot wire anemometer 线(性)化热线风速计
linearized potential field 线(性)化势流场
linearized theory 线(性)化理论
linearly dependent vector 线性相关的矢量
line-to-neutral 线与中性点间的
link 挂环；链环
link switch 联动开关
link system 联动系统
linkage 连杆机构；连接；连接方法
linkage section 连接段；连接节

linking rod 连杆；联杆
linkwork 铰接机构；联动装置
lintel 楣；过梁
lintel beam 水平横楣梁
liquid 液体
liquid column 液柱
liquid coolant 冷却液
liquid droplet 液滴
liquid penetrant examination 液体渗透探伤
liquid quenching 液体淬火处理
liquid sensor (LS) 液体传感器
liquid state 液态
listing 列表；横倾
lithometeor 大气尘粒
live axle 动轴；驱动轴；传动轴；主动轴
live conductor 带电导体
live frame of balance 天平动框
live load 动力负载
live spindle 旋转主轴
load 加载；负载
load angle 载荷角；负载角；功角
load bearing frame 承载构架
load bearing structure 承重结构
load bearing system 承重系统
load capacity 负荷容量负载能力；载重能力
load carrying ability 起(载)重能力
load carrying covering 受力蒙皮
load case 荷载工况
load cell 测力传感器；载荷天平；载荷传感器；重传感器；测压元件
load center 负荷中心
load characteristic 负载特性
load coefficient 负荷系数
load compression diagram 负载压缩图
load condition 负荷条件；载重状态；负载情况

load control 负荷调节；负载控制；荷载控制
load current 负载电流
load curve 负荷曲线
load cycling 负荷周期
load deformation curve 应力应变曲线
load estimation 负荷估算；负荷计算
load factor 荷载因子；负荷系数
load inducing mechanical yawing system 装载机械偏航系统；负荷诱导机械偏航系统
load leveling 负荷调整；负载均衡
load limit 负荷限度
load line 载重线
load ratio (LR) 载荷比；有效载荷
load range 负荷范围
load removal 负荷卸载
load saturation curve 负载饱和曲线
load shedding 甩负荷；减负荷
load spectrum 载荷谱
load test 负荷试验
load torque 负载转矩
load variation 负荷变化
loading 加载；装货；负荷；含量
loading capacity 承载能力
loading shift 载荷重新分布
local angle of latitude 当地方位角
local angle of attack 当地迎角
local battery 自给电池
local blade chord 当地叶片弦 (长)
local breeze 局部风
local building code 地方建筑规范
local carburization 局部渗碳
local circulation 局部环流
local climate 局部气候
local climate condition 局部气候条件
local concentration 当地浓度

local control 就地控制
local diffusivity 当地扩散率
local destabilization 局部扰动
local drag 当地阻力
local eddy 局地涡流；局部涡流
local elongation 局部伸长
local equilibrium 局部平衡
local flow 局部气流；当地流动
local fluctuating pressure 局部脉动压力
local friction drag 局部摩擦阻力
local gust 局部阵风
local heat treatment 局部热处理
local inversion 局部逆温
local isotropy 局部各向同性；局部均向性
local lift 当地升力
local loss 局部损失
local Mach number 当地马赫数
local microclimate 局部微气候
local panel 现场配电盘
local peak of pressure 当地峰压
local per unit surface 电位面积负荷
local pollution 局部污染
local power network 地方电网
local pressure 当地压力
local resistance 局部阻力
local resistance factor 局部阻力系数
local separation 局部分离
local similarity 局部相似率
local topography 局部地形
local turbulence 局部湍流 (度)
local wind 局部风
local wind characteristics 局部风特性
local wind profile 局部风廓线
local wind regime 局部风况
local wind resource 当地风力资源

localized damage 局部损失；地区性损失；局部损伤
localized length scale 当地积分长度尺度；局部尺度
location 位置
lock 加锁；锁紧；锁定；自动跟踪
lock bolt 锁紧螺栓；锁紧螺钉；防松螺栓
lock in 锁定
lock in band 锁定带
lock in effect 锁定效应
lock in excitation 锁定激励
lock in of vortex 锁涡
lock nut 锁紧螺母
lock on 锁定
lock washer 锁紧垫圈；止动垫圈；防松垫圈
locked in resonant oscillation 锁定共振
locked in stall 稳定失速
locked in vortex-induced response 锁涡响应
locked rotor 锁定转子；堵转
locked rotor torque 锁定转子转矩
locking 锁定
locking device 锁定装置
locking nut 自锁螺母；防松螺母；锁紧螺帽
lofting plume 屋脊型羽流；高耸的烟羽；高升空中的烟羽
log （运行）记录；（系统）日志
log file 日志文件
log linear distribution 对数线性分布
log linear wind profile 对数线性风廓线
log normal distribution 对数正态分布
log paper 对数坐标纸
logarithm 对数
logarithmic decrement 对数衰减
logarithmic demodulation 对数触调器；对数反调制器

logarithmic differentiation 对数微分
logarithmic distribution 对数分布
logarithmic distribution law 对数分布律
logarithmic frequency spectrum 对数频谱
logarithmic function 对数函数
logarithmic law 对数（变化）律
logarithmic scale 对数尺度
logarithmic velocity profile 对数速度廓线
logarithmic wind profile 对数风（速）廓线
logarithmic wind shear law 对数风切变律
London type smog 伦敦型烟雾
long acceleration 持续加速度
long dated 长期的
long distance control 远程控制
long lived 耐久的
long range control 大范围调节
long range prediction 长期预报
long range transport model 远程运输模式
long run test 连续试验
long separation bubble 长分离气泡；长泡分离
long service life 长使用寿命
long span cable 大跨度缆线
long span structure 大跨度结构
long span suspension bridge 大跨度悬索桥
long term biological hazard 长期生物危害
long term effect 长期效应
long term mean wind speed 长期平均风速
long term observation system 长期观测系统
long term prediction 长期预报
long term wind data 长期风资料
long term wind speed average 长期平均风速
long test section wind tunnel 长试验段（边界层型）风洞
long time diffusion factor 长期扩散因子
long time washout factor 长期冲洗因子

long track tornado 长迹陆龙卷
long wave radition 长波辐射
longitude 经度
longitudinal axis 纵轴
longitudinal crack 纵裂
longitudinal data 纵向数据
longitudinal diffusivity 纵向扩散率
longitudinal dispersion 纵向弥散
longitudinal dune 纵向沙丘；沙垄；纵沙丘；纵向沙垄；线性沙丘
longitudinal flapping 纵向挥舞；纵向拍打
longitudinal metacenter 纵稳心
longitudinal oscillation 纵向振荡
longitudinal pressure gradient 纵向压力梯度
longitudinal response 纵向响应
longitudinal stability 纵向稳定性
longitudinal turbulence component 纵向湍流分量
longitudinal turbulence spectrum 纵向湍流谱
longitudinal vortex 纵向涡
longitudinal wind 纵向风；径向风
loop 回路；观察孔
looping plume 波形羽流
loose ash 飞灰
Los Angeles smog 洛杉矶型烟雾
loss 损耗
loss of flow 流动损失
loss of head 压力头损失
loss of momentum 动量损失
loss of power 动力损失；功率损失
losses in transit 转运损耗
lost motion 空转
lot area 区域面积
louver 百叶窗；通风窗
low 低（气）压中心；低气压

low area storm 龙卷风
low aspect ratio 小展弦比
low battery voltage 电池欠压；低电池电压
low cloud 低云
low current 低强度电流
low density binder 低浓度黏合剂；低密度黏结剂
low density wind tunnel 低密度风洞
low drag airfoil 低阻翼型
low drag configuration 低阻构型
low drag cowl 低风阻外罩；减阻罩
low drag profile 低阻外形
low energy content wind 低能量风
low frequency band 低频带
low frequency end 低频端
low exhaust 低（污染）排气
low gauge 低速齿轮
low gear (LG) 低速齿轮
low incidence 小迎角
low level chassis 低底盘
low level jet 低空急流
low level nocturnal jet 夜间低空急流
low level wind 底层风
low noise 低噪声
low oil alarm 低油位警报
low pass filter 低通滤波器
low pitch roof 缓坡屋顶
low pollution energy source 低污染能源
low pressure 低压
low rise building 底层建筑
low rise structure 低矮结构
low sided ship 低弦船
low speed 低速
low speed aerodynamic 低速空气动力学
low speed aerofoil 低速翼型

low speed characteristics 低速特性

low speed high torque shaft 低速高扭矩轴

low speed operation 低速运行

low speed shaft 低速轴

low speed stability 低速稳定性

low speed WECS 低速风力机（额定叶尖速率比小于3的风力机）

low speed wind tunnel 低速风洞

low speed wind tunnel test 低速风洞试验

low speed winding 低速绕组

low temperature brittleness 低温脆性；冷脆性

low temperature resistance 耐低温性；低温阻力

low torque (LT) 低转矩

low turbulence wind tunnel 低湍流度风洞

low voltage apparatus 低压电器

low voltage ride through (LVRT) 低电压穿越

lower limit 下限

lower order harmonics 低阶谐波

LPS (lightning protection system) 防雷系统

LPZ (lightning protection zone) 防雷区

LR (load ratio) 载荷比；有效载荷

LS (liquid sensor) 液体传感器

LT (low torque) 低转矩

lubric 光滑的

lubric pump 润滑泵

lubricant 润滑剂

lubricant additive 润滑油添加剂

lubricate 润滑

lubricating gauge 润滑装置

lubrication 润滑；注油；润滑（作用）；润滑剂；润滑方式

lubrication groove 润滑油槽

lubrication nipple 润滑喷嘴

lubrication systems 润滑系统

lubricators 注油机

lucarne 老虎窗；屋顶窗

luffing 抢风；转向迎风；俯仰收放运动；升降运动

luffing the helm 迎风使舵

lug 接线柱；柄；把手；突起；吊耳

lull 风暴间歇；息静；暂时平静

luminescent pigment 荧光颜料；发光颜料

lumped mass aeroelastic model 集总质量气动弹性模型

lumped mass system 集总质量系统

LV (linear velocity) 线速度

LVE (linear vector equation) 线性矢量方程

LVF (linear vector function) 线性矢量函数

LVRT (low voltage ride through) 低电压穿越

LWECS (large scale wind energy conversion system) 大型风能转换系统

现代英汉风力发电工程

A Modern English-Chinese Dictionary of Wind Power Engineering

M

MA (maintenance analysis) 维护分析
MAC (maximum acceptable concentration) 最大容许浓度
MAC (maximum allowable concentration) 最大容许浓度
MAC (mean aerodynamic chord) 平均气动力弦
Mach number 马赫数
Mach number effect 马赫数效应
machine 机器；机械
machine efficiency 机械效率
machine erection 机器安装
machine fault 机械故障
machine tapered covering 机械加工变厚度蒙皮
machine tools 工作母机
machining 机械加工
machining allowance 加工余量
machining allowance of casting 铸件机械加工余量
machining precision 加工精度
macroclimate 大气候
macrometeorology 大尺度气象学
macrorelief 大起伏；大地形
macroscale 宏观尺度
macroscopic convection 宏观对流
macroturbulence 大尺度湍流
macroviscosity 宏观黏性
MAD (maximum allowable dose) 最大容许剂量
Madaras rotor 马达拉斯转子
magnetic 磁的；有磁性的；有吸引力的
magnetic amplifier 磁放大器
magnetic annular shock tube (MAST) （风动研究用）磁性环形激波管
magnetic arc stabilizer 电磁弧稳定器
magnetic circuit 磁路
magnetic clutch 电磁离合器
magnetic conductance 磁导
magnetic coupling 磁耦合；磁性联轴器；电磁联结器
magnetic crack detection 磁力探伤
magnetic creeping 磁滞现象
magnetic domain 磁域
magnetic field 磁场
magnetic field distribution 磁场分布
magnetic field indicator 磁场指示器
magnetic field intensity 磁场强度
magnetic field meter 磁场计
magnetic field strength 磁场强度

magnetic flux 磁通
magnetic flux density 磁通密度
magnetic flux wave 磁通波
magnetic force 磁力
magnetic hard steel 硬磁钢
magnetic history 磁化史
magnetic hysteresis 磁性滞后；磁滞现象
magnetic insulation 磁绝缘
magnetic leakage field 漏磁场
magnetic leakage flux 漏磁通
magnetic moment 磁矩
magnetic particle 磁粉
magnetic particle examination 磁粉检验
magnetic particle indication 磁痕；磁粉显示
magnetic particle inspection 磁力探伤；磁粒检测
magnetic particle test 磁力探伤；磁粉探伤；磁粒测试；磁化检验
magnetic path 磁路
magnetic permeability 磁导率
magnetic pole 磁极
magnetic powder flaw detector 磁气粉探伤机
magnetic pull 磁铁引力；磁引力
magnetic reluctance 磁阻
magnetic remanence 顽磁
magnetic resultant 磁力合力
magnetic saturation 磁饱和
magnetic screening action 磁屏蔽作用
magnetic steel 磁钢；磁性钢
magnetic storage 磁存储器
magnetic storage medium 磁存储介质
magnetic susceptibility 磁化率
magnetic test 磁探伤；磁试验
magnetic testing 磁力探伤

magnetic tools 磁性工具
magnetic torque 电磁转矩
magnetic valve 电磁阀
magnetic writing 磁写
magnetisation characteristic 磁化特性
magnetization 磁化强度
magnetizing current 励磁电流
magnetizing EMF 磁化电动势
magnetizing inductance 励磁电感
magnetizing power 磁化功率
magnetizing reactance 磁化电抗
magneto 永磁发电机；磁发电机
magneto alternator 永磁交流发电机
magneto motive force (MMF) 磁通势
magneto(-)fluid(-)dynamics 磁流体力学
magnetoaerodynamics 磁空气动力学
magnetoconductivity 磁导率；导磁性
magnetoelectric machine 永磁式电机
magnetoelectric(al) generator 永磁发电机
magnetogasdynamics 磁性气体动力学
magnetohydrodynamic turbulence scale 磁流体动力紊流度
magnetohydrodynamically driven vortex 磁流体动力驱动涡
magnetohydrodynamics 磁流体动力学
magnification factor 放大因子
magnifier 放大器
magnitude 振幅；量级
Magnus effect 马格努斯效应
Magnus effect rotor 马格努斯效应转子
Magnus force 马格努斯力
main 主要部分
main beam 主梁
main bearing 主轴承
main circuit 主电路

main contact 主触头；主触点
main deck 主甲板
main flow air 主气流
main frame 主框架
main hull 主船体
main line 干线
main parameter 主要参数
main part 主要部件
main pipe 主管道
main powerline 电力干线
main rotor 主风轮
main rotor blade 主旋翼叶片
main shaft 主轴
main shaft flange 主轴法兰
main shelterbelt 主防护林带
main span 主跨
main spar 主梁
main stream 主流
main vortex 主旋涡
mains 电力干线
mains supply 干线供给；交流电源
mainstream wind 主流风
maintainability 易保养性
maintenance 维护；日常维修；保养
maintenance analysis (MA) 维护分析
maintenance and repair 维护修理
maintenance charge 维护费用
maintenance effort 维护工作
maintenance engineer 维修工程师
maintenance equipment 维修设备
maintenance facility 维护设备
maintenance factor 维修系数
maintenance instruction 保养说明；技术维护规则
maintenance man 维修工，维修人员

maintenance manual 保养维修手册
maintenance operating instructions 维修操作细则
maintenance overhaul 日常维护
maintenance people 维修人员
maintenance system 维护系统
maintenance test 维护试验
major 主要的
major equipment 重要设备
major oscillation 主振荡
major overhaul 大修
major parameter 重要参数
major part 重要零部件
major repair 大修
major specification 主要规格
make allowance 留余量
maldistributed airflow 不均匀空气流
maldistribution 不均匀分布
male flange 凸缘法兰
male-female facing flange 凹凸面法兰
malfunction 故障；不正常工作
malfunction rate 故障率
mandrel 半导体阴极金属心；心轴
maneuvering performance 操纵性能
maneuvering test 操纵试验
manhole 人孔；检修孔
manifolds 集合管
manmade environment hazard 人为环境公害
manmade source 人工源
manometer 压力计；测压计；血压计
manometer pressure 表压力
manometric 压力计的；用压力计测量的
mansard roof 折线形屋顶
mantle 地幔
manual 手册；指南；人工；手控；手动

manual control 手动控制
manual friction brake 手动摩擦制动器
manufactory 制造厂
manufactory system 生产体系
manufacturer 制造商
margin 边缘
marginal adjustment 边界调整
marginal vortex 边界涡
marine 船舶的；海生的
marine aerosol 海洋气溶胶
marine air mass 海洋气团
marine atmosphere 海洋大气
marine environment 海洋环境
marine weather observation 海洋天气观测
marine wind regime 海洋风况
maritime air mass 海洋气团
marked capacity 额定容量；额定生产率
marketing assessment 市场评价
mash seam weld 滚压焊
Maskell method 马斯克尔（风洞阻塞）修正法
masonry 石工；石工行业；石造建筑；砌体
masonry drill 石钻
masonry nail 水泥钉
mass 块；团
mass air flow 空气质量流量
mass axis 质量轴
mass balance 质量平衡
mass center 质量中心
mass conservation 质量守恒
mass conservation equation 质量守恒方程
mass damper spring system 质量阻尼器弹簧系统
mass distribution 质量分布；群分布
mass doublet 质量双线
mass entrainment flux 卷挟质量通量

mass flow 质量流量
mass flow constant 质量流量常数
mass flow continuity 质量流量连续性
mass flow parameter 质量流量参数
mass flow rate 质量流量（单位时间内通过的物质质量）；质量流率
mass flow ratio 质量流量比
mass flux 质量通量
mass inertia 质量惯性
mass law 质量定律
mass law effect 质量定律效应
mass load 惯性力
mass loss 质量损失
mass loss rate 质量损耗率；质量损失率
mass manufacture 大量制造
mass median diameter 质量中位直径
mass moment of inertia 质量惯性矩
mass sink 质量汇
mass spectrometer 质谱仪
mass spread 质量分布
mass stiffness 质量刚度
mass transfer 质量输运
mass transfer boundary layer 质量输运边界层
mass transport 质量输运
mass unbalance 质量不平衡
mast 桅杆；塔桅
MAST (magnetic annular shock tube) （风动研究用）磁性环形激波管
master 征服；控制；精通；主导装置；主导的
master alloying 主合金；母合金
master body 标准样件
master characteristics 校准特性；主特性曲线
master controller 主控制器
master cylinder 主液压缸；主缸

master electronic control panel 主电子控制箱；主电子控制板
master frequency 主频
master instruments 校准用仪器
master processor 主处理器
master schedule 主要图表；综合图表；设计任务书；主要作业表
material 材料
material compatibility 材料相容性
material composite 复合材料
material damping 材料阻尼
material safety data sheet 材料安全性数据表
material test 材料试验
materials engineering laboratory (MEL) 材料工程实验室
mathematical 数学的
mathematical expression 数学表达式
mathematical model 数学模型
mathematical model of controlled plant 控制装置数学模型
mathematical stimulation 数学模拟
matrix 矩阵
matrix converter 矩阵变流器
matrix element 矩阵元
matrix equation 矩阵方程
matrix method 矩阵法
matrix reduction 矩阵化简
matt surface 亚光表面；无光泽面；粗面
matt surface finish 亚光表面光洁度
mature 成熟的
mature hurricane 成熟飓风
mature storm 成熟风暴
maxed output power 最大输出功率
maximal time between faults (MTBF) 最大无故障工作时间

maximal work 最大功
maximum 最大化；最大限度；极限
maximum acceptable concentration (MAC) 最大容许浓度
maximum admissible dimension 最大容许尺寸；最大容许尺度
maximum allowable concentration (MAC) 最大容许浓度
maximum allowable dose (MAD) 最大容许剂量
maximum allowable draft 最大容许风量
maximum allowable draft loss 最大容许风量损失；最大容许通风损失
maximum allowable level 最大容许水平
maximum allowable operating temperature 最大许可操作温度
maximum allowable operating time 最大容许运行时间
maximum allowable speed 最高容许转速；最大容许速度
maximum allowable temperature 最大容许温度
maximum allowable velocity 最大容许速度
maximum allowable working pressure 最大容许工作压力
maximum available power 最大可用功率
maximum available time 最大有效（工作）时间
maximum bare table acceleration 空载最大加速度
maximum blockage 最大阻塞
maximum capacity 最大容量
maximum capacity standards 最大容量标准
maximum continuous load 最大连续负荷
maximum continuous rating (MCR) 最大持

续功率

maximum conversion rating 最大转换率

maximum daily temperature 日最高温度

maximum demand 最大需要负荷[电]最大需量

maximum depth of frozen ground 最大冻土深度

maximum design pressure 最高设计压力

maximum designed wind speed 最大设计风速

maximum difference 最大差值

maximum dimension 最大尺寸

maximum discharge 最大流量

maximum effectiveness 最大有效性；最大效用

maximum error 最大误差

maximum flow 最大流量

maximum ground concentration 最大地面浓度

maximum gust lapse interval 阵风最大递减时段；阵风最大递减时距

maximum hourly outdoor temperature 小时最高室外温度

maximum instantaneous power 最大瞬时功率

maximum instantaneous wind speed 最大瞬时风速

maximum lift coefficient 最大升力系数

maximum limit 最大极限

maximum load 最大负荷

maximum mean temperature 最高平均温度

maximum measured power 最大测量功率

maximum mixing depth (MMD) 最大混合层厚度

maximum operating efficiency 最大运转效率

maximum payload ratio 最大有效载荷比

maximum peak 最大峰值

maximum permissible dissipation 最大允许损耗

maximum permissible error 最大允许误差

maximum permissible limit 最大允许极限

maximum permissible power 最大容许功率

maximum permissible release 最大容许排放

maximum permissible speed 最高允许转速

maximum permissible temperature 最高允许温度

maximum permitted power 最大允许功率

maximum power 最大功率

maximum power of wind turbine 风力机最大功率

maximum power output (MPO) 最大功率输出

maximum power point tracking (MPPT) 最大功率点跟踪；最大功率点追踪

maximum pressure 最高压力

maximum probable error 最大可能误差

maximum rate of flow 最大流量速率

maximum rated load 最大额定负荷

maximum rating 最大额定值

maximum rotational speed 最大转速

maximum rotor blade envelope 风轮叶片最大包络线

maximum safe capacity 最大安全容量

maximum safe speed 最大安全转速

maximum safety load 最大安全负荷

maximum safety temperature 最高容许温度；最高安全温度

maximum service life 最大工作寿命

maximum service pressure 最大使用压力

maximum shaft power 最大轴功率

maximum size 最大尺寸
maximum stress (MS) 最大应力
maximum suction point 最大吸力点
maximum theoretical torque 最大理论转矩
maximum torque (MT) 最大扭矩
maximum torque coefficient 最大力矩系数
maximum turning speed of rotor 风轮最高转速
maximum wind speed (MWS) 最大风速
maximum working load 最大工作负荷
maximum working temperature 最大工作温度
Maxwell 麦克斯韦（磁通量单位）
Maxwell distribution 麦克斯韦尔分布
Maxwell field equation 麦克斯韦场方程式
Maxwell's formula 麦克斯韦公式
Maxwell's hypothesis 麦克斯假设
MCP (measure correlate predict) 测试相关预测
MCR (maximum continuous rating) 最大持续功率
mean 平均值
mean aerodynamic center 平均空气动力中心
mean aerodynamic chord (MAC) 平均气动力弦
mean air temperature 平均气温
mean annual climate temperature 年平均气象温度
mean annual precipitation 平均年降水量
mean annual range of temperature 平均温度年较差
mean annual temperature 年平均温度
mean annual wind power density 年平均风能密度
mean blade chord 叶片平均弦长
mean camber line 中弧线
mean daily maximum temperature 日平均最高温度值
mean daily minimum temperature 日平均最低温度值
mean daily temperature 日平均温度
mean distance 平均距离
mean effective load 平均有效负荷
mean effective pressure (MEP) 平均有效压力
mean effort 平均作用
mean environmental wind 平均环境风
mean error 平均误差
mean external wind loading 外部平均风荷载
mean failure rate (MFR) 平均故障率
mean flow 平均流量
mean free error 平均无故障时间
mean free path （分子）平均自由程
mean geometric chord 平均几何弦
mean geometric chord of airfoil 平均几何弦长
mean grain size 平均粒径
mean highest temperature 平均最高温度
mean hourly wind speed 平均小时风速
mean internal wind loading 平均内部风荷载
mean life 平均寿命
mean line 中弧线
mean load 平均负荷
mean lowest temperature 平均最低温度
mean maximum air velocity 最大平均空气速度
mean monthly maximum temperature 月平均最高温度
mean monthly temperature 月平均温度
mean monthly wet-bulb temperature 月平均湿球温度
mean parameter 平均参数

mean power 平均功率
mean pressure 平均压力
mean response 平均响应
mean sea level 平均海平面
mean solar time 平均太阳时
mean speed 平均速率
mean square departure 均方偏差
mean square deviation 均方偏差
mean square error 均方差
mean square response 均方响应
mean square wind speed 均方风速
mean surface temperature 表面平均温度
mean temperature 平均温度
mean temperature difference 平均温差
mean time 平均时间
mean time between critical failures (MTBCF) 平均严重故障间隔时间
mean time between defects (MTBD) 平均故障间隔时间
mean time between degradation (MTBD) 平均衰变间隔时间
mean time between detection (MTBD) 平均（故障）检测时间
mean time between errors (MTBE) 平均错误间隔时间
mean time between failures (MTBF) 平均故障间隔时间
mean time between interruption (MTBI) 平均中断间隔时间
mean time between maintenance (MTBM) 平均维修间隔时间
mean time between malfunction (MTBM) 平均误动作间隔时间
mean time between overhauls (MTBO) 平均检修间隔时间
mean time between removals (MTBR) 平均拆换间隔时间
mean Time between repairs (MTBR) 平均修理间隔时间
mean value 平均值
mean velocity 平均速度
mean velocity defect 平均速度亏损
mean velocity profile 平均速度廓线
mean velocity scaling 平均速度换算
mean wetted length 平均浸湿长度
mean wind 平均风速
mean wind profile 平均风速廓线
mean wind speed 平均风速
mean wind velocity 平均风速
meandering plume model 蛇行羽流模式
measure 测量
measure correlate predict (MCP) 测试相关预测
measured 测量的；量过的；标准的；测定的
measured hole 测量孔；测压孔
measured power curve 测量功率曲线
measurement 测量；度量
measurement device 测量装置
measurement error 测量误差
measurement mast 测风杆
measurement of flow rate 流量测量
measurement parameters 测量参数
measurement period 测量周期
measurement seat 测量位置
measurement sector 测量扇区
measurement set 测量装置
measurement value 测定值
measuring 测量；衡量
measuring accuracy 测量精度
measuring element 测量元件

measuring equipment 测试设备
measuring error 测量误差
measuring instrument 测量仪器
measuring mast 测量桅杆；测风塔
measuring method 测量方法
measuring range 量程
measuring unit 测量装置
mechanical 机械的；力学的
mechanical actuator system 机械执行系统
mechanical admittance 机械导纳；力导纳
mechanical agitation 机械搅拌
mechanical anemometer 机械式风速仪
mechanical balance 机械平衡
mechanical braking system 机械制动系统
mechanical control & protection 机械控制和保护
mechanical convection 强制对流
mechanical cooling tower 机械通风冷却塔；机械冷却塔
mechanical coupling 机械耦合
mechanical damage 机械损伤
mechanical damper 机械阻尼器
mechanical dampling 机械阻尼
mechanical device 机械装置
mechanical draft 机械通风
mechanical drawing 机械图
mechanical efficient 机械效率
mechanical endurance 机械寿命
mechanical energy 机械能
mechanical equipment 机械设备
mechanical equivalent 机械功当量
mechanical erosion 机械磨损
mechanical failure 机械故障
mechanical features 机械性能
mechanical flywheel energy storage 机械飞轮储能
mechanical impedance 力阻抗
mechanical injury 机械损伤
mechanical joint 机械连接
mechanical loss 机械损失
mechanical magnification factor 机械放大因子
mechanical maintenance 机械维修保养
mechanical mixing 机械混合
mechanical movement 机械运动
mechanical noise 机械噪音
mechanical oscillation 机械振动
mechanical part 机械零件
mechanical pitching mechanism 机械调桨机构
mechanical power 机械功率
mechanical power control 机械功率控制
mechanical property 机械性能
mechanical property test 机械性能试验
mechanical rectifier 机械式整流器
mechanical rotor brake 机械风轮制动器
mechanical sand control 机械固沙；机械防砂
mechanical similarity 力学相似性
mechanical stability 机械稳定性
mechanical stirring 机械搅拌
mechanical strength 机械强度
mechanical stress 机械应力
mechanical torque 机械力矩
mechanical torque transient 机械转矩暂态
mechanical transfer function 力传递函数
mechanical turbulence 机械湍流
mechanical ventilation 机械通风
mechanical work 机械功
mechanically 机械地
mechanically switched capacitor 机械开关电

容器

mechanically switched shunt capacitor 机械投切并联电容器

mechanics 力学；结构

mechanics of materials 材料力学

mechanics of rigid bodies 刚体力学

mechanism 机理；机构

mechatronic 机电整合

mechatronic system 机电一体化系统

median 中值；中位数

median size 中值粒径

median wind speed 中值风速

medium 方法；媒介

medium carbonsteel 中碳钢

medium current 中强度电流

medium repair 中修

medium scale wind energy conversion system 中型风能转化系统

medium scale wind turbine 中型风电机组；中型风力机

medium sized wind turbine 中型风电机组；中型风力机

medium speed planetary gear 中速行星齿轮

medium speed rotor 中速风轮

megawatt 兆瓦特

megawatt hour 兆瓦时

megawatt size wind turbine 兆瓦级风轮机

MEL (materials engineering laboratory) 材料工程实验室

melon shaped dome 瓜形圆屋顶

membrane like structure 薄膜型结构

membrane roof 薄膜屋顶卷材防水屋面

memory effect 记忆效应

meniscus dune 新月形沙丘

metal backing 金属敷层

MEP (mean effective pressure) 平均有效压力

meridional wind 经向风

merry go round windmill 转塔式风车

mesa 平顶山

mesh cell 网眼

mesh point 网格点

mesh size 目径

meshing 啮合；建网

meshing gear 啮合正齿轮

meshing interference 啮合干涉

mesoclimate 中尺度气候

mesometeorology 中尺度气象学

mesopause 中层顶；散逸层顶

mesophase 中间期；中间相

mesorelief 中地形；中起伏

mesoscale 中尺度的

mesoscale circulation 中尺度环流

mesoscale terrain 中尺度地形

mesosphere 中间层；散逸层

MET (meteorological/meteorology) 气象学

MET mast 气象桅杆

metacenter 稳心；外心点；定倾中心

metacentric diagram 稳心曲线

metacentric height 稳心高度

metacentric radius 稳心半径

metal 金属

metal arc welding 金属电弧焊

metal covering 金属蒙皮

metal insulator materials (MIM) 金属绝缘体材料

metal mesh 金属网

metal oxide semiconductor field effect transistor (MOSFET) 金属氧化物半导体场效应晶体管

metallic character 金属特性

metallic ring 金属环

meteoric water 大气水

meteorological (MET) 气象的；气象学的

metorological convection 气象对流

meteorological data 气象资料

meteorological documentation 气象文件

meteorological element 气象要素

meteorological map 气象图

meteorological mast 气象桅杆

meteorological model 气象模型

meteorological observation 气象观测

meteorological parameter 气象参数

meteorological range(=standard visual range) 标准视距；气象视距

meteorological reference wind speed 气象参考风速

meteorological site 气象台(站)址

meteorological standard condition 标准气象条件

meteorological station 气象站

meteorological symbol 气象符号

meteorological tower 气象测量塔

meteorological wind tunnel 气象风洞

meteorology (MET) 气象学

meter 米；仪表

metering device 测量装置

metering error 测量误差

metering system 计量系统

method 方法

method of bins 比恩法

method of blade root attachment 叶根连接方式

method of image 镜像法

method of lattice 网格法

method of least square 最小二乘法

method of representative blade element 代表叶素法

method of electric analogy 电模拟法；流变电模拟法

method of singularities 奇点法；有限基本解法

method of small disturbance 小扰动法

method of superposition 叠加法

method of trail and error 尝试法

metropolitan atmosphere 大城市大气

Meyer hardness 迈耶硬度

MFR (mean failure rate) 平均故障率

micro mini wind turbine 微型或者迷你型风电机组

micro siting 微观选址

microburst 微猝发风微爆气流

microclimate 微气候

microclimate heat island 微气候热岛

microcomputer 微型计算机

micromanometer 微压计

micrometeorological data 微气象资料

micrometeorological environment 微气象环境

micrometeorology 微气象学

micrometer 千分尺

micrometer callipers 螺旋测径器

microphone 麦克风；微音器

microprocessor 微处理器

microrelief 小起伏；微地形

microscale 微尺度

microscale circulation 微尺度环流

microscale turbulence 微尺度湍流

microstrain 微应变

microtopography 小地形

microturbulence 微湍流

microviscosity 微黏性

microwave reflector 微波反射器
microwave telecommunication tower 微波通讯塔
microzonation 微区划
mid frequency band 中频带
mid wing 中翼
mid chord 弦线中点
middle tooth 主齿
mid span 中跨
mid span shrouded blade （有减震）凸台的叶片
mile of wind anemometer 英里风速计
mile wind speed 英里风速
mileage 英里数
mileage speed 英里风速
mill 碾磨；磨细；磨机；磨粉机
milling 轧齿边，磨；制粉
milling heads 铣头
milling machine 铣床
MIM (metal insulator materials) 金属绝缘体材料
miniature 微型的；小型的
miniature motor 微型电动机
minicomputer 微型计算机
minicomputer program 微机程序
minicomputer program control 微机程控
minimal value 极小值
minimum 最小的；最低限度
minimum area 最小面积
minimum decision limit 最低判定限
minimum detectable amount 最小检测量
minimum detection limit 最低探测限
minimum determination limit 最低测定限
minimum diameter 最小直径
minimum disturbance 最小干扰

minimum drag coefficient 最小阻力系数
minimum drag position 最小阻力姿势
minimum flow velocity 最小流速
minimum induced loss windmill 最小诱导损失风车
minimum load 最小负荷
minimum load operation 最低负荷运行
minimum power factor 最小功率因数
minimum requirement 最低要求
minimum response concentration 最低响应浓度
minimum sample length 最小取样长度
minimum spacing 最小间距
minimum speed 最低速度
minimum standard 最低标准
minimum temperature 最低温度
minimum value 最小值
minimum velocity 最低速度
minimum wind to yaw 最小调向风
minitype 微型；小型
minor determinant 子行列式
minor loading condition 次要载荷情况；小加载条件
minor repair 小修
minus tolerance 负公差
minute crack 发状裂缝；细裂缝
misaligned 线向不正的；方向偏离的；不重合的；未对准的
mismatch 失配
missile （风卷）飞掷物
missile impact effect 飞掷物撞击效应
mist （烟）雾
mist of fine sand 细沙雾
mistral (wind) 凛冽北风
mixed flow 混合流

mixed layer height 混合层高度
mixed sand 混合沙
mixing coefficient 混合系数；湍流扩散率
mixing depth 混合厚度
mixing height 混合高度
mixing layer 混合层
mixing length 混合长
mixing length theory 混合长理论
mixing ratio 混合比
mixing tube 混合管
mixture 混合物
MMD (maximum mixing depth) 最大混合层厚度
MMF (magneto motive force) 磁通势
mo(u)lding 塑造；铸型；造型；塑造物；铸造物
mobile 运动物体；可移动的
mobile anticyclone 移动性高气压
mobile load 活动载荷
mobile pollution source 运动污染源
mobile source emission 运动源排放
mobile wind speed unit 流动式风速测量车
modal 模型的
modal amplitude 模态幅值
modal analysis 模态分析
modal damping 模态阻尼
modal displacement 模态位移
modal force 模态力
modal frequency 模态频率
modal mass 模态质量
modal response 模态响应
modal stiffness 模态刚度
modal wind load 模态风载
mode 模态；振形
mode of operation 工作原理
mode of vibration 振型

mode shape 模态形状
mode velocity 模态速度
model 模型；模式
model analysis 模型分析
model basin 模型试验池
model blocking effect 模型阻塞效应
model law 模拟律；模型律
model similarity 模型相似性
model support 模型支架
model test 模型试验；模拟试验
model wake blocking 模型尾流阻塞
modelling criteria 模拟准则；建模标准
modem (modulator-demodulator) 调制解调器
moderate 稳健的；温和的
moderate breeze 和风
moderate gale 疾风 (七级风，风速 13.9～17.1 米/秒)
moderate gust 中等阵风
moderate tropical storm 中等热带风暴 (最大风速 34～47海里/时)
moderate wind 和风
moderately strong wind 中强风
mode-superposition procedure 振型叠加法
modification 修改；改型
modification of topography 地形改造
modification of wind 风场改造；风控制
modified shape 改型
modulator 调制器
module 模块；组件
modulus 系数；模数
modulus in shear 剪切模数
modulus in tension 拉伸模数
modulus of aeroelasticity 气动弹性模量
modulus of compressibility 压缩弹性模数

modulus of compression 抗压弹性模数
modulus of compressive elasticity 压缩弹性模数
modulus of continuity 连续模数
modulus of elasticity 杨氏模量；弹性模量；弹性模数
modulus of elasticity in direct stress 拉伸弹性模数
modulus of elasticity in shear 剪切弹性模数
modulus of elasticity in tension 抗拉弹性模数
modulus of flexibility 挠性模数
modulus of resilience 回弹模量
modulus of resistance 阻力系数
modulus of rigidity 刚性模数
modulus of rupture 断裂模量；破裂模数；弯曲极限强度；抗裂系数
modulus of rupture in bending 弯曲断裂模数
modulus of volume expansion 体积膨胀模数
Mohr's hardness 莫氏硬度
moist adiabatic lapse rate 湿绝热递减率
moist adiabatic process 湿绝热过程
moist plume 水汽羽流
moisture 湿气水分；湿度；潮湿；降雨量
moisture content 含水量
moisture equation 水汽量平衡方程
moisture repellent insulation 防潮绝缘层
moisture resistant coating 防潮涂层
moistureproof 防潮的；防湿的；不透水的
moistureproof barrier 防潮层；防湿层
moistureproof construction 防潮结构
moistureproof film 防潮膜
moistureproof liner 防潮垫层
moistureproof material 防潮（湿）材料
moistureproof type 防潮型

moistureproofness 耐湿性
mold and die components 模具单元
mold changing systems 换模系统
mold chillers 模具冷却器
mold core 模芯
mold heaters 模具加热器
mold polishing 模具打磨
mold rain 梅雨
mold repair 模具维修
mold texturing 模具磨纹
molded breadth 型宽
molded depth 型深
molds 模具
molecular conduction 分子传导
molecular diffusion 分子扩散
molecular dissipation rate 分子耗散率
molecular viscosity 分子黏性
moment 力矩
moment admittance 力矩导纳
moment arm 力矩臂
moment balance 力矩平衡
moment block 力矩块
moment coefficient 力矩系数
moment curve 弯矩图
moment derivative 力矩导数
moment diagram 力矩图
moment method 力矩法
moment of couple 力偶矩
moment of deflection 弯矩；挠矩
moment of friction 摩擦力矩
moment of inertia 惯性矩；转动惯量
moment of inertia in yaw 偏航惯性矩
moment of momentum 动量矩
moment of momentum equation 动量矩方程
moment of momentum theorem 动量矩定理

moment of spectral density function 谱密度函数

moment of stability 稳定矩

moment of torsion 扭矩；扭转力矩

moment reference 空气动力焦点；空气动力矩基准点

moment unbalance 力矩不平衡

moment vector 力矩矢量

momentum 动量；冲量

momentum boundary layer 动量边界层

momentum budget 动量收支

momentum conservation 动量守恒

momentum deficit 动量亏损；动量损失

momentum deficit thickness 动量亏损厚度；动量损失厚度

momentum equation 动量方程

momentum flow vector 动流量矢量

momentum flux 动量通量

momentum mixing length 动量混合长

momentum mixing theory 动量混合理论

momentum plume 动量羽流

momentum spectrum 动量分布

momentum theorem 动量定理

momentum thickness 动量厚度

momentum thickness of boundary layer 边界层动量厚度

momentum transfer 动量传递

momentumless buoyant plume 无动量浮力羽流

Monin coordinate 莫宁坐标

Monin-Obukhov (stability) length 莫宁-奥布霍夫（稳定）长度

Monin-Obukhov length scale 莫宁-奥布霍夫长度尺寸

Monin-Obukhov universal function 莫宁-奥布霍夫通用函数

monitor 监控器

monitored information 监视信息

monitoring 监控

monitoring apparatus 监控设备

monitoring device 监控装置

monitoring system 监控系统

monocoque construction 硬壳式结构

monocoque structure 硬壳结构

monodisperse aerosol 单分散气溶胶

monospar construction 单梁结构

monotonic function 单调函数

monsoon 季风；（印度等地的）雨季；季候风

monsoon wind 季风

Monte Carlo method 蒙特-卡罗法

monthly average wind speed 月平均风速

monthly lowest temperature 月最低温度

monthly maximum temperature 月最高气温

monthly minimum temperature 月最低气温

monthly variation of windspeed 月风速变化

mooring 系留；停泊

mooring stall 浮台

morphological 形态学的

morphological characteristics 地貌特征

most frequent wind direction 盛行风向

mote 微尘

motion 动作；移动

motion characteristic 运动特性

motion pattern 流线图

motion round wing 绕机翼的环流运动

motor 电动机；马达

motor capacity 电动机容量

motor drive 电动机驱动

motor dynamo 电动直流发电机

motor efficiency 电动机效率

motor exhaust 发动机排气
motor load 电动机负载；电机负荷
motor output 电机输出
motor speed 电机转速
motor starter 电动机起动器
motor switch 电动开关
motor type 电动机类型
motor valve 电动阀
motor valve actuator 电动阀执行机构
motoring 电动机驱动
motoring friction 空转摩擦力
moulding 装饰线条；造型，模塑；铸模成形；压条
mound 山丘；堆；高地
mount 安装；底座；山峰
mountain 山；山岳
mountain air 山地空气
mountain and valley breeze 山谷风
mountain barrier 山地障碍物
mountain chain 山脉
mountain effect 山地效应
mountain pass 山隘口；隘口；山口
mountain valley wind 山谷风
mountain wind 山风
mounting 安装；装配
mounting bolt 装配螺栓
mounting flange 安装用法兰盘
mounting height 安装高度
movable 可移动的
movable belt （风洞）移动带；活动地板
movable blade 可动叶片
movable flap 活动襟翼
movable load 动载
movable roof 活动屋顶
movement 运动；移动；运转

movement of air 空气流动；气体运动
moving 移动的
moving belt （风洞）移动带；活动地板
moving blade 动叶片
moving contact 活动接点；移动接触；动触头；活动触点
moving coordinate system 运动坐标系
moving gaseous medium 气流
moving ground （风洞）活动地板
moving load 动荷载
moving part 运动部件
moving reference system 运动参考系
moving tunnel 可移动风洞
moving urban pollution source 城市运动污染源
moving vane 动叶片
MPO (maximum power output) 最大功率输出
MPPT (maximum power point tracking) 最大功率点跟踪；最大功率点追踪
MS (maximum stress) 最大应力
MT (maximum torque) 最大扭矩
MTBCF (mean time between critical failures) 平均严重故障间隔时间
MTBD (mean time between defects) 平均故障间隔时间
MTBD (mean time between degradation) 平均衰变间隔时间
MTBD (mean time between detection) 平均（故障）检测时间
MTBE (mean time between errors) 平均错误间隔时间
MTBF (mean time between failures) 平均故障间隔时间
MTBI (mean time between interruption) 平均

中断间隔时间

MTBM (mean time between maintenance) 平均维修间隔时间

MTBM (mean time between malfunction) 平均误动作间隔时间

MTBO (mean time between overhauls) 平均检修间隔时间

MTBR (mean time between removals) 平均拆换间隔时间

MTBR (mean time between repairs) 平均修理间隔时间

MTMD (multiple tuned mass damper) 多重调谐质量阻尼器

mud baffle 挡泥板

mud flap 挡泥胶皮

mud wing 翼子板

muff coupling 套筒联轴节

multi manometer 多管压力计

multi bladed rotor 多叶片风轮

multi bladed windmill 多叶片风车

multi degree of freedom 多自由度

multi directional wind 多向风；不定向风

multi element model 多段模型

multi flue stack 多烟道烟囱

multi force transducer 多力传感器

multicomponent balance 多分量天平；多分力天平

multifinned body 多翅体

multigap arrester 多隙避雷器

multigap discharger 多隙放电器

multilayer sandwich 多层夹层结构

multilayered flow 多层流动

multilevel converter 多点式变流器

multi-megawatt wind turbine 多兆瓦级的风电机组

multimode 多模态

multinomial distribution 多项式分布

multioperating mode automatic conservation 多工况自动转换

multioperating mode control system 多工况控制系统

multi-phase inverter 多相逆变器

multiple box girder 多室箱梁

multiple connection 并联接法

multiple disc clutch 多片式离合器

multiple flue chimney 多烟道烟囱

multiple pass atmospheric braking 多次气动力减速

multiple sensor 复合传感器

multiple source 复合源

multiple stage planetary gear train 多级行星齿轮系

multiple streamtube model 多流管模型

multiple tuned mass damper (MTMD) 多重调谐质量阻尼器

multiply fabric 多层蒙皮

multiply wood 多层板

multipoint earthing 多点接地

multipolar electrical generator stator 多极发电机定子

multipolar external stator 多极外部定子

multipole direct drive generator 多极直接驱动型发电机

multipole direct drive variable speed synchronous generator 多极直接驱动变速同步发电机

multipole internal rotor 多极内转子

multipole large diameter stator 多极大直径定子

multipole permanent magnet synchronous

generator 多极永磁同步发电机
multipole ring generator 多极环式发电机
multipressure measuring system 多点测压系统
multirib construction 多肋结构
multisection construction 多段结构
multisection flap 多段襟翼
multispeed motor 多速电机
multistage axial-flow fan 多级轴流通风机
multistage compressor driven wind tunnel 多级压气机驱动风洞
multistory block 多层大厦
multistory box building 多层盒式建筑

multi-tube manometer 多管压力计
municipal environment 城市环境
municipal pollution 城市污染
municipal program 城市规划
mushroom roof 伞形屋顶
mutation 突变；转变
mutual 共同的；相互的
mutual action 相互作用
mutual effect 相互作用
mutual flux 互感磁通
mutual inductance 互感系数
mutual inductor 互感
MWS (maximum wind speed) 最大风速

现代英汉风力发电工程

A Modern English-Chinese Dictionary of Wind Power Engineering

N

NACA (National Advisory Committee for Aeronautics) 美国国家航空咨询委员会
NACA airfoil NACA 翼型
nacelle 机舱；发动机舱
nacelle canopy 机舱罩
nacelle cover 机舱盖
nacelle electronic controller 机舱电子控制器
nacelle housing 机舱外壳
nacelle main frame 机舱主框架；机舱主机架
nail 指甲；钉；钉子
nail making machines 造钉机
nameplate 铭牌；厂名牌
nameplate capacity 铭牌容量
nameplate pressure 铭牌压力
nameplate rating 铭牌定额
nameplate specification 铭牌说明
narrow band correlation 窄带相关
narrow band cross-correlation 窄带互相关
narrow band excitation 窄带激励
narrow band noise 窄带噪声
narrow band spectrum 窄带谱
NASA (National Aeronautics and Space Administration) 美国国家航空与航天局
National Advisory Committee for Aeronautics (NACA) 美国国家航空咨询委员会
National Aeronautics and Space Administration (NASA) 美国国家航空与航天局
national ambient air quality standard 国家环境空气质量标准
national electrical code 全国电气规程
National Environmental Policy Act 国家环境政策法
national grid 国家电网
National Renewable Energy Laboratory (NREL) 美国国家可再生能源实验室
National Science Board (NSB) （美国）国家科学委员会
National Society of Professional Engineers (NSPE) （美国）国家专业工程师协会
national standard 国家标准
National Standard Association (NSA) （美国）国家标准协会
National Standard Part Association (NSPA) 全国标准零件制造业协会
National Wind Power Association 全国风电协会
national wind resource 国家风力资源
natural 自然的
natural activity 天然放射性

natural aerodynamics 自然空气动力学
natural air cooling 自然通风冷却
natural atmosphere 天然大气
natural attenuation 自然衰减
natural background 天然本底
natural boundary layer wind 自然边界层风
natural circulation 自然循环
natural convection 自然对流
natural convection flow 自然对流流动
natural convection range 自然对流范围
natural cooling system 自然冷却系统
natural damping 自然阻尼
natural draft 自然通风
natural draft condensation tower 自然通风冷却塔
natural draft cooling tower 自然通风冷却塔
natural environment 自然环境
natural frequency 固有频率
natural gas 天然气
natural gusty wind 天然阵风
natural hardness 自然硬度
natural hazard 自然危害
natural mode shape 固有振形
natural oscillation 固有振荡
natural pollution 天然污染
natural purification 自净作用
natural radioactivity background 天然放射性本质
natural rolling 自然横摇
natural source of pollution 天然污染源
natural strain 自然应变
natural stress 自然应力
natural turbulence 自然湍流
natural turbulent wind 自然湍流风
natural undamped frequency 无阻尼固有频率
natural velocity 固有速度
natural ventilation 自然通风
natural vibration 固有振动
natural wind 自然风
natural wind boundary layer 自然风边界层
natural wind environment 自然风环境
natural wind field 自然风场
natural wind gust 自然阵风
natural wind hazard 自然灾害
Navier-Stokes equation 纳维-斯托克斯方程
navigating bridge 驾驶桥楼
near 近的；靠近
near field 近场
near gale 疾风（七级风，风速13.9~17.1米/秒）
near wake 近尾流
near wake region 近尾流区
near neutral regime 近中性状态
near point correction 近点校正
near surface layer 近地面层
near surface wind 近地风
needle file 针锉
needle valve 针形阀；针状活门
negative 负的；反面的；负数
negative aerodynamic damping 负气动阻尼
negative aerodynamic stiffness 负气动刚度
negative allowance 负公差
negative angle of attack 负迎角
negative buoyancy 负浮力
negative converter 反向变流器
negative correlation coefficient 负相关系数
negative deviation 反向偏差
negative direction 反向；逆向
negative feedback 负反馈
negative feedback pitch control 负反馈桨距控

制

negative frequency 负频率
negative lift device 负升力装置
negative lift wing 负升力翼板
negative mass 负质量
negative peak 负峰值
negative phase sequence 负相序
negative phase sequence impedance 负相序阻抗
negative phase sequence voltage 负相序电压
negative plume rise 羽流负抬升
negative pressure 负压
negative sequence component 负序分量
negative virtual mass 负虚质量
negatively buoyancy plume 负浮力羽流
neodymium-iron-boron (Nd-Fe-B) 钕-铁-硼
nephelometer 浊度计
Nestor number 奈斯特数
net 网
net area 净面积
net damping 净阻尼
net electric power output 净电功率输出
net gain 净增益
net incoming radition 净入射辐射
net load 净荷载
net loss 净损失
net mean effective pressure 净平均有效压力
net positive suction head 净压头
net power 净功率
net power output 净输出功率
net pressure loss 净压力损失
net radiation 净辐射
net radiation flux 净辐射通量
net radiometer 净辐射计

net sectional area 净截面积
net value 净值
net weight 净重
net work 净功；有效功
network 网络；管网；电网
network connection point 电网连接点
network impedance 电网阻抗
network impedance phase angle 电网阻抗相角
neutral 中相；中性的
neutral atmosphere 中性大气
neutral atmosphere boundary layer 中性大气边界层
neutral ballon 中性气球
neutral boundary layer 中性边界层
neutral conductor 中性线
neutral density plume 中性密度羽流
neutral drag coefficient 中性风阻系数
neutral equilibrium 中性平衡；随遇平衡
neutral flow 中性流动
neutral lapse rate 中性递减率
neutral point 中性点
neutral stratified flow 中性分层流
neutral turbulence flow 中性湍流流动
neutrally buoyancy plume 中性浮升羽流
neutrally stability 中性稳定性
neutrally steer 中性转向
neutrally stratification 中性层结
neutrally stratified air 中性分层大气
neutrally stratified flow 中性分层流
neutrally thermal stability 中性热稳定性
neutrally turbulent flow 中性湍流流动
neutrally stable atmosphere 中性稳定大气
neutropause 中性层顶
neutrosphere 中性层

new energy source 新能源
new jet 新生射流
new plume 新生羽流
Newtonian flow 牛顿流动
Newtonian fluid 牛顿流体
Newtonian friction law 牛顿摩擦定律
Newtonian mechanics 牛顿力学
Newtonian viscosity 牛顿黏性（系数）
Newton's law 牛顿定律
Newton's law of fluid resistance 牛顿流体阻力定律
nimbostratus 雨层云
nimbus 雨云
nimbus cumuliformis 积云状雨云
nipple 螺纹接头
no load 空载
no load mode 空载模式
no load operation 空载运行
no slip condition 无滑流条件
no wind period 无风期
nocturnal 夜的
nocturnal boundary layer 夜间边界层
nocturnal cooling 夜间冷却
nocturnal inversion breakup fumigation 夜间逆温破坏下熏
nocturnal radiation cooling 夜间辐射冷却
nocturnal stability 夜间稳定度
nocturnal thermal boundary layer 夜间温度边界层
nodding action 摆动
node 节点，叉点
node voltage 节点电压
nodular cast iron 球墨铸铁
noise 噪声
noise control 噪声控制

noise criteria 噪声标准
noise criteria curve 噪声标准曲线
noise damper 消声器
noise effect 噪音效应
noise intensity 噪声强度
noise level 噪声等级；噪音级；噪声位准
noise measurement 噪声测量
noise pollution 噪声污染
noise reduction 消声；声衰减；降噪
noise reduction device 消音装置
no-load 无载；空载；空负荷
no-load characteristic 空载特性
no-load cogging torque 空载齿槽转矩
no-load compensation 空载补偿
no-load current 空载电流
no-load loss 空载损耗
no-load operation 卸载运行
no-load operation test 卸载运行试验
no-load running 空载运行
no-load speed 空载转速
no-load starting 卸载起动
no-load test 空载试验
nomenclature 术语；专用名称
nominal 名义上的
nominal air flow rate 额定空气流速率；额定风量
nominal capacity 额定容量；公称容量
nominal current 额定电流
nominal dimension 公称尺寸；标称尺寸
nominal error 公称误差
nominal load 额定负荷；公称负荷
nominal output 额定输出；标称生产率
nominal power 额定功率
nominal power of wind turbine 风电机组的额定功率

nominal power wind speed 额定风速
nominal pressure 标称压力；名义压力
nominal rating 额定值；额定出力
nominal rotational speed 额定转速
nominal tip speed 额定叶尖速度
nominal torque 额定扭矩；额定转矩
nominal value 公称值
nominal voltage 额定电压
nominal volume 公称容积
nominal wind speed 额定风速
nominally 在名义上；标称
nomogram 诺模图；列线图
non-aeronautical aerodynamics 非航空空气动力学
non-articulated blade 非铰接叶片
non-binding 无约束力的
non-buoyant plume 非浮力羽流
non-conducting voltage 截止电压
nonconforming 非一致性的
noncorrodibility 耐腐蚀性；不腐蚀性
non-damage test 无损探伤试验
nondestructive examination 非破坏性检验
nondestructive test 无损探伤试验
non-dimension 无量纲
non-dimensional coefficient 无量纲系数
non-dimensional factor 无量纲因子
non-dimensional velocity 无量纲速度
non-dimensional form 无量纲形式
non-dimensional frequency 无量纲频率
non-dimensional wind shear 无量纲风切变
non-dimensionalization 无量纲化
nondisjunction 不分离
non-disperse aerosol 非分散气溶胶
non-ductile fracture 无塑性破坏
non-eddying flow 无旋流

non-elasticity 非弹性
non-equilibrium condition 非平衡条件
non-equilibrium state 非平衡状态
non-frequency dependent damper 非频率阻尼器
nongeostrophic flow 非地转流动
non-homogeneity 不同性质
non-homogeneous 不均匀的；不均衡的；非稳态的
noninertial coordinate system 非惯性坐标系
nonisotropic 各向异性的；非各向同性的
non-isotropic turbulence 非均匀性湍流
nonlinear 非线性的
nonlinear aerodynamic force 非线性气动力
nonlinear aerodynamics 非线性空气动力学
nonlinear extraction 非线性提取
nonlinear interaction 非线性相互作用
nonlinear oscillation equation 非线性振荡方程
nonlinear resonance 非线性共振
nonlinear vibration 非线性振动
non-load current 空载电流
non-Newtonian flow 非牛顿流体
non-Newtonian viscosity 非牛顿黏性
non-oscillatory divergence 非振荡发散；静力发散
non-perfect fluid 非理想流体
non-periodic overhaul 不定期检修
non-renewable buffering 不可再生缓冲
non-renewable energy source 不可再生能源
non-renewable source of energy 非再生能源
non-resonant chattering 非共振抖振
non-resonant response 非共振响应
non-return-flow water tunnel 开路式水洞
non-return-flow wind tunnel 开路式风洞

non-rigid coupling device 弹性连接装置
non-rotating part 不旋转部件
non-salient pole alternator 隐极同步发电机
non-separated flow 非分离流
non-shrinking 非收缩性
non-shrinking cement 不收缩水泥
non-slewing crane 非回转式吊机
non-stationary aerodynamic derivative 非定常气动导数
non-stationary flow 非稳定流动
non-steady aerodynamics 非定常空气动力学
non-steady lifting surface theory 非定常升力面理论
non-steady motion 非定常运动
non-steady state 非稳态
non-steady state model 非稳态模型
non-streamlined 非流线型的
non-structural element 非结构部件
non-structural module 非结构性模块
non-tapered aerofoil 等弦翼面
nonuniform air 不均匀空气
nonuniform airflow 不均匀气流
nonuniform beam 变截面梁
nonuniform flow 紊流
nonuniform lift 不均匀分布升力
nonuniform medium 不均匀介质
nonuniform motion 变速运动
nonuniform topography 不均匀地形
non-uniformity 不均匀性；异质性
non-uniformity coefficient 不均匀系数
non-working flank 非工作齿面
nonsymmetric bending 不对称弯曲
norm 标准；定额；规格
normal 正常的；正规的
normal air 标准空气

normal atmosphere 标准大气；常态大气；正常大气
normal axis 垂直轴
normal bend 法向弯管；法线弯管
normal braking system 正常制动系统
normal change 正常变化
normal close contact 常闭接点
normal condition 正常状态；正常状况；标准条件
normal displacement 法向位移
normal distribution 正态分布
normal force 法向力
normal load 正常负载
normal mode 正规模态
normal open contact 常开接点
normal operating life 正常工作寿命
normal operation 正常运行
normal output 额定输出
normal phenomenon 正常现象
normal pitch 法向齿距
normal plane 法面
normal pressure and temperature (NPT) 常温常压；标准温度与压力
normal pressure gradient 法向压力梯度
normal probability curve 正态概率曲线
normal rated load 标准额定负荷
normal rated power 额定功率
normal shutdown for wind turbine 正常关机
normal shutdown situation 正常停机状态
normal size 标准尺寸
normal speed 额定速度
normal state 标准状态
normal stress 正应力
normal surface 法面
normal wind year 正常风年

normal working condition 正常工况
normality 常态；标准态
normalization 标准化；规格化
normalization condition 归一化条件
normalization constant 归一化常数
normalized coefficient 归一化系数
normalized power spectrum function 归一化功率谱函数
normalized spectrum 归一化谱
normally 正常地；通常地
normally close 常闭的
normally close valve 常闭阀
normally closed contact 常闭触点
normally open 常开
normally open contact 常开触点
normally open valve 常开阀
north latitude 北纬
northeast trades 东北信风带
nose cone 整流罩；整流帽；鼻锥体；头锥体
nose down moment 低头力矩
nose fairing 头部整流（罩）
nose nacelle 前舱
nose of wing 翼前缘
nose slice 机头晃动
nose spar 前缘梁
nose up moment 抬头力矩
noxious gas 有害气体
noxious odor 有害气味
nozzle 喷嘴；（风洞）试验段
nozzle throat 喷管喉道
NPT (normal pressure and temperature) 常温常压；标准温度与压力
NREL (National Renewable Energy Laboratory) 美国国家可再生能源实验室

NSA (National Standard Association) （美国）国家标准协会
NSB (National Science Board) （美国）国家科学委员会
NSPA (National Standard Part Association) 全国标准零件制造业协会
NSPE (National Society of Professional Engineers) （美国）国家专业工程师协会
nuclear accident 核事故
nuclear environment 核环境
nuclear pollution 核污染
nuclear power station 核电站
nuclear reactor 核反应堆
nuclide 核素
nuisance 公害
number of blades 叶片数
number of poles 极数
number of stator turn 定子匝数
number of teeth 齿数
number of yaw drives 偏航驱动器数量
numerical control 数字控制
numerical control device 数控装置
numerical data 数字数据
numerical differentiation 数值微分
numerical forecast 数字预报
numerical method 数值法
numerical model 数值模型
numerical simulation 数值模拟
numerical value 数值
nut bolt 带帽螺栓
nut collar 螺母垫圈
nut lock washer 锁紧垫圈
n-year wind speed n年一遇风速

现代英汉风力发电工程

A Modern English-Chinese Dictionary of Wind Power Engineering

O

objective analysis 客观分析
oblique wind 斜风
oblong bolt hole 椭圆形螺栓孔；长圆螺栓孔
obnoxious gas 秽气；恶臭气体
obnoxiousness 讨厌；可恶
observability 视野性
observation 观察；检查
observation balloon 观测气球
observation door 观察门
observation error 观测误差
observation hole 观察孔
obstacle 障碍物
obstacle avoidance 排除故障；障碍规避；避障
obstacle indicator 故障指示器
obstruction to vision 视程障碍
obtainable accuracy 可达精度
obtuse 迟钝的
obtuse angle 钝角
obtuse angled triangle 钝角三角形
OC (oil cooler) 油冷却器
occluded 闭塞的；堵塞
occluded cyclone 锢囚气旋
occluded front 锢囚锋
occlusion 锢囚（作用）
occupant comfort 居住者舒适性；乘坐着舒适性
oce steel 无渗碳表面硬化用钢
ocean 海洋
ocean anticyclone 海洋反气旋
ocean circulation 海洋环流
ocean current 洋流
ocean wave field 洋浪场
ocean wind 海洋风
ocean wind power generation 海洋风能发电
octagon 八边形；八角的
octagon wind tunnel 八角形试验段风洞
ocular angle 视角
ocular estimation 目测
odd 奇怪的
odd harmonics 奇次谐波
odo(u)r 气味；臭气
odor dilution 臭气稀释
odor nuisance 恶臭公害
odor pollution 臭气污染
odor source 臭气源
odorousness 臭气浓度
off delay 断电延时；延迟断开
off delay relay 断电延迟继电器
off design condition 非设计工况
off emergency 紧急断开

off gas 废气
off grid 离网
off grid operation 离网运行
off line process 脱机处理
off load 无负荷
off peak 正常的；非高峰的；峰值外的
off peak load 非高峰负荷
off peak system 非高峰系统
off peak time 离峰时间
off the shelf 现货供应的；现成的；常备的；成品的；非专门设计的
off wind 顺风
offensive odor 恶臭
offset 抵消；补偿
offset limit 屈服极限
offshore 近海的
off shore area 近海区
off shore discharge 近海排放
off shore drilling 近海钻井
off shore engineering 近海工程
off shore environment 近海环境
off shore oil rig 近海石油平台
off shore structure 近海结构物
off shore wind 离岸风；陆风
offshore wind energy 海上风能
offshore wind power 海上风电
ogive 尖顶的
oil 油
oil ageing 润滑油的氧化
oil buffer 油压缓冲器
oil channel 油槽
oil chuck 液压卡盘
oil circulation 滑油循环
oil cooler (OC) 油冷却器
oil crisis 石油危机

oil dashpot 油阻尼器
oil duct 油路；油道；润滑油道；油导管
oil filled cable 充油式电缆
oil film technique 油膜（流动显示）技术
oil film visualization 油膜显示
oil filter 滤油器
oil flow picture 油流谱
oil flow visualisation 油流流动显示
oil fog 油雾
oil hardening 油淬
oil immersed type transformer 油浸式变压器
oil level 油位；油面
oil motor 液压马达
oil pump capacity 油泵流量
oil quenching 油淬火
oil radiator 滑油散热器
oil seal 油封
oil shock absorber 油压缓冲器；油压减震器
oil trough 油槽
oil tube cooler 管状油冷却器
oldham coupling 滑块连接；双转块机构；欧氏联轴节
OLTC (on line tap changing) 在线轴头变换；有载轴头变换
OLTC (on load tap changer) 加载抽头变换器；有载换接器
O&M (operation and maintenance) 运行与维修（费用）
omni 全方位
omni directional wind 全向风
omni directionality 不定向性；全向性
on line tap changing (OLTC) 在线抽头变换；有载抽头变换
on line tap changing transformer 在线抽头变换式变压器；有载抽头变换式变压器

on load indicator 有载指示器
on load tap changer (OLTC) 加载抽头变换器；带负载抽头变换开关
on load tap changing transformer 带负载抽头变换式变压器
on peak demand 在高峰时间的供电要求
on site 就地；当地
on site energy storage 就地储能
on site measurement 现场测量；实测
on site observation 现场观测；实测
onboard gantry crane 机载龙门起重机；车载龙门起重机
once in 50 year (return) wind 50年一遇的风
oncoming flow 来流；迎面流
oncoming turbulence 来流湍流（度）
oncoming wind 来流风；迎面风
one bladed wind turbine 单叶风电机组；单叶风力机
one caliper disc brake 一卡钳盘式制动器
one component balance 单分量天平
one dimensional flow 一维流动
one direction thrust bearing 单向推力轴承
one piece blade 整叶片
one sided spectral density function 单边谱密度函数
one to one control mode 一对一控制方式
one velocity diffusion theory 单速扩散理论
one way clutch 单向离合器
on-off action 通—断作用；开闭动作
on-off servo 继电伺服（随动）系统
onset galloping velocity 驰振初始速度
onset wind speed 起始风速
onshore 向陆地；在岸上；陆上
onshore area 海岸区
onshore wind 向岸风

on-state 开态；通路状态；导通状态
on-state loss 通态损耗
open 公开的；敞开的
open air 露天；户外
open air climate 露天气候
open and shut action 开闭动作
open area 开口面积；开阔地带
open circle 开圆
open circuit 断路；开路；断路；切断电路；开式回路
open circuit characteristic 开路特性
open circuit operation 开路运行
open circuit wind tunnel 开路式风洞
open collector 集电极开路输出门
open condition 开敞状态
open construction 无蒙皮结构；骨架
open country 开阔地
open countyard 开口场院
open ended wind tunnel 开路式风洞
open floor and roof wind tunnel 无上下壁风洞
open frame structure 开敞式框架结构
open gear lubricant 开式齿轮润滑油
open into 通向
open jet wind tunnel 开口试验段风洞
open loop circuit 开环电路
open loop gain 开环增益
open loop system 开环控制系统
open phase protection 断相保护
open ring wrench 开口扳手
open stadium 露天运动场
open terrain 开阔地形
open throat wind tunnel 开口风洞
open truss bridge deck 开式桁架桥面
open web girder 空腹梁

open working section 开口试验段
opened ended tunnel 直流式风洞
open-hearth furnace 平炉
opening 开洞；开裂
opening of tuyere 风口
operating 运行；控制；管理
operating apparatus 操作装置
operating capacity 运行容量
operating characteristic 运行特性；操作特性
operating characteristic curve 运行特性曲线
operating charge 运行费用
operating condition 工况；运行状态；运行条件
operating control 运行控制
operating cost 操作费用；运转费用
operating data 运行数据
operating device 操作装置
operating disturbance 运行干扰
operating duty 运行负荷；操作负荷
operating efficiency 运行效率
operating life 使用寿命；工作寿命
operating lifetime 运行寿命
operating load 运行负荷；工作负荷
operating maintenance 运行维护
operating margin 运行安全裕量
operating method 操作方法
operating mode 工作状态
operating parameter 运行参数
operating point 工作点
operating process 操作过程
operating program 操作程序
operating range 操作范围；作用半径
operating rate 运行速率
operating ratio 运转时间
operating record 运行记录
operating repairs 日常维护检修
operating rotational speed range 运行转速范围
operating routine 运行程序；操作过程
operating speed 运行速度
operating table 操作台
operating test 运行试验
operating time 工作时间；运行时间
operating wind speed 运行风速
operation 操作；工序；处理；管理
operation activity 维护；保养
operation and maintenance (O&M) 运行与维修（费用）
operation at idling condition 空载运行
operation life 工作寿命
operation management 运行管理
operation plane 运行计划
operation schedule 运行程序；操作日程
operational 操作的；运作的
operational calculus 算符演算；运算微积分
operational envelope 运行包络线
operational environment 工作环境
operational experience 操作经验
operational interval 转速变化范围
operational requirement 操作要求
operational voltage 运行电压
operational wind speed 运行风速
operator 操作人员；值班员；执行机构
opposing 反对的
opposing wind 逆风；逆面风
opposite 对立面
opposite stress 相反的压力
opposite vorticity 反向涡量
optical 光学的
optical air mass 大气光学质量比

optical depth 光学厚度
optical fiber 光纤
optical fibre cable 光纤电缆
optical fibre controller 光纤控制器
optical fibre signal 光纤信号
optical outline method 光学轮廓法
optical sensor 光学传感器
optimal 最佳的；最理想的
optimal rotational frequency 最佳旋转频率
optimization 优化；最佳化；最佳效率阶段
optimization method 优选法
optimum 最佳效果
optimum angle 最佳角
optimum condition 最佳条件
optimum efficiency 最佳效率
optimum flow 最佳流量
optimum method 最佳方法
optimum operation 最佳操作；最佳运行
optimum performance 最佳性能
optimum rotational speed 最优转速
optimum seeking method 优选法
optimum selection 最佳选择
optimum site 最佳场址
optimum slip 最佳滑差
optimum speed 最佳速度
optimum speed to windward 最佳迎风速度
optimum system design 系统优化设计
optimum value 最佳值
optimum velocity 最佳速度
ordering scheme 程序图
ordinance 法规
ordinance load 规定负荷
ordinary 普通的
ordinary differential equation 常微分方程
ordinary flow 正常流

ordinary loading 日常负荷
ordinary maintenance 日常维修
ordinate 纵坐标
organic chemical pollutant 有机化学污染物
organic sediment 有机沉淀物
organized vortex trail 规则涡迹
orientation 方向；方位
orientation control 调向；朝向控制
orientation mechanism 定位机构；定向机构；迎风机构
orifice 孔；洞口
orifice plate 孔板
orifice static tap 静压测量点
origin 原点
origin of coordinates 坐标原点
origin of force 力的作用点
original 原始的
original data 原始数据
original design 原设计；初始设计
original material 初始材料
original state 初始状态
O-ring joint O形垫圈
orographic 山岳的
orographic anticyclone 地形反气旋
orographic condition 地形条件
orographic effect 地形影响
orographic factor 地形因子
orographic lifting 地形抬升
orographic rain 地形雨
orographic upward wind 地形性上升风
Orsat gas analyzer 奥萨特气体分析器
orthodox construction 传统结构
orthogonality 正交性；相互垂直
orthogonality relation 正交关系
oscillating 振荡的

oscillating pressure field 振荡压力场
oscillation 振荡；振动
oscillation boundary 振荡边界
oscillation damping 摆动阻尼；振荡阻尼
oscillation frequency 振荡频率
oscillation orbit 振荡轨道
oscillation response 振荡响应
oscillation spectrum 振荡谱
oscillation test 振动试验
oscillator 振荡器
oscillatory 振荡的
oscillatory flow 振荡流
oscillatory instability 振荡不稳定性
oscillograph trace 示波图
osculating plane 接触面
Oseen flow 奥幸流
out of action 不再运转；失去效用
out of control 失控
out of door hydrodynamic 室外流体动力学
out of order 失控
out of phase 异相；异相位
out of plane bending 面外弯曲
out of step 不同步的
out rigger 舷外支架
out yawing 外偏航
outboard support 外伸支架；外侧支架
outdoor activity 户外活动
outdoor air load 室外空气负荷；新鲜空气负荷
outdoor area 户外区
outdoor environment 户外环境
outdoors 户外；野外
outer 外面的
outer coating 外层
outer layer 外层
outer race （滚动轴承）外圈；轴承外部轨道；

外环
outer zone 表面层
outflow 流出物
outgoing radiation 出射辐射
outlet 出口
outlet opening 排除口
outlet velocity 出口速度；排除速度
output 输出
output characteristic of WTGS 风力发电机组输出特性
output coupling 输出连接
output load 输出负荷
output nominal speed 额定输出转速
output power(for WTGS) 输出功率（风力发电机组）
output resistance 输出电阻
output shaft 输出轴
output signal 输出信号
output voltage 输出电压
outside appearance 外观
outside callipers 外卡尺
outskirt 市郊
outswinger 外弧球
outward 向外的
outward flange 外凸缘（法兰）
oval 椭圆形
oval nail 椭圆钉
oval orbit 椭圆轨道
over 越过
over aging 超龄的；老朽的；旧式的
over all 遍及
over allowance 尺寸上偏差
over beam 过梁
over capacity design 过容量设计
over current 过电流

over current protection 过电流保护
over current protective device 过电流保护装置
over current relay 过电流继电器
over frequency 过频率
over load 过载
over loading 过载；超载；重载
over power(for wind turbines) 过载功率
over speed 超速
over speed protection 过风速保护
over twisting 过度扭转
over voltage 过电压
over voltage alarm 过压报警
over voltage protection 过电压保护
overachieve 超过（预期目的等）
overall 全部的
overall accuracy 总精度
overall coefficient 总系数；综合系数
overall design 总体设计
overall dimension 外形尺寸；轮廓尺寸
overall drag 总阻力
overall energy 总能量
overall error 总误差
overall evaluation 综合评价
overall load 总负荷
overall span 全翼展
overcast sky 阴天
overcrossing 上跨交叉
overestimate 估算过高
overexpansion 过度膨胀
overflow cutting machines for aluminium wheels 铝轮冒口切断机
overhanging 悬伸的
overhanging beam 悬臂梁
overhanging eaves 飞檐

overhanging support 外伸支架；悬臂支架
overhaul 大检修
overhaul life 大修周期
overhauling 大检修
overhead 天花板；高架的
overhead cable 高架缆线
overhead inversion 高空逆温
overhead line 架空线
overhead pipe 高架管道
overhead power line 高架输电线
overhead railway 高架铁路
overhead side deck 外伸侧桥面
overhead stable layer 高空稳定层
overhead valve engine 顶阀发动机；顶置气门发动机
overhead weld 仰焊
overhead wire 高架电线
overlap 搭接（长度）；重叠
overload 过负荷；过载
overload allowance 容许超负荷
overload capacity 超负荷量；过载能量
overload characteristic 超负荷特性
overload imitation 超量模拟
overload operation 超载运行
overload protector 过载保护器
overload rating 额定过载
overload relay 过载继电器
overload relief valve 过载安全阀
overload safeguard 过载保护装置
overload test 超负荷试验
overloading 超载
overlying inversion layer 覆盖逆温层
overprediction 过高预计；过高预报
overshoot 超越
overshoot in angle of attack 迎角急增量

overspeed 超速
overspeed control 超速控制
overspeed device 超速保护装置
overspeed protection 超速保护
overspeed protection device 超速保护机构
overspeed spoiler 超速保护扰流板
overspeed trail 超速试验
overspeeding 超速；过速
overstaff 为…配备人员过多
overstretch 延伸过长
overspin 上旋
oversteer 过度转向
overstress 过应力

overturning 倾覆；倾翻
overturning moment 倾覆力矩；倾覆弯矩
overturning wind speed 倾覆风速
oxidation 氧化；氧化作用；氧化层
oxide skin 氧化层
oxyacetylene 氧乙炔的
oxyacetylene welding 氧乙炔焊
oxygen electrode 氧电极
oxygen propane cutting 氧气丙烷炬切割
ozone layer 臭氧层
ozonopause 臭氧层顶
ozonosphere 臭氧层

现代英汉风力发电工程

A Modern English-Chinese Dictionary of Wind Power Engineering

P

packaged electronic circuit (PEC) 电子线路程序包
packing 密封；填料；衬垫
packing collar 垫圈
packing grease 密封润滑脂
packing joint 密封连接
packing list 装箱单
packing plate 密封垫；垫板
packing ring 垫圈；填料环
packing washer 密封垫圈
pad 衬垫；护具；便笺簿；填补；滑板
paddle 踏板；桨叶板
padlock 挂锁；扣锁；关闭
padmount transformer 组合式变压器
paint film visualisation 涂膜流动显示
pair of vortices 涡对
pair of wheels 轮对
panel flutter 壁板颤振
panel method 板块法；面元法
panel wall 幕墙；嵌板墙幕墙；水冷壁
panemone 全风向阻力型竖轴风车
pans 盘状凹地（尤指盆地）
panting 挠振；拍击；冲击；脉动
parabola 抛物线；抛物面
parabolic diffusion equation 抛物线型扩散方程
paraboloid antenna 抛物面天线
paraboloid of revolution 回转抛物面
paraboloidal aerial 抛物面天线
parachute jumping 跳伞
parachuting 跳伞运动
parafoil 伞翼；翼型伞
parallel 平行的
parallel circuit 并联电路
parallel connection 并联
parallel operation 平行操作；并列运转
parallel section 等截面段
parallel slices （气流）平行层
parallel stream 平行流
parameter 参数；系数
parameter detection 参数检测
parameter optimization 参数优化
parametric equation 参数方程
parapet 女儿墙；护墙
parasite 寄生虫
parasite drag 寄生阻力
parasite load 寄生载荷
parasitic 寄生的
parcel of air 气球
parcel speed 气块速度

parent pollutant 原污染物
park 停放
park efficiency 停机效率
parked wind turbine 风力机停机
parking 停机
parking brake (for wind turbine) 停机制动（对于风电机组）
part depth simulation （大气边界层）部分厚度模拟
part(ial) span flap 部分翼展襟翼
partial 局部的
partial derivative 偏导数
partial differentiation 偏微分
partial earth 部分接地
partial load 部分负荷
partial oxidation 部分氧化
partial pressure 分压
partial similarity 部分相似
partial simulation 部分模拟
partial span pitch control 部分跨度桨距控制
partial throwing 部分倒伏状（V级植物风力指示）
particle 颗粒
particle size 粒度
particulate 微粒
particulate cloud 粒子云
particulate plume 粒子羽流
particulate pollutant 粒状污染物
partition column 分配柱
partly turbulent flow 部分湍流
Pascal's law 帕斯卡定律
Pasquill stability 帕斯奎尔稳定度
Pasquill-Gifford curve 帕斯奎尔-吉福特曲线
pass 垭口；山口；通道
passage 通道
passing 经过的

passive 被动的
passive circuit elements 被动电路元件
passive contaminant 被污染物
passive control 被动控制
passive diffusion 被动扩散
passive filter 无源滤波器
passive load 无源负载
passive method 被动法
passive plume 被动羽流
passive pollution 被动污染
passive resistive load control 无源电阻性负载控制
passive resistor 无源电阻器
passive safety 被动安全性
passive source 被动源
passive spread 被动扩散
passive yaw （风轮）被动调向；被动对风
passive yawing 被动偏航
pastagram 温高图
paste 张贴；粘贴
paste adhesive 糊状胶黏剂
patch board 配电板
path line 迹线；流线
path of devil 尘暴路径
path of plume 羽流路径
pattern allowance 模型余量
payback 偿付
payback period 还本期；归本年期；回收期
payload fraction 有效载荷部分
PBL (planetary boundary layer) 行星边界层
PCC (point of common coupling) 公共耦合点；公共连接点；公共供电点
PD (potential drop) 电压降
peak 峰值；波峰
peak amount 高峰量

peak broadening 峰加宽
peak capacity 高峰容量；最大容量
peak chopper 波巅限幅器
peak clipper circuit 削峰电路
peak clipping 峰值限幅器
peak contact 齿顶啮合
peak curve 尖顶曲线
peak discharge 高峰流量；最大流量
peak electrical load 高峰电力负荷
peak energy 峰值能量
peak factor 峰值因子
peak flood 高峰流量
peak frequency 高峰频率
peak gust 最大阵风
peak half width 峰半宽度
peak hour 高峰时间
peak load 峰值负荷
peak load operation 高峰负荷运行
peak load time 高峰负荷时间
peak of negative pressure 负压峰
peak output 最大输出
peak point 峰值点；最高点
peak point current 峰值电流
peak power (PP) 峰值功率
peak power output 峰值功率输出
peak pressure 峰压
peak rate of flow 高峰流量
peak sharing 调峰
peak shaving 高峰调节；高峰消减
peak to valley ratio 峰谷比
peak torque 峰值转矩
peak units 调峰机组
peak value 峰值
peak voltage 高峰电压
peak width 峰宽

peak wind speed 最大风速
peaked roof 尖屋顶
peaking 调峰；剧烈增加
peak-to-average ratio 最大值与平均值之比
peak-to-peak 峰间值；峰到峰；峰至峰值
peak-to-peak amplitude 峰间幅值
peak-to-peak fluctuation of amplitude 振幅峰—峰值起伏
peak-to-peak separation 峰间幅值
peak-to-peak value 峰间值；峰—峰差值
pebble 卵石
PEC (packaged electronic circuit) 电子线路程序包
PEC (photo electric cell) 光电管；光电池
PEC (power electronic converter) 电力电子变流器
PEC (printed electronic circuit) 印刷电路
PEC (program element code) 程序单元代码
PEC controlled variable external resistance PEC控制的外部可变电阻器
pediment 山形墙；三角墙
peg board 小钉板
penetrameter 透度计
penetrating capacity 渗透能力
penetrating inversion 渗透逆温
penetration coefficient 渗透系数
penetration flaw detector 渗透探伤
per unit value 标么值
percentage 百分比
percentage elongation 延伸率
percentage of block 阻塞度
percentage of turbulence intensity 湍流度
perceptible 可感觉的
perception threshold 可感觉阈
percussion action 冲击作用

percussion weld 冲击焊接
perfect 完美的
perfect black body 理想黑体
perfect fluid 理想流体
perfect gas 理想气体
perfect reflection 理想反射
perforated 穿孔的
perforated shrouds of chimney 多孔烟囱罩
perforated throat wind tunnel 开孔壁风洞（试验段壁有孔）
perforated wall 孔壁；有漏窗墙
perforated wing flap 穿孔襟翼
performance 性能；特性；生产力；实施
performance analysis 性能分析
performance build-up 性能提高
performance calculation 工况计算
performance characteristic 工作特性；运行特性
performance chart 工作特性图
performance coefficient 性能系数
performance curve 性能曲线
performance data 性能数据
performance diagram 工况图
performance envelope 性能包络线
performance estimation 性能估计
performance evaluate 性能估算
performance figure 质量指标
performance index 性能指数；性能指标
performance measurement 性能测定；工况测定
performance number 特性数；功率值
performance optimization 性能最佳化
performance parameter 性能参数
performance period 执行周期
performance report (PR) 性能报告

performance specification 运行特性
performance test 性能试验；运行试验
performance test center 性能测试中心
perimeter 周边
period 周期
period change 周期变化
period load 周期性负荷
period of duty 运行时间
period of oscillation 振荡周期
period of service 使用期限
periodic 周期的
periodic component 周期分量
periodic current 周期电流
periodic duty 周期性负荷
periodic fluctuating windspeed 周期性脉动风速
periodic function 周期函数
periodic inspection 定期检修
periodic load 周期负荷
periodic overhaul 定期检修
periodic turbulent fluctuation 周期性湍流脉动
periodic vibration 周期振动
periodic vortex shedding 周期性旋涡脱落
periodic vortex street 周期性涡街
periodic wake 周变尾流
periodicity 周期性
peripheral 周边的；外围的
peripheral component of velocity 切向分速度；圆周分速度
peripheral speed 圆周速度
peripheral velocity 圆周速度
permanent 永久的；永恒的
permanent deformation 永久变形
permanent flow 稳流；定常流动

permanent load 持久负荷

permanent magnet (PM) 永久磁铁；永磁体；永久性磁石

permanent magnet generator (PMG) 永磁发电机

permanent magnet synchronous generator (PMSG) 永磁同步发电机

permanent plant 固定设备

permanent site 永久场地

permeability 导磁率；导磁性；透磁率

permeance 磁导

permissible deviation 容许偏差

permissible error 容许误差

permissible limit 容许限度

permissible load 容许负荷

permissible speed 容许速度

permissible tolerance 可容许公差

permissible value 容许值

permissible variation 容许误差；容许偏差

permission flexibility 容许挠性

perpendicular 垂线；垂直；正交

perpendicular bisector 中垂线

perpendicular line 垂直线

perpendicular of velocity 速度变动

Persian windmill 波斯（竖轴）风车

persistence 持续

persistence of energy 能量守恒

persistent oscillation 持续振荡

persistent wind 持续风

perspex 有机玻璃

perturbation 小扰动；摄动

perturbation analysis 小扰动分析；摄动分析

perturbation method 小扰动法；摄动法

perturbed flow 受扰流动

PF (power factor) 功率因数

PFM (power factor meter) 功率因数计

phase 相；相位

phase angle 相位角

phase change 相变

phase change material 相变材料

phase coefficient 相位系数

phase compensation 相位补偿

phase compensation battery 相位补偿电池

phase conductor 相线

phase control 相位控制

phase controlled rectifier 相位可控整流器

phase controlled rectifying circuit 相控整流电路

phase conversion 相态转换

phase difference 相差；相位差

phase displacement 相位差

phase equality control 同相位控制

phase equilibrium 相平衡

phase imbalance 相间不平衡

phase inversion 相变化

phase inversion theory 相变理论

phase lag 相位滞后

phase regulated rectifier 可控相位整流器

phase reversal 反相

phase shift 相位位移；相位移转；相角程；相移

phasor 相矢量；相向量

phenomenological relation 表象关系

photo [词头] 与光有关的

photo electric cell (PEC) 光电管；光电池

photocathode 光电阴极

photochemical pollution 光化学污染

photochemical process 光化学过程

photochemical smog 光化学烟雾

photoelectric 光电的

photoelectric device 光电器件

photographic process 照相法
photographic recording 照相记录
photovoltaic panel 光伏电池板
photovoltaic (PV) 光伏
physical 物理的
physical atmosphere 物理大气压
physical climatology 物理气候学
physical dimension 物理尺寸
physical disturbance 物理扰动
physical incidence 物理迎角
physical model 物理模型
physical modelling 物理模拟
physical simulation 物理模拟
physical stack height 烟囱实际高度
physical time 物理时间；实际时间
physical verisimilitude 物理模拟；物理逼真
physiognomy 地貌
PI (proportional-integral) 比例积分
pibal 测风气球
pick off gear 可互换齿轮
picker 采摘者；挖掘者；采摘机；采摘工具
pickling 酸洗
pickup calibration 传感器校准
pickup current 始动电流
piezoelectric crystal balance 压电晶体天平
piezoelectric effect 压电效应
piezoelectric transducer 压电传感器
piezometer 流体压力计；测压管；压电计；压力表
piezometric ring 测压环
pile foundation 桩基础
piling up effect 堆积效应
pillar 桩；墩；支柱
pilot balloon 测风气球
pilot circuit 控制电路

pilot lamp 指示灯；信号灯
pilot plant facility 试验性设备
pilot tube 指示灯
pilot wind power plant 风力发电试验场
pin 销；栓；钉
pin down 把…固定住；使动弹不得
pinch 抢风夹点；箍缩；芯柱
pinion 小齿轮；柱销；轴齿轮
pinion gauge 小齿轮；游星齿轮
pipe & tube making machines 管筒制造机
pipe accessories 阀件
pipe arrangement 管配置
pipe stanchion 管支柱
pipeline with fins 带鳍片管道
pipette method 吸管法
piping 管道系统
piping system 管道系统
piping work 铺管工程
piston 活塞
piston cylinder arrangement 活塞气缸部件；活塞气缸排列
piston groove 活塞槽
piston pin 活塞销
piston pin boss 活塞销毂；活塞销孔；活塞销座；活塞销衬套
pit skin 金属表面气孔
pitch 变桨；调桨；桨距；齿距
pitch actuation 调桨驱动机构
pitch angle 桨距角
pitch angle controller 桨距角控制器
pitch bearing 调桨轴承
pitch change mechanism 变桨距机构
pitch changing 变桨距
pitch circle 节圆
pitch control mechanism 桨距控制机构；防纵

摇机构

pitch control system 调桨控制系统

pitch controller 调桨控制器

pitch curve 分度曲线

pitch cylinder 桨距调节气缸

pitch damping 俯仰阻尼；纵摇力矩

pitch diameter 节圆直径；分度圆直径

pitch gear 径节齿轮

pitch inertia 纵摇惯量；俯仰惯量

pitch motor 调桨马达

pitch of teeth 齿距

pitch play 齿隙

pitch point 节点

pitch regulated rotor 变距限速风轮

pitch regulated teetering rotor 调桨摇摆风轮

pitch regulated wind turbine 桨距调节风电机组

pitch regulating 桨距调节

pitch regulation 桨距调节

pitch serve motor 调节伺服马达

pitch setting 桨距调节

pitch system 调桨系统

pitch teeter coupling 变桨摇摆连接器

pitch to feather 顺风调桨

pitch to stall 调桨失速

pitch up 剧仰

pitch vibration 俯仰振动；纵摇振动

pitchable tip 可调桨叶尖

pitched roof 斜屋面

pitching 俯仰

pitching coefficient 调桨系数；俯仰系数

pitching equation 调桨方程

pitching gear 调桨齿轮

pitching lever 调桨杠杆

pitching moment 俯仰力矩

pitching moment balance 俯仰力矩平衡

pitching moment coefficient 俯仰力矩系数

pitching moment linearity 俯仰力矩的线性变化

pitching motion 俯仰运动

pitching resolution 调桨分辨率

pitching vibration 俯仰振动；纵摇振动

pitometer measurement 皮托管量测

pitot hole 总压孔

pitot line 皮托管；空速管；风速管

pitot loss 总压损失

pitot pressure 皮托管压力；皮托压力；全压

pitot pressure inlet port 皮托管压力进口

pitot probe 皮托管

pitot rake 梳状皮托管；全压管

pitot static tube 皮托静压管

pitot support rod 皮托管支杆

pitot tube 皮托管；总压管

pitot tube installed rod 皮托管安装杆

pitot tube method 皮托管法

pitot tube mouth 皮托管入口

pitot static difference 全—静压差；动压头；全压力与静压力差

pitot static drain 全静压管排水口

pitot static head 皮托静压水头；全静压探测管头；全静压头

pitot static rake 梳状动静压管

pitot static traverse 总静压游测

pitting 点蚀；点腐蚀；凹痕

pivot 支点；枢轴

pivot pin 枢轴销钉

plain 平的；简单的

plain conductor 普通导线

plain flap 简单襟翼

plane cell 翼组

plane flow 平面流动
plane source 平面源
planet 行星
planet carrier 行星齿轮架；行星架
planet gear 行星齿轮
planet gearbox 行星齿轮箱
planet (planetary) gauge 行星齿轮
planetary 行星的
planetary atmosphere 行星大气
planetary boundary layer (PBL) 行星边界层
planetary circulation 行星环流
planetary gear drive mechanism 行星齿轮传动机构
planetary gear train 行星齿轮系
planetary gearbox 行星齿轮箱
planetary gearbox yaw drive 行星齿轮偏航驱动器
planetary gearing 行星齿轮传动
planetary train 行星齿轮系
planetary wave 行星波
planetary wind 行星风
planing 滑行；刨的
planing boat 滑行艇
planing machines 刨床
planing machines vertical 立式刨床
planning chart 计划图
plant cover 植被
plant factor 设备利用率；发电厂利用率
plant metabolism 植物新陈代谢
plastic 塑料
plastic behaviour 塑性
plastic clamp 塑料夹具
plastic composite 复合塑料
plastic construction 塑料结构
plastic flow 塑性流动；塑性流；黏(滞)流(动)

plastic material 塑性材料
plasticized elastomer 增塑弹性材料
plate discharger 板形放电器
plate foundation 板式基础
plate thickness 筛板厚度；板块厚度；金属板厚度
plate web spar 腹板式(翼)梁
plateau 高原
platform 平台
platform hatch 平台短门；平台舱口
platinum 铂
platinum wire anemometer 铂丝风速计
platykurtic 低峰态
PLC (programmable logic controller) 可编程逻辑控制器
PLC failure 可编程逻辑控制器故障；PLC故障
plenum 充气；压力通风
plenum chamber 驻室；进气增压室
pliability 柔韧性
plot 曲线图；作图
plug 塞
plugging 反向制动；堵塞
plumb line 垂直线
plumber block 止推轴承；轴承台
plume 羽流；烟羽
plume advection 羽流平流
plume angle 羽流展角
plume bifurcation 羽流分岔
plume buoyancy 羽流浮力
plume centerline concentration 羽流轴线浓度
plume centerline height 羽流轴线高度
plume coalesce 羽流集聚
plume concentration 羽流浓度
plume depletion 羽流耗损
plume diffusion 羽流扩散

plume dispersion 羽流弥散

plume exhalation 羽流消散

plume exit velocity 羽流出口速度

plume expansion 羽流碰撞

plume geometry 羽流几何形状

plume growth 羽流增长

plume impingement 羽流拍撞

plume lifetime 羽流生存期

plume of bubbles 气泡羽流

plume of warm air 热空气羽流

plume optical depth 羽流光学厚度

plume oscillation 羽流振荡

plume passage 羽流行程

plume profile 羽流廓线

plume rise 羽流抬升

plume rise equation 羽流抬升方程

plume rise formula 羽流抬升公式

plume rise prediction 羽流抬升预报

plume smoke visualization 羽流烟显示

plume spread 羽流扩展

plume temperature 羽流温度

plume trajectory 羽流轨迹

plume transport wind speed 羽流输运风速

plume trapping 羽流陷落；羽流收集

plume travel distance 羽流运行距离

plume volume flux 羽流体积通量

plume width 羽流展宽

pluviograph 雨量计

pluviometer 雨量表

PM (permanent magnet) 永久磁铁；永磁体；永久性磁石

PMG (permanent magnet generator) 永磁发电机

PMSG (permanent magnet synchronous generator) 永磁同步发电机

pneumatic 气动的

pneumatic analog computer 气动模拟装置

pneumatic buffer 气压缓冲器

pneumatic building 充气建筑物

pneumatic control 风动控制

pneumatic conveyer 风力输送机

pneumatic cushioning 气压减震

pneumatic dashpot 风动消振器

pneumatic hydraulic clamps 气油压虎钳

pneumatic power tools 气动工具

pneumatic pressure 气压

pneumatic servo 气动伺服机构

pneumatic signal 气动信号

pneumatic stiffness 充气刚度

pneumatic storage 充气贮能

pneumatic structure 充气结构

pneumatic tool 充气工具

pneumatics 气动力学

pneumatology 气体力学

pneumodynamics 气动力学

podium 墩座；矮墙

POI (point of interconnection) 互联点

point doublet 点偶极子

point load 点荷载

point of common coupling (PCC) 公共耦合点；公共连接点；公共供电点

point of interconnection (POI) 互联点

point source 电源

point source model 点源模式

point value 点值

point vortex 点涡

point welding 点焊

point welding machine 点焊机

pointed wing tip 尖角翼尖

poising action 平衡作用

Poisson distribution 泊松分布
Poisson equation 泊松方程
Poisson ratio 泊松比
poke welding 手点焊；手动焊；钳点焊
polar 极地的
polar coordinate 极坐标
polar curve 极曲线
polar easterlies 极地东风带
polar front 极锋
polar moment 极矩
polar moment of inertia 极惯性矩
polarity 极性
polder 围海洼地；围圩
pole 电极
pole changing constant speed induction generator 换极恒速异步发电机
pole changing control 变极控制
pole changing generator 变极发电机
pole changing induction machine 变极异步电机
pole count 极数
pole damper 阻尼器
pole pair 极对数
pole piece 磁极片；极靴；极片
pole pitch 磁极距
pole shoe 极靴；极片
polling 轮询
polling controller 轮询控制器
polling list 轮询表
polling mode 轮询方式
polling program 轮询程序
pollutant 污染物
pollutant burden 污染物负荷
pollutant concentration 污染物浓度
pollutant diffusion 污染物扩散

pollutant dispersion 污染物弥散
pollutant downwash effect 污染物下洗效应
pollutant emission 污染物排放
pollutant flux 污染物通量
pollutant index 污染物指数
pollutant plume 污染物羽流
pollutant recirculation 污染物回流
pollutant release height 污染物释放高度
pollutant source 污染源
polluted air 受污染空气
polluted atmosphere 受污染大气
polluted environment 受污染环境
polluter pays principle (PPP) 污染肇事者付款原则
pollution 污染
pollution abatement 污染治理；减轻污染
pollution budge theory 污染预算理论
pollution control 污染防治
pollution cycle 污染循环
pollution deposit 污染沉积物
pollution environment 污染环境
pollution flux 污染通量
pollution index 污染指数
pollution intensity 污染强度
pollution level 污染程度
pollution load 污染负荷
pollution meteorology 污染气象学
pollution plume 污染烟羽
pollution zone 污染带
pollution-free energy source 无污染能源
poly urethane (PU) 聚氨酯
polyamide slide bearing 聚酰胺滑动轴承
polydisperse aerosol 多相分散气溶胶
polymeric material 聚合物材料
polynomial hill 多项式形山丘

polyphase 多相(的)
polyphase rectifier 多相整流器
polytropic atmosphere 多元大气
pontoon shape body 浮筒式车身
poppet 提升阀；垫架
populated area 居住区
population center 居住中心；人口中心
population density 人口密度
population dose 群体剂量
population exposure 群体照射
porosity 孔隙度；开比度
porous 多孔渗水的
porous shroud 多孔罩
porous wall wind tunnel 多孔壁风洞
porous wind screen 多孔风障
portable arc welding machine 便携式弧焊机
portable magnetic flaw detector 手提式磁力探伤仪
position 位置
position indicator 位置指示器
position transducer 位置传感器
position vector 位置矢量
position welding 定位焊
positioned weld 定位焊；暂焊
positive aerodynamic damping 正气动阻尼
positive allowance 正公差
positive approach 主动方法；积极方法
positive converter 正向变流器
positive draft 人工通风
positive drift 正阻力
positive elongation 正延性
positive feed forward control 正前馈控制
positive feedback 正反馈
positive gearing 直接传动
positive pressure 正压力

postcritical region 过临界区
postcritical vortex shedding 过临界旋涡脱落
post stall 过失速的
post stall regime 失速后状态
post strom inspection 风暴事后调查
postulated accident 假想事故
postweld heat treatment 焊后热处理
potential 潜在的
potential damage region 潜在危害区
potential drop (PD) 电压降
potential energy 势能
potential energy storage system 势能储能系统
potential evaporation 潜在蒸发
potential flow 势流
potential flow field 势流场
potential flow model 势流模型
potential flow theory 势流理论
potential function 势函数
potential gradient 位势梯度
potential heat 潜热
potential of wind energy 风能潜力
potential pollutant 潜在污染物
potential pressure 静压
potential site for wind machines 风力机潜场址
potential source 势源；潜在源
potential temperature 位温
potential transformer 电压互感器
potential vortex 位涡
potential wind power site 潜在风能场址
potentially 可能地；潜在地
potentially available wind power 潜在可用风能
poured in place structural foam 浇模发泡

结构

powder metallurgic forming machines 粉末冶金成型机

powder technology 颗粒技术；粉末工艺

powdery snow 粉状雪

power 功率

power allowance 功率余量

power amplifier 功率放大器

power angle 功角

power bogie 动力转向架

power bracket 功率的范围

power buggy 动力牵引车；电动小车

power cable 电力电缆

power capacity 功率容量

power coastdown 功率下降

power coefficient 功率系数

power collection 电力汇集系统

power collection system (for WTGS) 电力汇集系统（对于风力发电机组）

power consumption 功耗

power control 功率控制

power control principle 功率控制原理

power controller 功率调节器

power converter 功率变换器；功率整流器

power curve 功率曲线

power demand 能的需要量

power density 功率密度；风能密度

power diode 功率二极管

power dissipation 功率损耗

power distribution panel 配电盘

power drill 机械钻

power driven 动力传动

power duration curve 功率持续时间曲线

power efficiency 功率效率

power electronic circuit 电力电子电路

power electronic control 电力电子控制

power electronic converter (PEC) 电力电子变流器

power electronic equipment 电力电子器件

power electronic interface 电力电子接口

power electronic switch 电力电子开关

power element 动力元件

power extraction 获能

power extraction coefficient 获能系数；功率系数

power factor (PF) 功率因数

power factor correction 功率因数调整

power factor correction capacitor 功率因数矫正电容器

power factor meter (PFM) 功率因数计

power failure 断电

power flow 电力潮流

power flow control 潮流控制

power frequency 工频

power frequency breakdown 工频击穿特性

power frequency current 工频电流

power frequency overvoltage 工频过电压

power frequency recovery voltage 工频恢复电压

power frequency sparkover voltage 工频跳火电压

power frequency testing transformer 工频试验变压器

power frequency voltage test 工频电压实验

power frequency welding 工频焊

power gain 功率增益

power generation 发电

power grid 电网

power law 指数定律；幂次法则；幂次定律

power law exponent 幂律指数

power law for wind shear 风切变幂律
power law index 幂律指数
power law profile 幂律廓线
power level 功率级
power line 输电线
power line bundle 集束输电线
power mains 输电干线
power measurement terminal 电量测量端子；功率测量终端
power MOSFET 功率金属氧化物（半导体）场效应晶体管；功率场效应晶体管
power of motor 电动机功率
power on time 供电时间
power oscillation 功率波动
power output 功率输出
power park 发电厂区
power performance 功率系数；功率特性
power plant 发电厂
power plant emission 发电厂排放
power producer 动力源
power production 发电量
power purchase agreement (PPA) 购电协议
power quality 电力质量
power quality standard 电力质量标准
power rating 额定功率
power receptacle 电力插座
power regulation 功率调节
power regulator 功率调节器
power set 动力装置；发电机组
power shafting 传动轴系
power socket 电源插座
power source 电源；动力源
power spectral density 功率谱密度
power spectrum 功率谱
power supply 供电

power supply unit 供电设备
power switch 电源开关
power system 电力系统
power take-off 功力输出装置
power termination 功率负载
power train 动力传动系统
power train efficiency 功率传输效率
power transistor 功率晶体管
power transmission 电力传输
power triangle 功率三角形
power trimming rotor 功率调节风轮
power valve 增力阀
power wiring 电力布线
powered 有动力装置的
powered mechanism 传动机构
power-frontal area ratio 功率—迎风面积比
powerful 强大的
powerful spark 强火花
power-to-weight ratio 功率—重量比；动力—重量比；滑转率
PP (peak power) 峰值功率
PPA (power purchase agreement) 购电协议
PPP (polluter pays principle) 污染肇事者付款原则
PR (performance report) 性能报告
practical duty 实际功率
practicality 实用性
prairie 草原
Prandtl pitot tube 普朗特风速管；普朗特皮托管
Prandtl tube 普朗特管
Prandtl type wind tunnel 普朗特型回路式风洞普朗特型式风洞
Prandtl-Glauert correction 普朗特—葛劳渥修正

Prandtl's number 普朗特数
Prandtl's tube 普朗特管
pre cone angle 预锥角
pre door angle （桨叶）下垂角
pre emphasis 预加重
pre engineered building 预经工程设计建筑
pre rotation 预扭；预旋
pre rotation vane 预旋叶片
pre stall regime 失速前状态
pre tension load 预张力荷载
preassembly 预汇编；预组合
precast 预浇筑的；预制的
precast concrete tower 预制混凝土塔架
precipitation 降水
precipitation efficiency 降水效率；沉降效率
precipitation gage 雨量筒
precipitation rate 降水率；沉降率
precipitation wind rose 降水风玫瑰
precise 精确的
precise regulation 精密调节
precision 精密度
precision accuracy 精度
precision adjustment 精调
precision brazing 精密钎焊
precision finishing 精密加工
precision in products 产品精度
precision instrument 精密仪器
precision measurement 精确测定
precision of analysis 分析的准确性
precision prescribed 要求精度
precision selection 精度选择
precision standard 精度标准
precision work 精密加工；精密工作
preconditioning length （风洞气流）预处理长度

predominant 主要的
predominant diameter 主要粒径
predominant frequency 主频率
predominant wind direction 主风向
preliminary 准备；初步的
preliminary adjustment 预调
preliminary calibration 初步校准
preliminary checkout 初步测试
preliminary correction 初校正
preliminary design 初步设计
preliminary dimensions 预定尺寸
preliminary elongation 预拉伸
preliminary estimation 初步估计
preliminary operation 试行运转
preliminary test 初步试验
preload spring 预紧弹簧
preload tension 预载张力
premature separation 提前分离
premature transition 提前转捩
preparative treatment 预加工
prepreg lay-up 预浸布铺层
preprodution test 正式投入运行前的试验
preprototype 预样机；预原型；预制样品
preset adjustment 预调准
preset counter 预置计数器
preset position 预调位置
press bonding 压焊
press forming 模压成型
pressed out boss 压制毂
presses cold forging 冷锻冲压机
presses crank 曲柄压力机
presses eccentric 离心压力机
presses forging 锻压机
presses hydraulic 液压冲床
presses knuckle joint 肘杆式压力机

presses pneumatic 气动冲床
presses servo 伺服冲床
presses transfer 自动压力机
pressing dies 压模
pressure 压力
pressure anemometer 压力风速计
pressure angle 压力角
pressure belt 受压带
pressure cell 压力传感器
pressure center 压力中心
pressure coefficient 压力系数
pressure control valve 压力控制阀
pressure die away 压力消失
pressure difference 压差
pressure discontinue 压力不连续
pressure distribution 压力分布
pressure drag 压差阻力
pressure drop 压力降
pressure energy 压力能
pressure exponent 压力指数
pressure field 压力场
pressure fluctuation 压力脉动
pressure gauge 压力表
pressure governor 压力调节器
pressure gradient 压力梯度
pressure gradient force 气压梯度力
pressure head 压力头；压力传感器
pressure hole 通气孔
pressure instrument 压力仪表；压力计
pressure jump 气压涌升；压力跃变
pressure load 压力荷载
pressure lubrication 压力润滑；强制润滑
pressure orifice 测压孔
pressure pattern 气压分布型式
pressure plate anemometer 压板风速计

pressure plotting 压力分布图
pressure port 测压孔
pressure profile 压力廓线
pressure recovery 压力恢复
pressure reducing valve 减压阀
pressure regulation valve 调压阀
pressure relief valve 泄压阀
pressure resistance 压差阻力
pressure scanning switch 压力扫描开关
pressure selector switch 测压选择开关
pressure sensor 测压传感器
pressure side 受压侧；受压面
pressure slope 压力曲线斜率
pressure sphere anemometer 压力球风速计
pressure spike 压力尖峰
pressure stress 压应力
pressure surge 气压涌升；压力跃变
pressure switch 压力继电器
pressure system 气压系统
pressure tap 测压孔
pressure test 压力试验
pressure traverse 压力游测
pressure trough 气压槽
pressure tube anemometer 压力管风速计
pressure tubing 测压管
pressure wind tunnel 增压风洞
pressured gauge 压力表
pressurized wind tunnel 增压风洞
Preston tube （壁面剪应力测量用）普雷斯顿管
prestressed bridge 预应力桥
prestressed concrete cable stayed bridge 预应力混凝土斜张桥
pretwist 预扭
pretwist angle （桨叶）预扭角
prevailing 流行的

prevailing power direction 风能盛行方向
prevailing westerlies 盛行西风带
prevailing wind 盛行风；主风
prevailing wind direction 主风向
preventive 预防的
preventive maintenance 预防性维修；预防性保养
preventive routine maintenance 预防性维修；常规维修
primary 主要的
primary air 主气流
primary airflow 主气流
primary cell 原生电池
primary coat 底漆
primary colour 底漆色
primary coolant 一次冷却剂
primary current 一次电流
primary dune 初期沙丘
primary energy resource 一次能源
primary environment 原生环境
primary frequency 主频率
primary members 主要构件
primary pollutant 原生污染物
primary settling 初级沉降
primary source 初始源
primary voltage 一次电压
prime 主要的
prime lacquer 底漆
prime material 底层材料；打底材料
prime motor 原动机
prime mover 原动机
prime paint 底层漆
primer coat 底漆层
priming painting 底漆
priming valve 起动注油阀；起动注水阀；起动阀
principal 原理；定律；因素
principal accumulator 主蓄能器
principal frequency 主频率
principal mode 主模态；主振型
principal of image 镜像原理
principal of least work 最小功原理
principal of mass conservation 质量守恒定律
principal of momentum and energy 动量与能量原理
principal of operation 工作原理；操作原理
principal of superposition 迭加原理
principal optimality 最优原理
principal stress 主应力
principle diagonal 主对角线
principle dimension 基本尺寸；主尺度
printed electronic circuit (PEC) 印刷电路
prismatic building 棱柱形建筑
prismatic structure 棱柱形结构
private line 专用线路
probability 概率
probability density function 概率密度函数
probability distribution 概率分布
probability of occurrence 出现概率
probability paper 概率坐标纸
probability theory 概率论
probable error 概差；概然误差；可能错误
probe 探头；探针
probe gas 示踪气体；探头气体；探测气体
probe material 示踪物质
procedure 程序
process of self-excitation 自励过程
processing 加工；处理
processing behaviour 加工性能
processing condition 加工条件

processing equipment 加工设备
processing method 处理方法；加工方法
processing of data 数据处理；资料整理
product 产品
production prototype 投产样机
production type wind tunnel 生产性风洞
profile 翼型
profile angle 齿形角
profile chord 翼弦
profile component 翼剖面分量
profile correction 齿廓修行
profile drag 翼型阻力
profile error 齿形误差
profile flow 翼型绕流
profile machine 仿形机床；仿形铣床
profile meter 表面测量仪；（表面粗糙度）轮廓仪
profile modelling 仿形；靠模
profile modification 齿形修整
profile of teeth 齿形
profile resistance 翼型阻力
profile set 翼型系列
profile shape 翼型
profile shifted gear 变位齿轮
profile steel 型钢
profile thickness 翼剖面（最大）厚度
profile tracer 靠模
profiled outline 外形轮廓
program 程序
program board 程序控制台
program control 程序控制
program controller 程序控制器
program element 程序单元
program element code (PEC) 程序单元代码
program exception code (PEC) 程序异常代码

program regulation system 程序调节系统
program relay 程序继电器
programmable 可编程的
programmable by software 软件设置
programmable control 可编程控制
programmable logic controller (PLC) 可编程逻辑控制器
programming 设计；规划
programming device 程序编制装置
programming system 程序设计系统
project reporting manual (PRM) 工程报告手册
projected 投影的
projected area 投影面积
projected area of blade 叶片投影面积
pronounced stall 严重失速
proof 证明；证据
proof of concept machine 方案验证机
proof voltage 耐电压
propagate 传导；传播
propeller 螺旋桨；推进器
propeller advance 螺旋桨进程
propeller blade 螺旋桨式叶片
propeller boss 桨毂
propeller discontinue 桨叶旋转面上气流不连续性
propeller hub 螺旋桨毂；螺旋桨桨毂
propeller type anemometer 螺旋桨风速计
propeller type wind machine 螺旋桨型风力机
propeller vane 螺旋式叶片
propeller vane anemometer 螺旋桨风速计；螺旋桨叶片风速计
proper alignment 同心度
proper oscillations 固有振动

prophylatic repair 预防检修
proportional-integral (PI) 比例积分
proportional-integral controller 比例积分控制器；PI控制器
proportioning valve 比例调节阀
propulsion wind tunnel (PWT) 推进试验风洞
protect 保护；防止
protect relay 保护继电器
protected 受保护的
protected against dropping water 防滴
protected against splashing 防溅
protected against the effects of immersion 防浸水
protection 保护
protection circuit 保护电路
protection coat 防护层
protection earthing 保护接地
protection lever 保护等级
protection relay 保护继电器
protection spectacles 护目镜
protection status 保护现状
protection system (for WTGS) 保护系统（对于风力发电机组）
protective 防护的
protective circuit breaker 保护断路器
protective earthing 保护接地
protective gap 保护性间隙放电
protective glove 防护手套
protective interlock 保安互锁装置
protective relay 保护继电器
protocol 协议
prototype 样机；原型物
prototype atmospheric wind 原型大气风
prototype Reynolds number 原型雷诺数

prototype wind speed 原型风速
protuberance 突出物；凸度
proximity sensor 接近度（防撞）传感器；近距离传感器
pseudo adiabatic convection 假绝热对流
pseudo adiabatic expansion 假绝热膨胀
pseudo adiabatic lapse rate 假绝热直减率
pseudo adiabatic process 假绝热过程
pseudo potential flow 假位流；赝势流
pseudo random force 假随机力
pto coupling 动力输出轴联轴节
pto driven 动力输出轴驱动的
pto power 动力输出轴功率
p-to-p 从峰值到峰值
PU (poly urethane) 聚氨酯
public acceptance 公众接受性
public hazard 公害
public place 公共场所
public utility 公用事业；公共设施
public utility network 公用电网
puff 喷团；烟团
puff diffusion 喷团扩散
puff model 喷团模式
puff of smoke 烟团
puff release rate 烟团释放率
puff rope 拉绳
puff source 喷团源
puff superposition model 喷团叠加模式
puff trajectory 喷团轨迹
puffing plume 喷团型羽流
pull 拉
pull up torque 最小起动扭矩；最低起动转矩
pulley 滑轮
pullin torque 牵入扭矩
pullout torque 牵出扭矩；失步转矩

pulsating current 脉动电流
pulsating torque 脉动转矩
pulsating voltage 脉动电压
pulsation loss 脉动损耗
pulsation welding 脉冲焊接
pulse 脉冲
pulse frequency 脉冲频率
pulsed 脉冲的
pulsed wire anemometer 脉冲热线
pulse-width modulated (PWM) 脉冲宽度调制；脉宽调制
pulse-width modulated converter 脉宽调制变流器
pultruded aluminium blade 挤拉铝叶片
pultrusion 拉挤成型
pump 泵
pump circulation 强制循环
pumped 用泵送
pumped hydro energy storage 抽水储能
pumped storage 泵储能（抽水或压气）
pumping appliance 排水设备
puncture 刺穿

puncture voltage 击穿电压
pure 纯的
pure bending instability 纯弯不稳定性
pure circulation round wing 纯机翼环流
pure oscillation 纯振荡；正弦振荡
pure shear 纯剪（切）
pure stress 单向应力
pure torsion flutter 纯扭颤振
pure wave 正弦波
purification constant 净化常数
purlin 桁条
push button actuator 按钮开关
push pull rectifier 推挽式整流器
push welding 手点焊；钳点焊
PV (photovoltaic) 光伏
PWM (pulse-width modulated) 脉冲宽度调制；脉宽调制
PWM switching circuit 脉冲宽度调制开关电路；脉宽调制开关电路；PWM开关电路
PWT (propulsion wind tunnel) 推进试验风洞
pyramid balance 塔式天平

现代英汉风力发电工程

A Modern English-Chinese Dictionary of Wind Power Engineering

Q

quadratic constraints 二次约束
quadratic equation 二次方程
quadratic expression 二次式
quadratic form 二次型
quadrature spectrum 正交谱
qualitative description 定性描述
quality 质量
quality assessment 质量评定
quality assurance 质量保证
quality assurance acceptance standards 质量保证验收标准
quality assurance planning 质量保证计划
quality certificate 质量证明书
quality change 质量变化
quality character 质量特性
quality check 质量检查
quality coefficient 质量系数
quality control 质量控制；品质管理
quality criteria 质量标准
quality defect 质量缺陷
quality evaluation 质量评定
quality factor 品质因数
quality inspection 质量检验
quality loss 质量损失
quality measurement 质量测定
quality of air environment 空气环境质量
quality of tolerance 公差等级
quality requirement 质量要求
quality specification 质量标准；技术规格
quantitative 定量的
quantitative analysis 定量分析
quantitative description 定量描述
quantitative determination 定量测定
quantitative estimation 定量估算
quantitative test 定量试验
quantity 数量；定量
quantity of current 流量
quantity of energy 能量
quantity of flow 流量
quantity of vegetative cover 植被覆盖量
quarry dust 采石场粉尘
quarter 四分之一
quarter chord point 四分之一弦点
quarter deck 后甲板
quartering wind 斜风；后侧风；尾舷风；侧风
quasi-attached flow 准附着流
quasi-geostrophic approximation 准地转近似
quasi-harmonic oscillation 准谐振
quasi-homogeneous turbulence 准均匀湍流
quasi-incompressible fluid 准不可压流体

quasi-periodic vibration 准周期振动
quasi-state cycling loading 准静态周期载荷
quasi-static 准静态的
quasi-static assumption 准静态假设
quasi-static deflection 准静态挠度
quasi-static derivative 准静导数
quasi-static response 准静态响应
quasi-static wind load 准静态风载
quasi-steady aerodynamics 准定常空气动力学
quasi-steady approach 准定常方法
quasi-steady flow 准定常流
quefrency 类频率；拟频率；逆频
quench 淬火
quench aging 淬火（后自然）时效
quench alloy steel 淬硬合金钢
quench hardening 淬硬化
quench hot 高温淬火
quenched and tempered steel 调制钢
quenched steel 淬硬钢
quenching 淬火
quenching bath 淬火浴
quenching crack 淬火裂纹
quenching degree 淬透性
quick 快速的
quick response 快速响应
quill shaft 挠性短轴；套筒轴；中空轴

现代英汉风力发电工程

A Modern English-Chinese Dictionary of Wind Power Engineering

R

race running 空转
race way 电缆管道
rack angle 机架角
rack railway 齿轨铁道
radar aerial 雷达天线
radar antenna 雷达天线
radar sounding 雷达探空
radar wind sounding 雷达测风
radial 半径的；放射状的
radial acceleration 径向加速度
radial flow model 径向流模型
radial flux permanent magnet synchronous generator (RF-PMSG) 径向磁通永磁同步发电机
radial journal bearing 径向轴承；径向轴承
radial model 径向模态
radial pin coupling 径向销连接
radial pressure gradient 径向压力梯度
radial thrust bearing 径向止推轴承
radial vector 径向向量
radial velocity 径向速度
radial whirl 径向涡流
radial wind sounding 径向风测深
radiance 辐射率
radiant 辐射的
radiant flux 辐射通量
radiating 辐射的
radiating fin 散热片
radiating gill 散热片
radiating rib 散热片
radiation 辐射
radiation balance 辐射平衡
radiation budget 辐射收支
radiation cooling 辐射冷却
radiation fog 辐射雾
radiation heating 辐射增温
radiation intensity 辐射强度
radiation inversion 辐射逆温
radiation protection 辐射防护
radiation transfer 辐射传递
radiative diffusivity 辐射扩散率
radiative heat exchange 辐射热交换
radiator 散热器
radiator false front 散热器护栅板；散热器前护栅
radiator gill 散热器散热片
radiator shutter 散热器百叶窗
radio antenna 无线电天线
radioactive aerosol 放射性气溶胶
radioactive cloud 放射性（烟）云

radioactive concentration 放射性浓度
radioactive contaminate 放射性污染物
radioactive deposit 放射性沉降物
radioactive dust 放射性尘埃
radioactive emission 放射性排放
radioactive fallout 放射性散落物；放射性尘降物
radioactive indicater 放射性示踪物；放射性指示器
radioactive isotope 放射性同位素
radioactive pollution 放射性污染物
radioactive rain 放射性雨
radioactive source 放射源
radioactive tracer 放射性示踪物
radioactivity 放射性
radioaerosol 放射性气溶胶
radiocontrast agent 放射性对比剂
radiographic inspection 射线照相探伤法
radiographic test 放射线探伤
radioisotope tracer 放射性同位素示踪物
radiosonde 无线电探空仪
radiowind 无线电测风仪
radius 半径
radius of curvature 曲率半径
radius of gyration 回转半径；转动惯量半径
radius of inertia 惯性半径
radius of plume 羽流半径
radius of rounding 圆角半径
radius vector 矢径
radome 雷达罩
rain 雨
rain erosion 雨冲蚀
rain erosion damage 雨蚀损伤
rain gush(gust) 暴雨
rain gutter 雨水槽
rain in torrents 暴雨
rain visor 遮雨板
rainfall frequency 降雨频率
rainfall intensity 降雨强度
rainfall rate 降雨率
raingauge 雨量计
rainmaking 人工降雨
rainout 雨散落物；雨沉降物；雨涤
rake angle 桨叶倾斜角
rake forward 前斜
rake of blade 桨叶斜角；桨叶倾斜度
rake of bow 斜度
rake of sampling tubes 梳状取样排管
rake probe 排管；梳状探针
raked wing tip 斜翼尖
random 随机的
random analysis 随机分析
random array 随机阵列
random distribution 随机分布
random error 随机误差
random event 随机事件
random excitation 随机激励
random fluctuation 随机脉动；随机涨落
random gust 随机阵风
random inspection 随机抽查
random load 随机荷载
random modulated signal 随机调制信号
random process 随机过程
random radiography 随机射线照相检查
random response 随机响应
random sampling 随机取样
random turbulence 随机湍流
random variable 随机变量
random vibration 随机振动
random vortex shedding 随机旋涡脱落

random walk model 随机走动模式
range 范围；山脉；较差
range of load 负荷范围
range of measurement 测量范围
range of regulation 调节范围
range of temperature 温度范围
rank smell 难闻气温
Rankine scale 兰氏温标
Rankine's vortex 兰金涡流
rapid 迅速的；急促的
rapid analysis 快速分析
rapid cooling 快速冷却
rapid current 急流
rapid flow 急流；湍流
rapid fluctuation 急剧波动
rapping allowance 铸型尺寸增量
RAPS (remote area power supply) 偏远地区供电
rare earth 稀土
rare earth metal 稀土金属
ratchet spanner 齿轮扳手
rate 比率
rate of curving 弯曲率
rate of dilution 稀释比
rate of discharge 排出率
rate of dissipation 耗散率
rate of distortion 畸变率
rate of divergence 发散速率
rate of emission 排放率
rate of fall 沉降速率
rate of fall-out 衰减率
rate of flow 流速；流量率
rate of loading 负荷率
rate of loss 损耗率
rate of rainfall 降雨率

rate of rise 曲线上升斜率
rate of sand movement 移沙率
rate of strain 应变率
rate of volume flow 体积流量；体积流动速率
rate of wear 磨损率
rated 额定的
rated apparent power 额定视在功率
rated boost pressure 额定升压
rated brake power 额定制动功率
rated capacity 额定容量
rated condition 额定工况
rated consumed power 额定消耗功率
rated current 额定电流
rated energy 额定能量
rated flow 额定流量
rated frequency 额定频率
rated input 额定输入；额定输入功率
rated life 额定寿命；额定使用期限
rated load 额定负载
rated load operation 额定负荷运行
rated load test 额定负荷试验
rated load torque 额定转矩
rated operational current 额定工作电流
rated operational voltage 额定工作电压
rated output 额定输出
rated output capacity 额定输出容量
rated output power 额定输出功率
rated power 额定功率（风力发电机组）
rated power coefficient 额定力矩系数
rated quantity 额定量
rated reactive power 额定无功功率
rated slip 额定滑差率
rated speed 额定转速
rated tip speed ratio 额定叶尖速度比（标准高速性系数）

rated torque coefficient 额定力矩系数
rated turning speed of rotor 风轮额定转速
rated value 额定值
rated voltage (RV) 额定电压
rated wind speed (for wind turbines) 额定风速（风力机）
rating 额定值；等级；参数；定额；测定
rating data 额定数据
rating method 额定法；鉴定法
rating output 额定输出
rating power 额定功率
rating value 额定值；标称值；标准值
ratio 比率；比例
ratio of over load 过载度
ratio of tip-section chord to root-section chord 叶片根梢比
ratio of transmission 传动比
ratio of tunnel nozzle 风洞喷管面积比
ratio of viscosity 黏度比
ravine 谷；沟
ravine wind 峡谷风
raw data 原始数据
rawinsonde 无线电探空测风仪
Rayleigh criterion 瑞利判据
Rayleigh distribution 瑞利分布
Rayleigh number 瑞利数
Rayleigh scattering 瑞利散射
RCC (rotor current controller) 转子电流控制器
reach 达到；延伸
reactance 电抗
reactance capacity 无功功率
reactance voltage 电抗电压
reacting bar 承力杆
reaction 反应；感应

reaction balance 反作用平衡
reaction force 反作用力
reactive 反应的
reactive component 无功分量
reactive current 无功电流
reactive force 反作用力
reactive in respect to 相对……呈感性
reactive load compensation equipment 无功功率补偿设备
reactive loss 无功损耗；无功损失
reactive power 无功功率；无功
reactive power capacity 无功容量
reactive power compensation 无功补偿
reactive power compensation system 无功补偿系统
reactive power consumption 无功消耗
reactive power control 无功控制
reactive power factor 无功功率因数
reactive power set-point 无功功率设定点
reactive thrust 反作用力
reactive value 无功量
reactive volt-ampere 无功伏安
reactor control station 反应堆操纵器
reak in relay 插入继电器
real 实际的；真实的
real fluid 实际流体
real gas 实际气体
real load 有效负荷；实际负荷
real load curve 实际负荷曲线
real operating time 实际运转（操作）时间
real part 实部
real source 实际电源；实声源
real time 实时
real time data 实时数据
real time counter (RTC) 实时计数器

real time monitoring 实时监控
real value 有效值
real velocity 实际速度
real work 有效功
real-valued process 真实过程
rear edge 后缘
rear hub 后轮轴承
rear mounted aerofoil 后翼板
rear span 外形后宽
rear spar 后（翼）梁
rear stagnation point 后驻点
reattachment 再附（着）
reattachment line 再附线
reattachment point 再附点
reattachment zone 再附区
rebound 回弹；回跳
rebound elasticity 弹性回复
receptacle 插座
receptor 接闪器
recirculating jet 环形射流
recirculating wake flow 回流尾流
recirculation 回流；再循环
recirculation cavity 回流空穴
recirculation region 回流区
rectified action 整流作用
rectified feedback 整流反馈
rectified output 整流输出
rectifier bridge 整流器电桥
rectifier cell 整流元件
rectifier circuit 整流电路
rectifier inverter 整流换流器
reclaimed land 开垦地
recorder 记录器
recording 记录
recording anemograph 自记风速计

recovery 恢复
rectangular 矩形的
rectangular array 矩阵列
rectangular axis 直交轴
rectangular bar blade 矩形条叶片
rectangular bus bar 矩形母线
rectangular coordinates 直角坐标
rectangular equation 直角坐标方程
rectangular vortex ring 矩形涡环
rectangular wind tunnel 矩形试验段风洞
rectangular wing 矩形翼
rectangular wing tip 矩形翼尖
rectangular working section （风洞）矩形试验段
rectification 改正；校正
rectification characteristic 整流特性曲线
rectified action 整流作用
rectifier 整流器
rectifying device 整流装置
rectilinear flow 直匀流
rectilinear vorticity 直线涡流
rectilinear vortex 直线涡流；直涡丝
recurrence interval 重现间隔
recurrence period 重现周期
redistribution 再分布
reduced 减少的；简化的
reduced damping 折算阻尼
reduced frequency 折算频率
reduced mass 折算质量
reduced pressure coefficient 折算压力系数
reduced scale 缩尺
reduced spectrum 折算谱
reduced wind speed 折算风速
reducing 减低
reducing flange 异径法兰；缩口法兰

reducing valve	减压阀
reduction	减少；下降
reduction gear (RG)	减速齿轮
reduction of area	断面收缩率
reduction ratio	减速比
redundancy	冗余
redundant	多余的
redundant system	备用系统
reed fence	芦苇栅
reefing	缩帆
reference	参考；参照
reference area	基准面
reference axes	参考轴
reference chord	基准弦
reference coordinate	基准坐标
reference data	参考数据
reference dimension	参考尺寸
reference distance	基准距离
reference height	基准高度
reference length	参考长度
reference line	基准线
reference man	参考人
reference manual	参考手册
reference organ	参考器官
reference plane	基准面
reference point	基准点；参考点
reference pressure	参考压力
reference roughness length	基准粗糙长度
reference source	标准源；参考源
reference standard	参考标准
reference test	基准试验
reference value	参考值
reference voltage	基准电压
reference wind	参考风
reference wind speed	参考风速
reflecting layer	反射层
reflection factor	反射因子
reflection plane	反射板；反射平面
reflector bowl	抛物面形反射器
reflow	逆流
reflux	逆流；回流；反流；倒流
refluxing	回流
refrigerated	冷冻的；冷却的
refrigerated wind tunnel	低温风洞冷却风洞
refrigeration	制冷；冷冻
regeneration	再生；后反馈放大
regenerative	再生的；更生的
regenerative braking	回馈制动；再生制动
regenerative coolant	再生式冷却剂
regenerator	再生器；蓄热器
region of convection	对流层
regional climate	区域气候
regional climatology	区域气候学
regional damage index	区域性损失指数
regional development	地区开发
regional diffusion	区域扩散
regional pollution	地区性污染
regional scale storm	区域性风暴
regional wind	地区性风
regional wind climate	地区性风气候
regional wind regime	地区风况
regional wind resource	地区风力资源
regionalism	区域主义；地方主义
regular	定期的；有规律的；合格的；整齐的
regular check	定期检查
regular maintenance	日常维修；定期检查
regular pressure distribution	有规律的压力分布
regular service condition	正常工作条件
regular vortex shedding	规则旋涡脱落

regular vortex street 规则涡街
regular vortex trail 规则涡迹
regulate 调节
regulate flow 调节流量
regulating 调节；控制
regulating characteristics 调节特性
regulating mechanism 调速机构
regulating mechanism by adjusting the pitch of blade 变桨距调节机构
regulating mechanism of rotor out of the wind sideward 风轮偏测式调速机构（使风轮轴线偏离气流方向的调速机构）
regulating mechanism of turning wind rotor out of the wind sideward 变桨距调节机构
regulator 调节器
reinforced 加强的；加固的
reinforced concrete base 钢筋混凝土地基
reinforced elastomer 增强的弹性材料
reinforced flange fitting 加强的法兰管接件
reinforced plastic 增强塑料
reinforced rib 加强肋
reinforced seam 加强焊缝
reinforced square set 加固方框
reinforced structure 加固结构
reinforcement 加固
reinforcement of weld 加强焊缝
reinforcing fiber 增强纤维
reinforcing girder 加力梁
reinforcing material 加强材料
reinforcing pad 补强垫；增强衬板
reinforcing rib 加强肋
reinforcing rod 钢筋条
reinforcing steel 钢筋
rejection 抛弃；拒绝
rejection of heat 散热

relative 相对的
relative acceleration 相对加速度
relative accuracy 相对精度
relative airstream 相对气流
relative angle of attack 相对迎角
relative density 相对密度
relative diffusion 相对扩散
relative discharge 相对流量
relative eddy 相对涡流
relative efficiency 相对效率
relative elongation 相对伸长
relative error 相对误差
relative flow 相对流量
relative frequency distribution 相对频率分布
relative humidity 相对湿度
relative permeability 相对透气性
relative pressure 相对压力
relative risk 相对危险度
relative roughness 相对粗糙度
relative stability 相对稳定性
relative strength 相对强度
relative thickness of airfoil 翼型相对厚度
relative velocity 相对速度
relative wind 相对风
relative wind direction 相对风向
relative wind speed 相对风速
relativity 相对论；相关性
relativity principle 相对性原理
relaxation oscillation 松弛振荡
relaxation period 松弛周期
relaxation time 松弛时间
relay 继电器
relay setting 继保整定；继电保护；继电汽定；整定计算
release current 释放电流

release line of tracer 示踪物释放线
release point 释放点
reliability 可靠性
reliability determination test 可靠性测定试验
relief 泄压；卸载；溢流；地势；起伏
relief by-pass valve 安全旁通阀
relief device 减压装置
relief port 放气口
relief valve 安全阀
relieving 减轻
relieving of internal stress 消除内应力
religion building 宗教建筑
reloading 重复荷载
reloading curve 重新加载曲线
reluctance 磁阻；阻抗
reluctance generator 磁阻发电机
reluctance motor 磁阻电动机
reluctancy 磁阻；阻抗
remanence 剩磁；顽磁
remanent 剩余的；残余的
remanent core 带剩磁铁心
remanent field 剩余磁场
remanent induction 剩余磁感应
remanent magnetism 剩余磁性
remanent magnetization 剩磁；顽磁性；残留磁气
remanent strain 残余应变
remote 遥远的；偏僻的
remote area 边远地区
remote area power supply (RAPS) 偏远地区供电
remote control 远程控制；遥控
remote display 远程显示
remote indicator 遥控指示器
remote measurement 遥测

remote monitor 远距离监测器
remote monitoring system 远距离监控系统
remote reading 遥测显示；遥控读数
remote terminal unit (RTU) 远程终端设备；远程终端装置
remote velocity 远方速度
remotely 遥远地；偏僻地
remotely operated circuit 遥控电路
removal 移动；排除；搬迁
removal of faults 排除故障
removal of internal stress 消除内应力
renewable energy 可再生能源
renewable energy source 可再生能源
repair 修复
repair sleeve 补修管
repair time 修复时间
repairability 可修理性
repeatability 重复性
repeated 再三的；反复的
repeated addition 叠加
repeated bending test 弯曲疲劳试验
repeated compression test 压缩疲劳试验
repeated differentiation 多次微分
repeated fluctuating stress 反复应力
repeated hardening 多次淬火
repeated impact 反复冲击
repeated impact test 冲击疲劳试验
repeated load 交变荷载
repeated shock 反复冲击
repeated strain 疲劳应变
repeated stress 反复应力
repeated stress failure 疲劳断裂
repeated tempering 多次回火
repeated tensile stress test 反复拉伸试验
repeated tension and compression test 拉压

疲劳试验
repeated torsion test 扭曲疲劳试验
repetitive 重复的
repetitive error 重复误差
repetitive loading 反复载荷
repetitive process 迭代法
repetitive stress 重复应力
replenishable source of energy 能量补给源
replenishable sources 利用可再生能源
replica 复制品；复制物
replica aeroelastic model 仿样气动弹性模型
reposition 复位；回位
repowering 改造增容
representative 代表；典型
representative area 特征面积
representative length 特征长度
representative method of sampling 抽样表示法
representative model 表现模
representative sample 代表性试样
representative sampling 特征取样
representative scale 代表性比例尺
representative section 等效截面
representative speed 特征速度
representative value 代表值
representative year for wind energy resource assessment 风能资源评估代表年
reproduction 重现
required 必需的
required power 需用功率
reserve 储备；储存
reserve capacity 备用容量
reserve generator turbine 备用发电机
reserve unit 备用机组
reserve energy 储备能量

reservoir 水库；蓄水池
reset 复位
reset action 重新调整动作
reset bar 复位杆
reset condition 原始状态；复原状态；复位状态；重设状态
reset control 复位控制
reset spring 复位弹簧
reset valve 再调阀
residence time 停留时间
residential area 居住区
residential quarter 住宅区
residual 剩余的；残留的
residual current 残余电流
residual current circuit breaker 漏电断路器；剩余电流断路器
residual deformation 残余变形
residual induction 残余感应；剩磁电感；剩余磁感应；剩余电感
residual magnetic field 剩磁场
residual magnetization 剩余磁化强度
residual permeability 剩余导磁率
residual resistance 余阻力
residual strain 残余应变
residual voltage 残压
resilience 回弹
resilient 弹回的
resilient coupling 弹性联轴节
resilient material 弹性材料
resistance 电阻
resistance bulb 测温电阻
resistance coefficient 阻力系数
resistance coefficient for wind 风阻力系数
resistance erodibility 抗蚀性
resistance factor 阻力因数

resistance of an earthed conductor (earth resistance) 接地电阻
resistance of fluid friction 流体摩擦阻力
resistance of materials 材料力学
resistance to air flow 气流阻力
resistance to compression 抗压缩性；压缩变形阻力
resistance to overflow （对气、液体的）流出阻力
resistance voltage 电阻电压
resisted rolling 阻尼横摇
resisting 抵抗；忍住
resisting force 阻力
resisting medium 黏性介质；阻尼介质
resisting moment 抗力矩
resisting shear 抗剪力
resisting torque 抗转矩
resistive 有抵抗力的
resistive load 电阻性负载
resistivity 电阻率
resistor 电阻器
resolve 分解
resolution 解决；分辨力
resolution of force 力的分解
resolution of polar to cartesian 极坐标—直角坐标转换
resolution of vector 矢量的分解
resonance 共振；谐振
resonance oscillation 谐振；共振荡
resonance phenomena 谐振现象
resonant 共振的
resonant eddy 共振涡旋
resonant excitation 共振激励
resonant frequency 谐振频率
resonant pitching 共振俯仰；共振纵摇
resonant response 共振响应
resonant vortex excitation 共振旋涡激励
resonant vortex shedding 共振旋涡脱落
response 响应
response bandwidth 响应带宽
response frequency 响应频率
response gust factor 响应阵风因子
response spectrum 响应谱
response time 响应时间
rest condition 静止状态
restitution 回弹；恢复
restoration 恢复；复位
restoration force 恢复力
restorer 恢复器；复位器；还原器
restoring 恢复
restoring force 恢复力
restoring moment 恢复力矩
restrained beam 约束梁；固端梁
restraint 约束；限制；限制器；阻尼器
restrictor valve 限流阀
restriking 电弧再触发
resultant 合力
resultant action 总作用
resultant couple 合成力偶
resultant drag 总阻力
resultant force 合力
resultant gear ratio 总传动比
resultant of velocities 速度的合成
resultant strain 合成应变
resultant stress 合成应力
resultant vector 合矢量
resultant wind 合成风
resultant wind direction 合成风向
resulting 致使；产生
resulting stress 合成应力

resuspension coefficient 再悬浮系数
resuspension model 再悬浮模式
resuspension rate 再悬浮率
resynchronize 再同步
retaining valve 单向阀
retardation 阻滞；迟延
retardation of wind 风力减弱
retarding 迟滞；减速
retarding force 制动力
retarding torque 制动转矩
retention 贮留；保持
retrofit 改型
return 返回
return air 回流空气
return circuit wind tunnel 回路式风洞
return current 反流
return flow 回流
return flow wind tunnel 回流式风洞
return information 返回信息
return period 重现期；回复周期
return spring 恢复弹簧
return time 回复时间
return type wind tunnel 回路式风洞
return wash 回转冲刷
returning air mass 回归气团
reverse 背面
reverse bias 偏压
reverse feedback 负反馈
reverse flow 反向流；逆流；倒流
reverse flow region 逆流区
reverse power 逆功率
reverse slip-face 反滑落面
reverse torsion machine 扭转疲劳试验机
reverse torsion test 反复扭转试验
reverse voltage 反向电压

reversed stress 交变应力；反向应力
reversible deformation 变形回复；可逆变形
reversing gear 换向齿轮
revolution 回转运动；转动
revolution of polar to cartesian 极坐标—直角坐标转换
revolutions per minute 转/分
revolutions per second 转/秒
revolving storm 热带风暴；旋转风暴
Reynolds analogy 雷诺相似
Reynolds approach 雷诺方法
Reynolds averaging 雷诺平均
Reynolds criteria 雷诺准则
Reynolds equation 雷诺方程
Reynolds factor 雷诺因子
Reynolds law of similarity 雷诺相似律
Reynolds number (RN) 雷诺数
Reynolds number based on chord 弦长雷诺数
Reynolds number based on diameter 直径雷诺数
Reynolds number correction 雷诺数修正
Reynolds number effect 雷诺数效应
Reynolds number of turbulence 湍流雷诺数
Reynolds number range 雷诺数范围
Reynolds number similarity 雷诺数相似
Reynolds averaged Navier-Stokes equations (RANS) 雷诺平均纳维埃托克斯方法
Reynolds stress 雷诺应力
Reynolds stress tensor 雷诺应力张量
RF-PMSG (radial flux permanent magnet synchronous generator) 径向磁通永磁同步发电机
RG (reduction gear) 减速齿轮
RH (Rockwell hardness) 洛氏硬度
rheostat 变阻器

rib 加强肋；肋状物
rib reinforcement 加强肋；加强筋
rib roughness of cooling tower 冷却塔加强肋粗糙度
rib web 加强肋腹板
ribbon 带；饰条
ridge 山脊
ridge crest 脊顶
ridge roof 有脊屋顶
riding comfort 乘坐舒适性
right 右
right hand rule 右手定则
rigid 严格的
rigid alignment 精确对准
rigid axle 刚性轴
rigid body 刚体
rigid boundaries working section （风洞）刚性壁试验段
rigid bracing 刚性撑杆
rigid casing 刚性外壳
rigid composite material 硬质复合材料
rigid conduits 刚性导管
rigid coupling 刚性联轴节
rigid distortion 刚性扭曲
rigid element aeroelastic model 刚性组件气动弹性模型
rigid fixing 刚性固定
rigid frame 刚性架
rigid material 刚性材料
rigid member 刚性构件
rigid metal girder 刚度大的金属梁
rigid model 刚性模型
rigid model on elastic base 弹性底座刚性模型
rigid plastic 硬质塑料
rigid steel conduit 钢制电线管；硬钢管

rigid structure 刚性结构
rigid tunnel boundary 刚性风洞壁
rigid wake model 刚性尾流模型
rigid wall （风洞）刚性壁
rigidity 刚度
rigidity coefficient 刚性系数
rigidity condition 刚性条件
rigidity gear 刚性齿轮
rigidity modulus 刚性模量
rigidly mounted blade 刚性安装的叶片
rigorous similarity 严格相似性
rime 霜凇；白霜
rime fog 雾凇雾
rime ice 霜冰
riming 结凇
ring earth external 环形接地体
ring gear (annulus gear) 内齿圈
ring generator 环式发动机
ring joint 环接；围缘接合
ring shielding ring 屏蔽环
ring stiffened semi-monocoque construction 加筋半硬壳结构
ring vortex 环形涡；涡环
rinsing 清洗
ripple wave length 纹波的波长
rise 上升
riser 立管
riser characteristic 上升特性
rising 上升的
rising air 上升气流
rising gust 上升阵风
rising height of plume 羽流抬升高度
rising plume 抬升羽流
rivet 铆接
rivet allowance 铆孔留量

rivet(t)ing 铆；铆接
RMS (root-mean-square) 均方根
RMS error 均方根误差
RMS pressure coefficient 均方根压力系数
RMS values 均方根值
RMSE (root mean square error) 均方根误差
RN (Reynolds number) 雷诺数
robustness 鲁棒性；强度；坚固性
rocker arm 摇臂
rocker arm of gearbox 齿轮箱摇臂
Rockwell a scale 洛氏硬度A标
Rockwell apparatus 洛氏硬度计
Rockwell hardness (RH) 洛氏硬度
Rockwell hardness scale 洛氏硬度标尺；洛氏硬度值
Rockwell hardometer 洛氏硬度计
Rockwell number 洛氏硬度值
Rockwell test 洛氏硬度试验
Rockwell tester 洛氏硬度计
rockwool 矿毛绝缘纤维矿；石棉
roll 滚转
roll angle 滚转角
roll damping 滚转阻尼
roll oversteering 倾侧过度转向
roll rate 倾侧速率
roll resonance 横摇共振
roll steer 倾侧转向
roll welding 滚压焊
rolled up vortex 卷起涡
roller 轧辊；滚筒；滚筒
roller actuator 转子开关
roller bearing 滚柱轴承
rolling 旋转的
rolling country 丘陵区
rolling couple 倾侧力偶；横摇力偶

rolling flow wind tunnel 旋转流风洞
rolling friction 滚动摩擦
rolling friction resistance 滚动摩擦阻力
rolling ground 起伏地面
rolling instability 倾侧不稳定性；横摇不稳定性
rolling moment 滚转力矩；倾侧力矩
rolling resistance 滚动阻力
rolling speed 滚动速度
rolling terrain 丘陵地带
rolling topography 起伏地形
roofing plume 屋脊型羽流
room temperature 室温
root angle 安装角
root chord 翼根弦
root chord length 翼根弦长
root circle 齿根圆
root diameter 齿根圆直径
root distance 齿根距
root face （焊缝）根部面；齿根面；钝边
root mean square (RMS) 均方根
root mean square error (RMSE) 均方根误差
root mean square response 均方根响应
root of a weld 焊缝根部
root of blade 叶根
root radius 齿根半径
rope lay conductor 多股绞合电缆；复绞导线
Rossby number 罗斯贝数
Rossby parameter 罗斯贝参数
rotary transducer 转速传感器
rotary valve 回转阀
rotating 旋转的
rotating annulus 转环转盘
rotating commutator 旋转换向器；旋转整流子
rotating convection 旋转对流

rotating core 转动核心
rotating dishpan 转盘（全球气候物理模型）
rotating electrical machine 旋转电机
rotating equipment data 转动设备数据
rotating flow 有旋流
rotating flux 旋转磁通
rotating inertia 转动惯量
rotating magnetic field 旋转磁场
rotating mass 旋转质量
rotating paddle 旋转叶片
rotating rectifier 旋转整流器
rotating stall 旋转分离
rotating stiffness 旋转刚度
rotating torque 旋转力矩
rotating union 旋转接头
rotating valve 回转阀
rotating variable resistor 旋转式变阻器
rotating vector 旋转矢量
rotation 旋转；循环；轮流
rotation angle 旋转角
rotation axis 旋转轴
rotation cooling 循环冷却
rotation moment 转矩
rotation source 旋转源
rotation vector 旋转矢量
rotational 转动的
rotational balance 转动平衡
rotational component 角位移分量；旋转分量
rotational direction 旋转方向
rotational flow 有旋流
rotational inertia 转动惯量
rotational stiffness 旋转刚度
rotationally sampled wind velocity 旋转采样风矢量
rotative 回转的；循环的
rotative moment 转矩
rotor 风轮；转子；旋翼
rotor angular momentum 转子角动量
rotor balancing 转子平衡
rotor bar 转子条
rotor bearing 转子轴承
rotor blade 风轮叶片
rotor brake 转子刹车
rotor cage bar 转子笼条
rotor circuit 转子电路
rotor column 风轮柱
rotor conductor 转子导体
rotor copper loss 转子铜耗
rotor core 转子铁心
rotor current 转子电流
rotor current controller (RCC) 转子电流控制器
rotor diameter 风轮直径
rotor disc 风轮桨盘
rotor efficiency 风轮效率；转子效率
rotor hub 风轮桨毂
rotor induced EMF 转子感应电动势
rotor interface 风轮界面
rotor leakage resistance 转子漏抗
rotor lock 风轮锁
rotor locking disc 转子锁止盘
rotor locking ring 风轮锁定环
rotor magnetic field 转子磁场
rotor nominal speed 额定转速
rotor nominal speed for constant wind turbines 恒速风电机组风轮的额定转速
rotor nominal speed for dual speed wind turbines 双速风电机组风轮的额定转速
rotor nominal speed for variable speed wind turbines 变速风电机组风轮的额定转速

rotor overhang 风轮悬突体
rotor pole 转子极
rotor power coefficient 风能利用系数
rotor pre-cone 风轮预锥角
rotor resistance 转子电阻
rotor shaft tilt angle 主轴倾斜角度
rotor side converter 转子侧变流器
rotor slip 风轮滑流转子转差率
rotor solidity 风轮实度
rotor speed 风轮转速
rotor speed range 转速范围
rotor swept area 风轮扫风面积
rotor thrust 风轮推力
rotor tip vortex 旋翼叶尖旋涡
rotor torque 风轮转矩
rotor voltage 转子电压
rotor wake 风轮尾流
rotor winding 转子绕组；转子线圈
rough country 丘陵地带；崎岖地区
rough terrain diffusion model 粗糙地形扩散模式
roughness 粗糙度；粗糙元
roughness band 粗糙带
roughness class 粗糙度等级
roughness concentration 粗糙度
roughness elements 粗糙元
roughness exposure 粗糙地貌开敞度
roughness factor 粗糙因数；粗糙因素；粗糙度系数
roughness height 粗糙度高；粗糙度值
roughness layer 粗糙层
roughness length 粗糙长度
roughness of terrain 地表粗糙度
roughness parameter 粗糙度参数
roughness Reynolds number 糙率雷诺数（即卡门数）
roughness rose 粗糙度玫瑰图
roughness similitude 粗糙度模拟
round leading edge airfoil section 顿前缘翼型
round nosed airfoil 圆头叶形
round off accumulating 舍入误差的累加
Routh criterion 劳斯准则；劳斯判据
routine 日常工作
routine cleaning 定期清理；常规清理
routine control 常规管理
routine determination 常规检测
routine inspection 定期检查
routine maintenance 日常维修
routine quality control 日常质量管理
routine repair 日常修理
routine test 常规试验
RTC (real time counter) 实时计数器
RTU (remote terminal unit) 远程终端设备；远程终端装置
rubber 橡胶
rubber buffer 橡皮减震垫
rubber bump 减震垫；橡皮缓冲器
rugged duty 磨损量过大的工作状态
rugged terrain 畸形地形
rule and procedure 规则和程序
rule of thumb 经验法则
run 运行
run down 撞倒；走下坡路；（使）虚弱
run duration 风速持续时间
run idle 空转
run of the wind 风程
runaway speed 飞逸转速；发电机超速
running 运转；工作；流动；行程；控制
running balance 动平衡
running condition 运行工况

running cost 运转费
running gear 运行装置；行走机构
running period （设备）运转周期
running program 运行程序
running repair 日常修理
running resistance 运行阻力
running sand 流沙
running test 运行试验
running time 运转时间；操作时间
running trail 航行试验
running voltage 工作电压
running without load 空转
rupture life 持久强度
rupture test 破坏试验
rural area 乡村
rural boundary layer 乡村边界层
rural diffusion coefficient 乡村扩散系数
rush current 冲击电流
rust 锈斑锈；生锈
rust grease 防锈脂
rust inhibitor 防锈剂
rust preventing agent 防锈剂
rust preventive 防锈漆
rust protection 防锈
rustic brick 粗面砖
rustiness 生锈
rustproof 防锈
rusty surface 锈蚀表面
RV (rated voltage) 额定电压

现代英汉风力发电工程

A Modern English-Chinese Dictionary of Wind Power Engineering

S

S type rotor (Savonius type rotor) S型转子
saddle 鞍；鞍状物
saddle back roof 鞍形屋顶
safe 安全的；可靠的
safe allowable load 安全容许负载；安全负荷
safe carrying capacity 安全载流量
safe coefficient 安全系数
safe concentration 安全浓度
safe distance 安全距离
safe guard construction 安全防护结构
safe in operation 安全操作
safe life 安全寿命
safe load 安全负荷；容许负荷
safe measure 安全措施
safe operating area (SOA) 安全工作区
safe operation 安全操作
safe range of stress 应力安全范围
safe running 安全运行
safe standstill 安全停机
safe stress 容许应力
safety 安全；保险；安全设备；保险装置
safety action 防护作用；保护作用；屏蔽作用
safety aid 安全设备；安全援助
safety alarm 安全报警器
safety alarm device 安全报警装置

safety and control scheme 安全控制策略
safety apparatus 安全设备
safety appliance 安全设备
safety belt 安全带
safety belt buckle 安全带扣环
safety breaker 安全开关；安全断路器
safety chain 安全链
safety clearance 安全间隙
safety code 安全规程
safety coefficient (SC) 安全系数
safety color 安全色
safety concept 安全方案
safety control 安全控制
safety control equipment 安全控制设备
safety coupling 保险连接器；安全联轴节
safety device 安全装置
safety distance 安全距离
safety factor 安全因数；安全系数；安全率；安全因素
safety feature 安全装置
safety for overspeeding 过速安全性
safety fuse 安全熔断器
safety helmet 安全帽
safety impedance 安全阻抗
safety isolating transformer 安全隔离变压器

safety margin 安全限度
safety marking 安全标志
safety performance 安全性能
safety precaution 安全保护；预防措施
safety procedure 安全规程
safety regulation 安全规则；安全条例
safety signal 安全信号
safety switch 安全开关
safety system 安全系统
safety testing 安全试验
safety valve 安全阀
safety wire 保险丝
sail 帆片；风帆
sail fabric 风帆蒙布
sail handing 帆操纵
sail tip 帆尖
sail wind generator 帆翼风力发电机
sail wing rotor 帆翼风轮
sail wing windmill 帆翼风车
sailer 帆船
sailing boat 帆船
sailing ship 帆船
sailing vessel 帆船
sailing yacht 帆艇
salient 凸角
salient edge 凸缘；凸角
salient pole electric rotor 凸极转子
salient pole stator frame 凸极定子框架
salient pole stator winding frame 凸极定子绕组框架
salient pole synchronous generator 凸极同步发电机
salient pole 凸极
salt 盐
salt bath brazing 盐浴钎焊

salt fog 盐雾
salt haze 盐霾
salt spray 盐雾
salt water stratified towing tank 分层盐水拖槽
salt wind 盐风
salt windmill 盐场风车
saltation 跃移；跳跃
saltation friction velocity 跃移摩擦速度
saltation Froude number 跃移弗劳德数
saltation mode 跃移模型
saltation phase 跃移相位
saltation velocity threshold 跃移起动速度
sample 取样；样品
sample car 样车
sample size 样本大小
sample statistics 样本统计(学)
sampling 取样；抽样
sampling action 采样作用；采样动作；取样动作
sampling error 取样误差
sampling interval 取样间隔
sampling point 取样点
sampling port 取样孔
sampling probe 采样探针
sampling rake 取样排管
sampling test 抽样试验
sampling time 取样时间
sand 沙
sand bank 沙坝
sand bearing wind 含沙风
sand bed 沙层
sand blasting 吹沙
sand cloud 沙云
sand control dam 防沙堤

sand desert 纯沙沙漠
sand devil 沙旋风；沙暴
sand drift 风沙
sand drifting 流沙
sand driving wind 挟沙风
sand dune fixation 沙丘固定
sand grain 沙粒
sand haze 沙霾
sand hill 沙冈
sand levee 沙堤
sand mist 沙雾
sand mound 沙堆
sand moving wind 起沙风
sand protecting plantation 防沙林
sand ridge 沙脊
sand shadow 沙影；背风积沙区
sand sheet 沙片
sand snow 沙性雪
sand soil 沙土
sand storm 沙暴
sand stream 沙流
sand strip 沙带
sand tornado 沙龙卷
sand trap 集沙器；分沙器；沙坑障碍
sand wave 沙浪
sand whirl 沙旋
sandwich 夹层结构；多层的
sandwich panel 夹层板
sandwich plate 夹层板
sandwich beam 多层组合梁
saphe 拟位相；同态相位
satisfactory stall 良好失速
saturated 使渗透；使饱和
saturated adiabatic process 饱和绝热过程
saturated soil 饱和土

saturated zone 饱和区
saturation 饱和；渗透
saturation adiabatic lapse rate 饱和绝热递减率
saturation characteristic 饱和特性
saturation curve 饱和曲线
saturation deficit 饱和差
saturation effect 饱和效应
saturation mixing ratio 饱和混合比
saturation point 饱和点
saturation pressure 饱和压力
saturation specific humidity 饱和比湿
saturation temperature 饱和温度
Savonius type rotor (S type rotor) S型转子
saw blades 锯片；锯刀
sawdust 锯末
sawing machines 锯床
sawing machines-band 带锯床
saws band 带锯锯带
saws horizontal band 卧式带锯
saws vertical band 立式带锯
saw-tooth fin 锯齿形鳍板
saw-tooth roof 锯齿形屋顶
saw-tooth trip 锯齿形挡板
SC (safety coefficient) 安全系数
SC (shift counter) 移位计数器
SC (stratocumulus) 层积云
SC cas (stratocumulus castellanus) 堡状层积云
SC cug (stratocumulus cumulogenitus) 积云性层积云
SC op (stratocumulus opacus) 遮光层积云
SC ra (stratocumulus radiatus) 辐状层积云
SC tr (stratocumulus translucidus) 透光层积云

SCADA (supervisory control and data acquisition) 数据采集与监视控制系统
SCADA interface SCADA接口
scaffold 脚手架
scaffolding 脚手架材料；搭脚手架的材料
scalar 无向量；标量
scalar control 标量控制
scalar potential 标电位；标量势；标量位；标势
scalar product 点积
scale 氧化层；氧化皮；轧制鳞皮
scale down model 缩尺模型
scale effect 尺度效应
scale height 均质大气高度
scale model 缩尺模型
scale of eddies 涡旋尺度
scale of turbulence 湍流尺度
scale of wind damage 风害规模
scale removal 去除氧化皮；除垢；除锈；除鳞
scaling 缩放比例
scaling circuit 计数电路
scaling factor 缩尺因子；缩尺比
scaling parameter 缩尺参数
scanivalve 扫描阀
scarf welding 斜面焊接
scarp 悬崖
scattering 散射；散布
scavenge port 换气口；扫气口
scavenging 清除；打扫；排除废气
scavenging coefficient 清除系数
scheduled maintenance 进度维护；计划维修；预定检修；进度维护
scheduled repair 计划修理；定期修理
schematic diagram 原理图；示意图
schlieren method 纹影法
schlieren photograph 纹影照片

Schmidt number 施密特数
SCIG (squirrel cage induction generator) 鼠笼异步发电机
scintillometer 闪烁计数器
scler. (scleroscope hardness) 肖氏硬度；回跳硬度
sclerometer 硬度计
scleroscope 回跳硬度计；肖氏硬度计
scleroscope hardness (scler.) 肖氏硬度；回跳硬度
scleroscope hardness test 肖氏硬度试验
scoop 通风斗；掘；舀取；水斗
SCR (silicon controlled rectifier) 可控硅整流器
scraping 刮痕；擦伤
screen 格网；(风洞) 阻尼网
screen action 屏蔽作用
screen display 屏幕显示
screen mesh 网眼；丝网；筛孔；筛眼
screening 筛分；遮蔽
screening function 屏蔽功能
screw 螺丝钉；螺旋；螺杆；螺孔；螺旋桨
screw driver 螺丝刀
screw extractor 拉马；断螺钉联出器；螺杆旋出器；螺旋挤压机；起螺丝器
screw jack 螺旋千斤顶
screw pitch gauge 螺距规；螺纹样板
screw setting 定位螺旋
screw tap 螺丝攻
screw thread 螺纹
screw thread lubricant 螺纹润滑剂
screwed 螺丝状的
screwed joint 螺纹连接
screwed pipe joint 螺纹管连接
screwed piping joints 螺丝状的管接头

scriber 划线器；描绘标记的用具
sea 海
sea breeze 海风
sea breeze front 海风前锋
sea going capability 适航性
sea line 海岸线
seal 密封；封口；阀座；密封面
seal fitting 密封接头；密封配件
seal oil 密封用油
seal welding 密封焊接
sealing 密封
sealing gland 密封盖
sealing steam box 汽封蒸汽室
seam 缝；接缝
seam weld 滚焊
seam welding 滚（缝）焊
seam welding machine 缝焊机
seashore 海滨
seashore breeze 海岸风
seasonal wind 季风
seawind 海风
second shelterbelt 副防护林带
secondary action 副作用；二次作用
secondary air 二次气流
secondary circulation 二次环流
secondary cold front 副冷锋
secondary current 二次电流
secondary cyclone 副气旋
secondary depression 副低压；次（生）低压
secondary drag 二次流阻力
secondary effect 二次效应
secondary flow 二次流
secondary front 副锋
secondary inflow 二次入流
secondary member （建筑物）次要部件

secondary order correlation 二介相关
secondary pollutant 二次污染物
secondary pollution 二次污染
secondary stream 二次流
secondary voltage 二次电压
secondary vorticity 二次涡量
secondary wind direction 次级风向
secondary wind flow effect 次级风效应
section 截面
section chord 剖面弦；叶片剖面弦长
section chord technique 截面弦法
section contour 剖面外形；截面轮廓
section drag 截面阻力；翼型阻力
section lift 截面升力；翼型升力
section model 截段模型
section modulus 截段模量
sectional construction 预制构件拼装结构
sector box model 扇形箱模式
security coupling 安全联轴器
security wind speed 安全风速
sediment 沉积物
sediment diffusion 泥沙扩散
sediment runoff 输沙量
sediment transport 泥沙运输
sedimentary resistance 沉积阻力
sedimentated dust 降尘
sedimentation 沉淀；沉积
seesaw motion 跷板式运动
seesaw rotor 跷板式风轮
segment 段；节
selective absorption 选择性吸附
selective assembly 选配
selective headstock 变速箱
self acting 自动式
self acting control 自动控制

self acting valve 自动阀
self adjusting 自调整
self aligning roller bearing 调心滚子轴承；球面滚柱轴承
self balancing device 自动平衡装置
self bias resistor 自偏置电阻
self blocking 自动联锁
self braking 自动制动
self checking 自动检验
self closing 自动关闭
self closing stop valve 自动断流阀
self closing valve 自闭阀
self commutated inverter 自换流逆变器；自动换向逆变器
self compensation balance 自补偿天平
self control 自动控制
self damping 自阻尼
self damping property 自阻尼特性；固有减振性能
self excitation process 自励过程
self excited 自励
self excited aerodynamic moment 自激励空气动力矩
self excited dynamo 自激电机
self excited force 自激力
self excited oscillation 自激振荡
self excited vortex shedding 自激涡脱落
self exciting 自励的；励磁的
self ignition 自发火；自点火；自燃
self induced oscillation 自激振荡
self inductance 自感系数
self inductor 自感
self locking 自锁
self lubrication 自润滑
self operated controller 自动控制器

self operated measuring unit 自动测量装置
self orientating （风轮）自动调向；自动对风
self propulsion test 自航试验
self purification 自净作用
self recording anemometer 自计风速计
self recording barometer 自计气压计
self registering anemometer 自计风速计
self regulating windmill 自动调节风车
self standing tower 独立塔架
self starting 自起动
self starting property 自起动特性
self supporting 自支持；自支承；自支撑
self supporting tower 独立塔架
selfsustained oscillation 自激振动；自持振荡
semi aerodynamic shape 半气动形状；半流线型
semi axis 半轴
semi closed jet wind tunnel 半闭口式风洞
semi controlled device 半控型器件
semi empirical formula 半经验公式
semi finished material 半成品；半制品
semi geared wind turbine 半齿轮驱动风电机组
semi girder 悬臂梁
semi infinite cloud model 半无限烟云模式；半无限云模式
semi length scale 半长尺度
semi log paper 半对数（坐标）纸
semi logarithmic coordinate 半对数坐标
semi logarithmic paper 半对数（坐标）纸
semi major axis 长半轴
semi matt 半亚光
semi minor axis 短半轴
semi mobile dune 半流动沙丘
semi mobile sand 半流动沙

semi open jet wind tunnel 半开口式风洞
semi rigid approach 半刚性方法
semi rigid model 半刚性模型
semiautomatic control 半自动控制
semiconductor 半导体器件
semiconductor device 半导体器件
semifinished product 半成品
semilunar dune 新月形沙丘
semimanufactures 半成品
semimonocoque construction 半硬壳式结构
semispan 半翼展
semistall （气流）局部分离
semiturbulent 半紊流的
semiuniform 半均匀的
semi-urban environment 半城市环境
sending allowance 传输损耗
sensible heat 显热
sensible heat storage 显热贮能
sensing component 敏感元件
sensing transducer 传感器
sensitive element 敏感元件
sensitive sensor 传感器
sensitivity 灵敏度
sensitivity of following wind 调向灵敏性（表示随风向的变化，风轮迎风是否灵敏的属性）
sensor 传感器
SEP (standard engineering practice) 标准工程惯例
separated 分开；隔开
separated boundary layer 分离边界层
separated flow 分离流动
separated flow vortex 分离流的旋涡
separated shear layer 分离剪切层
separately excited 他励的
separately excited generator 他励磁发电机
separately excited rotor winding 他励磁转子绕组
separation 分离
separation angle 分离角
separation bubble 气流分离
separation cavity 分离空穴
separation coupling 可拆联轴节
separation flow 分离流
separation line 分离线
separation point 分离点
separation region 分离区
separation streamline 分离流线
separation vortex 分离涡
separation wake 分离尾流
sequence 顺序
sequence control 程序控制
sequence controller 程序控制器
sequence of operation 操作程序
sequence test 联锁顺序试验
sequential 连续的
sequential order of the phase 相序
SERI (Solar Energy Research Institute) 美国太阳能研究所
serial line internet protocol (SLIP) 串行线路网际协议
series 串联；系列；批
series circuits 串联电路
series connection 串联
series development 级数展开
series excited 串励
series parallel 混联
series parallel connection 串并联连接
series parallel control 串并联控制
series production 批量生产
service 业务；技术维护；检修；服务；操作

service aid 维修工具
service behavior 工作性能；服务行为；使用状态；运转情况
service conditions 使用条件
service crane 维护吊车
service equipment 维修设备
service hatch 维修孔
service hoist 维修吊车
service life 使用期限
service lifetime 使用寿命
service load 操作负荷
service manual 维修手册；维修说明书
service property 使用特性；服役性能
service reliable 使用可靠性
service requirement 运行要求
service speed 运行转速
service test 使用试验；运转试验
service tool 维修工具
serviceability 适用性
serviceability limit state 正常使用极限状态
servo 伺服
servo action 伺服作用
servo brake 伺服制动（器）
servo drive 伺服传动装置
servodyne 伺服系统的动力传动装置；随动系统的动力传动装置
servomechanism 伺服机构
servomotor 伺服电动机
servorudder 伺服舵
servosystem 伺服系统；随动系统
S.E.S. (Society of Engineering Science) （美国）工程科学学会
set 设定；机组；套
set deformation 永久变形
set up procedure 装配程序

set value 设定值
setting 校准；装置；支座
setting angle 安装角
setting angle of blade 叶片安装角
setting clearance 调整间隙
setting pressure 设定压力；整定压力
setting range 调整范围
settled dust 沉降尘埃
settling 安置；固定
settling chamber （风洞）稳定段；沉降室
settling rate 沉降速度
settling ratio 沉降系数
settling time 建立时间；过渡时间；安装时间
severe 剧烈的
severe gale 厉风
severe gust 强阵风
severe storm 强风暴
severe tropical storm 强烈热带风暴（最大风速48～63海里/时）
severity 严重；严格
severity factor 硬度系数
severity stall 失速严重
sextic equation 六次方程
SFC (specific fuel consumption) 比燃耗油
SH 硬度；肖氏硬度
shackle U形挂环；卸扣；锁扣
shade deck 遮阳甲板
shadow 阴影；遮蔽
shadow area 阴影区；遮蔽区
shadow method 阴影法
shaft 轴；杆
shaft bearing 轴承
shaft concentricity 轴同心度
shaft coupling 联轴节
shaft fatigue 轴疲劳

shaft flange 轴法兰	shear layer 剪切层；切变层
shaft horsepower 轴输出功率	shear layer reattachment 附面层再附着
shaft installing sleeve （密封用）轴套	shear loading 剪切荷载
shaft mounted unit 轴安装设备	shear modulus 抗剪弹性模量；切变模量
shaft packing 轴封填料	shear of wind 风切度
shaft power 轴功率	shear reinforcement 抗剪钢筋
shaft seal 轴封；轴封装置	shear stiffness 抗剪刚度
shaft torque 轴扭矩	shear strength 抗剪强度
shaft work 轴功	shear stress 剪应力
shaitan 尘旋	shear test 剪切试验；抗剪试验
shallow 浅的	shear turbulence 剪切湍流
shallow foundation 浅基础	shear velocity 剪切速度
shape coefficient 体型系数	shear viscosity coefficient 切变黏滞系数
shape factor 形状因子；体型系数	shear wall structure 剪切墙结构
shape parameter 形状参数	shear wind 切变风
shapers 牛头刨床	shear(ing) strain 剪应变
sharp angle 锐角	shear(ing) strength 抗剪强度
sharp draft 强力通风	shearing stress 剪切应力
sharp edged body 尖锐物体	shed vortex 脱体涡；脱出涡
sharp edged building 尖缘建筑	shed vortices 尾迹涡系
sharp edged gust 突发阵风	shed wake vortex 脱落涡
sharp edged separation 尖缘分离	sheet aluminium 铝板
sharp edged strip 锐缘条	sheet erosion 表面侵蚀
sharp nosed blade 尖头叶片	sheet gasket 密封垫圈
Shaw hardness 肖氏硬度	sheet metal forming machines 金属板成型机
shear 剪切；切变	sheet metal working machines 金属板加工机
shear beam 抗剪梁；剪切梁	sheet of vorticity 涡旋面
shear bolt 安全螺栓；保险螺栓；剪力栓	sheet resistance 表面电阻
shear center 剪切中心	sheet stringer construction 板桁结构
shear core structure 剪力核心（抗风）结构	sheeted roof 薄板屋顶
shear deformation 剪切变形；切变	shell construction 壳式结构
shear effect 剪切作用；剪切效应	shell structure roof 壳体屋顶
shear flow 剪切流	shell vibration 蒙皮振动；外壳振动
shear flow turbulence 剪切流紊流度	shelter deck 遮蔽甲板
shear generator （风）切变发生器；剪切发生器	shelter forest 防护林

shelterbelt 防风林带
sheltering effect 遮蔽效应；防护作用
shield gas 保护气体
shield jig 保护夹具
shielding 遮蔽的；防护的
shielding coefficient 遮蔽系数
shielding effect 遮蔽效应
shielding factor 遮蔽因子
shielding ring 屏蔽环
shift counter (SC) 移位计数器
shifting beam 活动梁
shifting bearing 活动支座
shifting center 稳心
shifting sand 流沙
shimmy 摆振；横摆
ship loading dock 装船码头
ship shape line 船舶型线
shipyard gantry crane 船坞门吊；船厂龙门起重机
shock 冲击
shock absorber 减震器；震动吸收器
shock current 触电电流
shock generated vorticity 激波产生的涡量
shock load 冲击荷载
shock loss 激波损失
shock tunnel simulation 激波风洞模型试验
shock wave 激波
shooting 射击
shop fabrication 车间制造
shore wind 海岸风
shoreline 海岸线
shoreline fumigation 海岸线熏烟
shoreward mass transport 向岸质量运输
short 短的
short circuit 短路
short circuit characteristic 短路特性
short circuit current 短路电流
short circuit current level 短路电流水平
short circuit operation 短路运行
short circuit power 短路功率
short circuit ratio 短路比
short circuited rotor 短路式转子；鼠笼式转子
short circuiting ring 短路环
short duration 短时
short duration gust 短期阵风
short duration loading 短期荷载
short footed 短柄的
short period overspeed 短期超速
short separation bubble 分离短气泡；短泡分离
short term 短期
short term cut-out wind speed 短时切出风速
short term effect 短时效应；瞬时效应；近期效益
short term wind data 短期风资料
short test section wind tunnel 短试验段风洞
short time diffusion factor 短期扩散因子
short time washout factor 短期冲洗因子
short wave radition 短波辐射
shortness 脆性
shovel 铲；铁铲
shrink disk 缩紧盘；收缩盘
shrinking ratio 收缩比；收缩率
shroud 罩；(风轮) 集风罩
shrouded rotor 有罩风轮
shrub 灌木
shunt 并励；分路器；分流
shunt capacitor 并联电容器
shunt capacitor banks 并联电容器组
shunt displacement current 旁路位移电流
shunt excited 并励

shunt field 并励磁场
shunt reactor 并联电抗器
shut 关闭；停止
shut down (for wind turbine) 关机；切机（风电机组）
shut down wind speed 刹车风速；停机风速
shut off 停止；关掉；隔绝；使不进入
shut off wind speed 刹车风速
shutter 百叶窗
SI (system international of units) 国际单位制
side 方面；侧面
side force 侧力
side gust 侧阵风
side headwind 侧逆风
side leading wind 侧顺风
side rake 侧斜角；侧前角
side span 边跨
side spin 侧旋
side sway 侧摆；侧移
side vane 侧翼（在风轮侧面利用风压使风轮偏离风向的机构）
side wall （风洞）侧壁
sideslip 侧滑
sideslip angle 侧滑角
sideslip derivative 侧滑导数
sidewalk 人行道
sidewall insert （风洞）侧壁插件
sidewash （气流）侧洗
sidewind 侧风；横风
sign board 标价牌；路标
signal 信号
signal alarm 信号报警
signal amplifier 信号放大器
signal anemometer 信号风速计
signal circuit 信号电路

signal equipment 信号设备
signal generator 信号发生器
signal graph 信号流图
signal indicator 信号指示灯
signal light 信号灯
signal to noise ratio 信噪比
signalling alarm equipment 信号报警设备
silding vane rotary flowmeter 刮板流量计
silica 硅石；二氧化硅
silicon 硅
silicon carbide 碳化硅；金刚砂；碳化矽
silicon controlled rectifier (SCR) 可控硅整流器
silk streamer （气流观察）丝线带
silt 泥沙；粉砂
similar geometry wind tunnel 几何相似风洞
similar number 相似准则数
similarity 类似；相似点
similarity condition 相似条件
similarity coordinate 相似坐标
similarity criteria 相似准则
similarity law 相似定律；相似性定律
similarity level 相似性水平
similarity measure 相似性测度
similarity parameter 相似参数；相似准数
similarity rules 相似率
similarity theory 相似理论
similitude 相似物
simple 简单的
simple harmonic motion 简谐运动
simple harmonic vibration 简谐振动
simple retrofit 简单改装
simplex transmission 单工传输；单向传输
simplicity 简易性
simulate 模拟
simulated 模拟的；模仿的

simulated condition	模拟条件
simulated flow	模拟流动
simulated snow	模拟雪
simulated wind	模拟风
simulating material	模拟材料
simulation	模拟
simulation chamber	模拟室
simulation hardware	模拟硬件；模拟装置
simulation program	模拟程序
simulation system	模拟系统
simulation technique	模拟技术
simulation test	模拟试验
simulator	模拟装置
sine mode	正弦模态
sine weighted PWM	正弦加权 PMW
singing	嗡鸣
singing sand	鸣沙
single acting	单作用的；单动的
signal anemometer	信号风速计
single bearing	单轴承
single blade adjustment	单叶片调整
single blade windmill	单叶片风车
single clamp	单卡头
single component balance	单分量天平
single disk	单个刹车盘
single disk brake	单圆盘制动器
single disk clutch	单片式离合器
single fed induction generator	单馈异步发电机
single fed asynchronous generator	单馈异步发电机
single frequency excitation	单频率激励
single gearbox with multiple electrical generators	带有多台发电机的单齿轮箱
single helical gear	单线螺旋齿轮
single line diagram	单线图
single load	集中荷载
single phase	单相
single phase bridge rectifier	单相桥式整流器
single planetary gear train	单级行星齿轮系
single point earthing	单点接地
single rectangular box girder bridge	单矩形箱桥梁
single return wind tunnel	单回路风洞
single side band (SSB)	单边带
single source model	单源模式
single spar construction	单梁结构
single tower	单柱塔；独塔
single trapezoidal box girder	单梯形箱梁
single wave operation	单波运行
single-wave rectifier	半波整流器
singular point	奇点
sink	汇点，汇集
sink streamline	汇流线
sink strength	汇强
sintering	烧结
sinuous	蜿蜒的；弯曲的
sinuous flow	蛇形流动；乱流
sinuous plume	蛇形羽流
sinuous vortex	蛇形涡
sinusoidal	正弦的；正弦曲线的
sinusoidal cyclic load	正弦周期载荷
sinusoidal density wave	正弦磁密度
sinusoidal excitation	正弦型激励
sinusoidal lock-in excitation	正弦锁定激励
sinusoidal mode shape	正弦模态形状
sinusoidal oscillation	正弦振荡
sinusoidal time function	正弦时间函数
site	地点；位置
site analysis	场址分析

site assessment 场址评价
site electrical facilities 风场电器设备
site evaluation 场址评价
site measurement 现场实测
site selection 选址
site survey 现场勘测
siting 定址
siting criteria 定址准则
six component balance 六分量天平
six poles dual feed generator 六极双馈发电机
six series airfoil NACA 六位数系列翼型
size analysis 粒度分析
size coefficient 粒度系数
size distribution 粒径分布
size reduction factor 缩尺因子
skew 斜的，歪的；歪斜，扭曲
skew bevel gear 斜齿锥齿轮
skew bridge 斜桥
skew direction 斜向
skew inclination 倾斜
skew wind 斜风
skewed boundary layer 扭曲边界层
skewed streamline 扭曲流线
skewing allowance 歪扭余量
skewness 斜度；偏差度，偏度；偏斜度
skidding behavior 滑动特性
skin 蒙皮；皮肤；外壳
skin bonded construction 蒙皮胶合结构
skin dose 皮肤剂量
skin friction 表面摩擦力
skin friction tests 蒙皮磨损阻力试验
skin lamination 蒙皮表层；表皮分层
skin material 蒙皮材料
skin patch 蒙皮补片
skin reinforcement 蒙皮加强

skin resistance 表面摩擦阻力
skin rivet 蒙皮铆钉
skin streamline 表面流线
skin stretch forming 蒙皮拉伸成形
skin stringer construction 蒙皮桁条结构
skin test 蒙皮试验
sky condition 天空状况
sky radiation 天空辐射
slapping 拍击
slat 板条；狭板
sleet 雨凇；雨夹雪；冻雨
sleeve 轴套
slender 细长的
slender body theory 细长体理论
slender tower 细高塔架；细高大厦
slenderness 细长度
slewing 快速定向
slewing ring 齿环，轴承齿环，回转支承；回转环
slide 滑动
slide bearing 滑动轴承
slide calliper 游标卡尺
sliding 滑行的
sliding calliper 游标卡尺
sliding clutch 滑动离合器
sliding friction 滑动摩擦
sliding pin 滑销
sliding pin system 滑销系统
sliding shoes 滑动制动器；滑靴
slight 轻微的；少量的
slight breeze 轻风（二级风）
slight flagging 轻微旗状（II级植物指示风力）
sling psychrometer 手摇干湿表
sling thermometer 手摇温度表
slinger 吊索

slip 转差率；滑动的
SLIP (serial line internet protocol) 串行线路网际协议
slip angle 横偏角；侧滑角
slip clutch 滑动离合器
slip flow 滑流
slip frequency 滑差频率；转差频率
slip power 滑差功率
slip ratio 转差率；滑动比
slip ring 滑环；集电环；滑动环；集电环
slip speed 滑差速度
slippage 滑动；下跌；滑动量；下跌量
slipping clutch 安全摩擦离合器；可调极限扭矩摩擦离合器；滑动离合器
slipstream 滑流
slope 斜率；坡度；山坡
slope angle 坡度角
slope down-wind 下坡风
slope flow 山坡气流；坡面流
slope ground 斜坡；倾斜地表
slope of lift curve 升力曲线斜率
slope up-wind 上坡风
slope wind 山坡风；坡风
slot 槽口；开缝；裂缝；切口
slot leakage flux 槽漏磁通
slot wall 槽壁
slot wedge 槽楔
slot weld 切口焊缝
slotless air-gap winding 无槽气隙绕组
slotted 有沟槽的
slotted aerofoil 开缝机翼
slotted stator 开槽定子
slotted teeth 开槽齿
slotted working section 开壁槽试验段；开缝壁试验段

slow 慢的
slow convection 弱对流
slow death 老化
slow down 减速
slow release action 缓释动作
slow speed main shaft 低速主轴
slow speed rotor 低速风轮
sludge 污泥
slug 子弹；金属块；斯勒格
slug flow (汽) 团状流动；活塞流
slug flow model 团状流动模型
slug of air 空气团
small 小的
small (amplitude) disturbance 小（振幅）扰动
small capacity wind turbine 小容量风电机组
small disturbances method 小扰动方法
small pane 小镶板
small perturbance theory 小扰动理论
small repair 小修
small scale gust 小尺度阵风
small scale model 小尺度模型
small scale turbulence 小尺度湍流
small scale WECS 小型风能转换装置
small scale wind energy conversion system (SWECS) 小型风能转换系统
small scale wind turbine 小型风电机组
small size eddy 小尺度涡流
small sized wind machine 小型风力机
small sized wind tunnel 小型风洞
smart grid 智能电网
smaze (=smog+haze) 烟霾
Smith-Putnam wind turbine 史密斯-普特南大型风轮机
smock mill 裙形古风车
smog (=smoke+fog) 烟雾

smog aerosol 烟雾气溶胶
smog control 烟雾控制
smog horizon 烟雾顶层
smog index 烟雾指数
smog injury 烟雾危害
smog warning 烟雾警报
smoke agent 发烟剂
smoke aloft 高空烟
smoke and soot 烟尘；烟灰
smoke black 烟黑
smoke cloud 烟云
smoke concentration 排烟浓度；烟浓度
smoke damage 烟害
smoke density 排烟密度
smoke dust 烟尘
smoke emission 排烟
smoke emission standard 排烟标准
smoke eye 烟雾报警器；烟眼
smoke flow visualization 烟流流动显示；烟流显形
smoke fog 烟雾
smoke funnel 烟囱
smoke generator 发烟器
smoke grenade 烟幕弹
smoke haze 烟霾
smoke horizon 烟层顶
smoke indicator 烟气浓度计
smoke injury 烟害
smoke laden air 含烟空气
smoke nozzle 喷烟嘴
smoke pall 厚烟层
smoke particle 烟粒
smoke pattern 烟流谱
smoke pipe 排烟管
smoke plume 烟羽

smoke pollution 烟污染
smoke probe 烟雾探测器
smoke puff 烟迹
smoke screen 烟幕
smoke spot 烟斑
smoke stack 烟囱；烟筒；烟囱帽
smoke stack outlet 烟囱出口
smoke trail 烟迹
smoke visualization 烟流流动显示
smoke wind tunnel 烟风洞
smoke wire method 烟丝法
smokeless zone 无烟区
smooth 光滑的
smooth flow 平滑流
smooth wind 平滑风
smoothness 光滑度；平滑度
snow 雪
snow and rain 雨夹雪
snow avalanche 雪崩
snow banner 雪旗
snow barrier 雪障
snow bearing wind 裹雪风；风花雪
snow belt 雪带
snow blockage 积雪阻塞
snow break 雪障
snow cave 雪洞
snow climate 雪原气候
snow cloud 雪云
snow cover 积雪层；雪被
snow crystal 雪晶
snow density meter 积雪密度计
snow deposit 积雪
snow devil 雪卷风
snow drift 风吹雪；雪堆；吹雪
snow drift control 吹雪控制

snow drift density 雪堆密度；雪空间密度
snow drift field 堆雪场
snow drift intensity 吹雪强度
snow drift station 堆雪站
snow dune 雪丘
snow fence 积雪栅；防雪栅
snow field 雪原
snow flake 雪花
snow forest climate 雪林气候
snow garland 雪花环
snow gauge 量雪器
snow grain 粒雪
snow guard 防雪栅
snow gully 雪沟
snow layer 雪层
snow line 雪线
snow load 雪载
snow loading 雪载
snow management 堆雪控制
snow mat 雪席
snow measuring plate 积雪板；量雪板
snow niche 雪蚀龛
snow particle 雪粒
snow pellet 雪丸
snow plume 雪羽流
snow precipitation line 降雪线
snow protection plantation 防雪林
snow region 雪区
snow retention 积雪
snow roller 雪卷
snow scale 雪尺
snow screen 防雪栅
snow shield 防雪设施
snow simulator 雪模拟器
snow slide 雪崩

snow slip 雪崩
snow slope 雪坡
snow squall 雪飑；暴风雪
snow stake 测雪桩
snow storm 雪暴
snow storm center 雪暴中心
snow storm meter 雪暴测定仪
snow storm path 雪暴路径
snow trap 集雪器；雪阱
snow tube 集雪管
snowberg 雪山
snowfall 雪量；降雪
snowfall totalizer 累计雪量计
snowsampler 雪取样器；取雪器
snubber 缓冲器；减震器
SOA (safe operating area) 安全工作区
soaring wind 上升气流
Society of Engineering Science (S.E.S) （美国）工程科学学会
socket 插座
socket spanner 套筒扳手
socket weld 承插焊接
socle girder 悬臂梁
SODAR (sonic detection and ranging) 声波探测与测距
soft 柔软的
soft and free expansion sheet making plant 软板（片）及自由发泡板机组
soft braking 软制动
soft cushioning 软式减震
soft cut in 软切入
soft damping 软阻尼
soft design 软件设计
soft driven system 软传动系统
soft driven train 软传动系

soft keyboard 软键盘
soft start 软起动；软并网
soft starter 软起动器
soft tower 软性塔架
soft transmission system 软传动系统
software 软件
software compatibility 软件兼容性
software cost 软件成本
software design procedure 软件设计过程
software development library 软件开发库
software development plan 软件开发计划
software development process 软件开发过程
software documentation 软件文件
software engineering 软件工程
software environment 软件环境
software library 软件库
software maintenance 软件维护
software monitor 软件监督程序
software package 软件包
software package of computer aided design 计算机辅助设计软件包
software platform 软件平台
software portability 软件可移植性
software product 软件产品
software psychology 软件心理学
software quality 软件质量
software reactive power droop setting 软件无功调差
software reliability 软件可靠性
software testing 软件测试
software testing plan 软件测试计划
software tool 软件工具
soil 土壤
soil climatology 土壤气候学
soil conservation 土壤保持

soil damping 基土阻尼
soil drifting 土壤风蚀
soil erosion 土壤侵蚀
soil evaporimeter 土壤蒸发仪
soil flexibility 基土柔度
soil moisture 土壤水分
soil moisture content analyser 土壤水分测定仪
soil texture 土壤粗密度；土质
soil water conservation 水土保持
solar 太阳的
solar altitude 太阳高度
solar climate 太阳气候
solar collector 太阳能集热器
solar constant 太阳常数
solar elevation angle 太阳高度角
solar energy 太阳能
solar energy collector 太阳能收集器
Solar Energy Research Institute (SERI) 美国太阳能研究所
solar energy system 太阳能系统
solar flux 太阳能通量
solar insolation 太阳辐射
solar irradiance 太阳照度
solar power plant 太阳能发电站
solar radiant energy 太阳辐射能量
solar radiation 太阳辐射
solar radiation flux 太阳辐射通量
solar spectrum 太阳光谱
solar terms 节气
solar wind energy hybrid system 太阳能风能混合系统
solar zenith angle 太阳天顶角；太阳高度角
solarigraph 总日射计
solarimeter 太阳辐射强度计

solder dipping 浸焊
solenoid 螺线管；电磁阀；电控气流阀
solenoid coil 电磁线圈
solenoid operated switch 电磁控制开关
solenoid operated valve 电测控制阀；螺线管阀
solenoid relay 螺管式继电器
solenoid switch 电磁开关
solenoid valve 螺线管阀；电磁阀
solid 固体的
solid blockage 固体阻塞
solid boundary （风洞）实壁边界
solid content(s) 容积；固体含量
solid coupling 刚性联轴器
solid debris 固体碎屑
solid draw pipe 无缝钢管
solid draw tube 无缝钢管
solid earth 固定接地；完全接地
solid matter 固体物质
solid measure 体积
solid mechanics 固体力学
solid state 固体状态
solid wall test section 实壁试验段
solid web girder 空腹梁
solid wind break 实壁风障；固体防风
solidity （风轮的）实度；固化性
solidity losses 实度损失
solidity ratio 实度比
solt 槽
solum 风化层
sonde 探测气球；探测装置；探测器
sonic 音速的；声音的
sonic anemometer 声波风速计
sonic detection and ranging (SODAR) 声波探测与测距

soot 烟灰
soot and dust 烟尘
soot blower 烟灰吹除机；吹灰机
soot deposit 积灰
sootfall 烟灰沉降
sound absorption mat 声音吸收垫
sound deadening 消音
sound field 声场
sound level 声级
sound level meter 噪声计
sound pressure 声压
sound pressure level 声压级；声级
sound proof 隔音的
sounding balloon 探空气球
source 源
source and sink method 源汇法
source density 源密度
source depletion model 源耗减模式
source distribution 源分布
source flow 源流动
source of air pollution 空气污染源
source of buoyancy 浮力源
source of energy 能源
source of pollution （大气）污染源
source streamline 源流线
source strength 源强
source superposition 源叠加
source term 源项
southeast trade 东南信风带
space 空间
space coordinate 空间坐标
space correlation 空间相关
space derivative 空间导数
space polar coordinate 空间极坐标
spacer 间隔棒；逆电流器

spacer flange 中间法兰；过渡法兰；对接法兰
spacing 间距
span 展长；展宽；跨度
span chord ratio 展弦比
span efficiency factor 翼展效率因数
span factor 翼展因数
span length 翼展长
span load distribution 展向载荷分布；跨向荷载分布
span loading 翼展负载
span of foil 叶片展长
spanner 扳手
spanwise constant thickness ratio 展向恒定厚度比
spanwise correlation 展向相关；跨向相关
spanwise distance 展向翼展距离
spanwise flow 展向流动；跨向流动
spanwise load change 展向载荷变化
spanwise spar 展向梁
spanwise velocity component 展向分速
spanwise vortex 展向涡；跨向涡
spar 翼梁
spar buoy （海上风车的）圆柱式浮台；柱形浮标；圆柱浮标
spare part 备件
spare power 备用电力
spare units 备用机组
spare valve 备用阀
spark arrestor 火花避雷器
spark gap 避雷器；火花放电隙
spasmodic variation 间歇性的变化
spatial 空间的
spatial autocorrelation function 空间自相关函数
spatial correlation 空间相关

spatial covariance 空间协方差
spatial flow visualization 空间流动显示
spatial gust 空间阵风
spatial non-uniformity 空间不均匀性
spatial randomness 空间随机性
spatial spectral density function 空间谱密度函数
spatial structure 空间结构
spatial vorticity distribution 空间涡量分布
spatial waveform 空间波形
special 特别的
special appliance 特种设备；专用设备
special purpose spanner 专用扳手
special steel 特种钢
specific 特殊的；特定的
specific activity 比放射性（强度）
specific conductance 电导率
specific density 比重
specific elongation 伸长率
specific emission （污染）比排放量
specific enthalpy 比焓
specific entropy 比熵
specific extinction coefficient 比消光系数
specific fuel consumption (SFC) 比燃耗油
specific gravity 比重
specific heat 比热
specific humidity 比湿度
specific load 单位负载；负荷率
specific loading 单位载荷
specific mass 密度
specific output 比输出
specific power 比功率
specific reluctance 磁阻率
specific surface area 比表面积
specific volume 比容

specified 规定的

specified capacity 额定容量；额定气体量

specified condition 额定工况；额定参数

specified horsepower 额定功率；铭牌功率

specifies breakaway torque 规定的最初起动转矩

specimen 样品；样机；样本

spectral density function 谱密度函数

spectral gap 谱隙

spectral shift 谱移

spectral window 谱窗

spectrum 谱；范围；余象

spectrum of horizontal gustiness 水平阵风谱

spectrum of turbulence 湍流谱

spectrum of turbulence energy 湍流能谱

spectrum of velocity fluctuation 脉动谱

spectrum of vertical gustiness 垂直阵风谱

speed brake 气动力减速装置；减速板

speed change gear 变速齿轮

speed characteristic curve 速度特性曲线

speed control 速度控制

speed control device 调速装置

speed controller 调速器

speed controller with inverter 变频调速器

speed factor 速度因子

speed increasing gear 增速齿轮；增速器

speed increasing gear pair 增速齿轮副

speed increasing gear train 增速齿轮系；增速齿轮副

speed increasing ratio 增速比

speed isopleth figure 等速线图

speed measurement 速度测量

speed monitor 转速监视器

speed of cooling 冷却速度

speed of flow 流速

speed of ratio 速度比

speed of reducer 减速器

speed of revolution 旋转速度

speed of rotation 转速

speed of variation 速率变化；变速

speed of wind 风速

speed orifice （风洞）测速孔

speed polar 速度极线图

speed range 转速范围

speed reducing gear train 减速齿轮系

speed reduction gear 减速齿轮；减速器

speed regulation 速度调节

speed torque characteristic 速度转矩特性

speed torque curve 转速力矩特性曲线

speed triangle 速度三角形

speed up effect 加速加速效应

speed envelope 超速包络线

speedometer 测速计

speedup factor 加速因子

spent gas 废气

sphere gap 球状避雷器；球隙避雷器

spherical 球形的

spherical roller bearing 球形滚柱轴承

spherical vortex 球形涡

spheroidal graphite cast iron 球墨铸铁

spider gear assembly 星形齿轮

spilling flap 阻尼板

spilling wind 溢出风能（限速法）

spin 自旋；旋转

spin tunnel model tests 螺旋风洞模型试验

spin vector 角速度矢量

spin(ing) wind tunnel 螺旋试验风洞（垂直风洞）

spinner 整流罩；螺旋桨整流罩；桨毂盖；旋转器；自旋体

spinning ball 旋转球

spinning mode 自转模式

spinning reserve 运转备用；旋转备用

spiral 螺旋

spiral bevel gear 螺旋伞齿轮；弧齿锥齿轮

spiral casing 蜗壳

spiral cooling 螺旋形冷却

spiral housing 蜗壳

spiral turbine 蜗壳式叶轮机；蜗壳式水轮机

spire 尖塔（旋涡发生器）

spire roughness simulation technique 尖塔粗糙元模拟技术

splash 溅

splash apron 挡泥板

splash type lubricating system 飞溅润滑系统

splash wing 挡泥板；翼子板

splashing 飞溅

splicing 拼接

splicing sleeve 接续管

spline 方栓；齿条；止转楔；花键

spline model 仿样模型；样条模型

splined coupling 花键连接

splined shaft 有齿轴；花键轴

splint 夹板；托板

split 分离；分解；分裂

split bearing 对开式滑动轴承；剖分轴承

split flap 分裂式襟翼

split flow 分流

split gear 拼合齿轮

split pin 开口销

split Savonius wind turbine 开裂式 S 型风力机

splitter plate 分割（稳定）板

spoiler 扰流器；扰流板

spoiler brake 阻流减速板；扰流板刹车

spoiling 损坏；破坏

spoiling flap 阻尼板（随风速的变化用来阻止风轮转数增加的构件）

spontaneous oscillation 自发振动

spool gear 长齿轮

spot corrosion 点腐蚀

spot repair 现场修理

spot weld 焊点

spot welding 点焊

spread 传播；伸展

spread of stall (area) 失速（气流分离）区的扩展

spread of wing 翼展

spring 弹簧

spring abutment 弹簧支座

spring buffer 弹簧缓冲器

spring coupler damper 弹簧防震器

spring deflection 弹簧挠度

spring loaded activator 弹簧加载执行机构

spring loaded brake 加载弹簧制动器；弹簧闸；弹簧制动装置

spring loaded centrifugal latch 加载弹簧离心闩锁

spring stiffness 弹簧刚度

spring viscous damper 弹簧黏滞阻尼器

sprocket 链轮齿；扣链齿；扣链齿轮

sprocket belt 链轮齿皮带

sprocket belt pitch drive 链轮齿皮带调桨驱动

sprung mass 悬挂质量；簧载质量

spun glass 玻璃纤维

spur gear 正齿轮；直齿圆柱齿轮

spur gearing 正齿轮传动装置

spur wheel 正齿轮

squall 飑

square 平方；方块；正方形的；四方的

square building 方形建筑

square cube law 平方立方原则
square head bolt 方头螺栓
square measure 面积
square nut 四方螺母；螺帽
square thread 矩形螺纹
square vortex ring 方形涡环
square wave 方波
square wind tunnel 方形试验段风洞
squared off tip 方形翼尖
squirrel cage 鼠笼；鼠笼式的
squirrel cage induction generator (SCIG) 鼠笼异步发电机
squirrel cage rotor 鼠笼式转子
squirrel cage winding 鼠笼式绕组
SSB (single side band) 单边带
SS (suspended solid) 悬浮固体
SSC (static series compensators) 静态串联补偿器
SSR (subsynchronous resonance) 次同步谐振
ST. (stratus) 层云
ST.-CU. (stratocumulus) 层积云
stability 稳定性；稳定度
stability against tipping 抗倾翻稳定性
stability boundary 稳定性边界
stability category 稳定性类别
stability class 稳定度等级
stability coefficient 温度系数
stability criterion 稳定性判据
stability derivative 稳定性导数
stability index 稳定度指数
stability of atmosphere 大气温度度
stability of following wind 调向稳定性
stability parameter 稳定性参数
stability range 稳定范围
stability test 机械稳定性试验

stabilivolt 稳压器；稳压管
stabilization 稳定作用
stabilization factor 稳定因素
stabilization network 稳定网络
stabilization of super conducting magnet 超导磁体稳定化
stabilizator 稳压器
stabilized 稳定的
stabilized operating temperature 稳定工作温度
stabilized plume height 羽流稳定高度
stabilizer 稳定器
stabilizing fin 稳定鳍；稳定翼板
stabilizing transformer 稳定变压器
stable 稳定的
stable air 稳定空气
stable air mass 稳定气团
stable atmosphere 稳定大气
stable condition 温度工况
stable density stratification 稳定密度层结
stable equilibrium 稳定平衡
stable lapse rate 稳定递减率
stable operation 稳定运行；稳定操作
stable oversteer 稳定过渡转向
stable stratification 稳定层结
stable stratified flow 稳定分层流
stable tunnel speed 风洞稳定风速
stack dilution factor 烟囱稀释因子
stack draft 烟囱排烟；烟囱通风
stack effect 烟囱效应；抽吸效应
stack effluent 烟囱排放物
stack emission 烟囱排放
stack exit velocity 烟囱出口速度
stack factor 烟囱因素
stack gas 烟道气

stack guy 烟囱风缆
stack height 烟囱高度
stack loss 烟囱损失
stack induced downwash 烟囱诱导下洗
stack nozzle 烟囱口
stack outlet 烟囱出烟孔
stack parameter 烟囱参数
stack plume 烟羽
stack velocity 出烟速度
stack vent 烟囱排烟道
stacked Darrieus rotor 重叠式达里厄转子
stadium 体育场
stagger angle 叶片扭角
staggered 错列的；交错的
staggered vortex 交错旋涡
staggered vortex street 交替涡街；交错涡街
staging area 集合场；分级区；集结待命地区
stagnant 停滞的；不景气的
stagnant air 停滞空气
stagnant area 滞流区
stagnant inversion 停滞逆温
stagnant layer 滞止层
stagnant wake 静区；死区；滞流区
stagnant water 死水
stagnated air mass 停滞气团
stagnating tunnel flow 风洞滞止气流
stagnation 滞止；停滞
stagnation area 滞止区
stagnation boundary layer 滞止边界层
stagnation flow 滞止流动
stagnation line 滞止线
stagnation point 驻点，临界点
stagnation pressure 滞止压力；驻点压力；临界压力
stagnation properties 滞止参数

stagnation state 滞止状态；停滞状态
stagnation streamline 滞止流线
stagnation surface 驻面
stagnation temperature 滞止温度；驻点温度；临界温度
stagnation value 滞止值
stagnation zone 滞止区
staining 锈蚀；污染
stainless steel 不锈钢
stall 失速
stall angle 失速角
stall angle of attack 失速攻角
stall condition 零速工况（转速比为零时的工况）；失速状态
stall control 失速控制
stall flutter 失速颤振
stall hub 失速轮毂
stall hysteresis excitation 失速迟滞激励
stall line 失速线
stall margin 失速裕度
stall performance 失速性能
stall point 失速点
stall profile blade 失速型叶片
stall regulated 失速调节
stall regulated rotor 失速式限速风轮
stall regulated wind turbine 失速调节风电机组
stall regulation 失速式限速
stall strip 失速条
stalled area 失速区
stalled blade 失速叶
stalled flow 失速气流
stalling action 失速行为
stalling angle of attack 失速迎角；临界迎角
stalling incidence 失速迎角

stalling moment 失速力矩
stalling speed 失速速度
stalling test 失速试验
stallout 失速；气流分离
stamping 冲压
stamping parts 冲压机
stand alone 独立运行
standard 标准；规范；准则
standard air 标准大气
standard air pressure 标准大气压
standard atmosphere 标准大气
standard atmosphere pressure 标准大气压
standard atmospheric state 标准大气状态
standard block 标准试件；标准块
standard capacity 标准容量
standard component 标准部件
standard curve 标准曲线
standard density altitude 标准密度高度
standard design 标准设计
standard deviation 标准差
standard dimension 标准尺寸
standard duty 标准负荷；标准功率
standard engineering practice (SEP) 标准工程惯例
standard equation 标准方程
standard error 标准误差
standard insulation class 标准绝缘等级
standard international unit 国际标准单位制
standard internet web browser 标准互联网浏览器
standard laboratory atmosphere 标准实验室（大）气压
standard laboratory temperature 标准实验室温度
standard layout 标准配置
standard load 标准负荷
standard mass 标准质量
standard mean chord 标准平均弦（等于机翼总面积除以全展长）；平均几何弦
standard meter 标准仪表
standard of USA 美国标准
standard part 标准（零）件
standard power 标准功率
standard rated output 标准额定功率；标准额定出力
standard rated power 标准额定功率
standard specification 标准技术规范；标准规范
standard state 标准状态
standard thyristor 标准晶闸管
standard uncertainty 标准误差
standard unit 标准单位
standard value 标准值
standard velocity 标准速度
standard visual range 标准视程
standard weight 标准重量
standard wire ga(u)ge 标准线规
standardized wind speed 标准风速
standby 备用品；待机
stand-by 后备；备用设备；准备
stand-by battery 备用电池
stand-by equipment 备用设备
stand-by plant 备用装置
standby source of power 备用动力源
standby time 待命时间
standing 长期的；直立的
standing balance 静平衡
standing bubble 驻泡
standing vortex 驻涡
standing wave 驻波

standing wave ratio 驻波系数；驻波比
standstill 静止；停机
Stanton number 斯坦顿数
Stanton tube 斯坦顿管
stator winding 定子绕组
star connection 星形连接
star coupling 万向联轴节
star point 星点
star type cable 星型电缆
star-delta connection 星形—三角形连接
star-delta switching 星形—三角形切换
star-delta switching starter 星形—三角形切换启动器
start up wind speed 启动风速
starting and ending time 起止时间
starting current 起动电流
starting device 起动装置
starting duty 起动功率
starting load 起动负荷
starting motor 起动电动机
starting performance test 起动性能试验
starting plume 初始羽流
starting power 起动动力；起动功率
starting signal 起动信号
starting test 起动试验
starting time 起动时间
starting torque 起动力矩
starting torque coefficient 起动力矩系数
starting up equipment 起动设备
starting vortex 起动涡
starting wind speed 起动风速
start-up 起动
start-up device 起动装置
start-up wind speed 起动风速
start-up torque 起动力矩；起动转矩

STATCOM (static synchronous compensator) 静止同步补偿器
state 状态；态；状况
state balance 状态平衡
state derivative 状态导数
state diagram 状态图
state equation 状态方程
state function 状态函数
state information 状态信息
state of alarm 报警状态
state of control 控制状态
state of runtime machine 机器运行状态
state of starting operation 起动工况
state parameter 状态参数
state space 状态空间
state vector 态矢量
states superior limit 上限
static 静态的
static accuracy 静态偏差；稳态精度
static aerodynamics 定常流空气动力学
static aeroelasticity 静气动弹性力学
static aerothermoelastic behaviour 空气热弹性静力特性
static air pressure 空气静压
static balance 静平衡；静力天平
static calibration 静态校准
static characteristic 静态特性
static condition 静态条件
static control 静态控制
static conversion 静态变换
static deflection 静挠度
static derivative 静导数
static (force) divergent 静（力）发散
static dynamic pressure technique 静动压技术

static efficiency 静态效率
static equilibrium 静（态力）平衡
static fatigue 静力疲劳
static friction 静摩擦
static hardness test 布氏硬度试验
static head 静压头；静压传感器
static hole 静压测量孔
static indentation test 球硬度试验静凹痕试验
static lift 静升力
static load 静载（负）荷；恒载
static loading 静态载荷
static model 静态模型
static of fluids 流体静力学
static performance 静态特性
static pressure 静压
static pressure coefficient 静压系数
static pressure compensation 静压力补偿
static pressure compensation joint 静态压力补偿接头
static pressure correction vent 静态压力校正孔
static pressure difference 静压差
static pressure distribution 静压分布
static pressure error 静压误差
static pressure gradient 静压力梯度
static pressure hole 静压孔
static pressure loss 静压损失
static pressure orifice 静压测量孔
static pressure pitot tube 静压皮托管
static pressure ratio 静压比
static pressure tap 静压孔
static pressure vent hole 静压力测定孔
static series compensators (SSC) 静态串联补偿器
static slot （风洞）静压补偿缝；静压缝口

static stability 静态稳定性
static stiffness 静态刚度
static switch 静态开关
static synchronous compensator (STATCOM) 静止同步补偿器
static test 静力试验
static transfer switch (STS) 静态开关；静态转换开关
static tube 静压管
static unbalance 静不平衡
static var compensator (SVC) 静止无功补偿器
static variable 静态变量
static voltage stability margin (SVSM) 静态电压稳定裕度
static wind load 静态风载
statical head 静压头；静压传感器
statical load 静载
statical model 静态模型
statics 静力学
statics of fluid 流体静力学
stationary 固定的；静止的
stationary blade 固定叶片
stationary condition 固定条件；恒定条件
stationary current 稳定流
stationary equilibrium 稳态均衡
stationary field 恒定场
stationary flow 稳态流
stationary flow field 平稳流场
stationary installation 固定装置
stationary instrument 固定仪表
stationary layer 平稳层
stationary load 固定负荷
stationary motion 常定运动
stationary optical fibre controller 静态光导纤维控制器

stationary process 平稳过程
stationary random load 平稳随机荷载
stationary state 平稳状态
stationary wave ratio 驻波比
statistical 统计的；统计学的
statistical approach 统计方法
statistical correlation 统计相关
statistical data 统计数据
statistical dependence 统计相关
statistical discrete gust method 孤立阵风统计方法
statistical distribution 统计分布
statistical factor 统计因子
statistical fluctuation 统计涨落
statistical mechanics 统计力学
statistical processing 统计（数据）处理
statistical rule 统计规律；统计相似率
statistical theory 统计理论
statistical weight factor 统计权重因数
stator 定子
stator blade 导向器叶片；静子叶片；定子叶片；固定叶片
stator coil overhang 定子线圈悬
stator current 定子电流
stator flux 定子磁通
stator inductance 定子电感
stator leakage reactance 定子漏抗；定子漏阻
stator permeance 定子磁导
stator resistance 定子电阻
stator slot 定子槽
stator tooth 定子齿
stator turns 定子匝数
stator voltage 定子电压
stator voltage phasor diagram 定子电压相量图

stator winding 定子绕组；定子线圈
stator yoke 定子磁轭；定子轭
staubosphere 尘圈
stay 支撑；拉线
stay bolt 支撑螺栓
stay cable 斜拉索
steadiness 稳定度
steady 稳定的
steady acceleration 等加速度
steady condition 稳定工况
steady direct current 恒稳直流电
steady emission rate 定常排放率
steady equation 稳态方程
steady flow 稳流
steady flow aerodynamics 定常流空气动力学
steady flow field 定常流场
steady flow system 稳定流系统
steady galloping response 定常驰振响应
steady gas dynamics 定常流气体动力学
steady load 静载荷
steady operation 稳定运行
steady rain 连绵雨
steady rate 定常速率
steady running 稳定运转
steady running condition 稳定运转工况
steady solution 定常解
steady speed 稳定速度；稳定转速
steady state 稳态
steady state behavior 稳态工况性能
steady state boundary condition 稳态边界条件
steady state characteristics 稳态特性（曲线）
steady state compressible flow 稳态可压缩流
steady state condition 稳态条件
steady state constant 稳态常数

steady state flow 稳态流
steady state incompressible flow 稳定不可压缩流
steady state performance 稳定工况性能；稳态特性
steady state plume 定常羽流
steady state process 稳态过程
steady state response 定常响应
steady state speed 稳态速率
steady stress 静应力
steady unsaturated flow 非饱和稳定流
steady vibration 定常振动
steady wind 定常风
steady wind load 定常风载
steady(-state) oscillations 稳态振荡
steam accumulator 蒸汽蓄热器；蒸汽蓄；蒸汽蓄力器
steam condenser 蒸汽冷凝器
steel 钢
steel channel 槽钢
steel conical tubular tower 锥形钢筒塔架
steel flats 扁钢
steel I-beam 工字钢
steel pit 钢表面凹坑
steel sheet 薄钢板
steel spike 钢针
steep 陡峭的；急剧升降的
steep descending gradient 陡下坡道
steep down grade 陡下坡
steep grade 陡坡
steep slope 陡坡
steepen （使）变得陡峭
steeply 险峻地；大坡度地
steering 操纵；转向器
steering angle 转向角

steering box 转向机构箱；操舵箱
steering computer 操纵系统计算机
steering current 引导气流
steering gear 转向机构；转向装置；轮向齿轮
steering resistance 转向阻力
steering response 操纵灵敏性；转向响应
stellate snow flake 星状雪花
step 步
step by step control 进步式控制
step by step procedure 逐步法
step by step switch 进步式开关
step by step variable gear 多级变速齿轮
step change 阶跃变化
step control 分步控制
step controller 分级控制器
step function 阶跃函数
step motor 进步式电动机
step up gear 增速传动装置
step up gearbox 增速齿轮箱
step up ratio 增速比；升压比率
step up transformer 升压变压器
step voltage 跨步电压；阶跃电压
stepless speed change device 无级变速装置
stepper motor 步进电机
stepwise cut-out control 逐步切出控制；分段断路器控制
stiction 静摩擦
stiff 坚硬的
stiff building 刚性建筑
stiff design 硬件设计
stiff driven train 硬传动系统
stiff tower 刚性塔架
stiff transmission system 硬传动系统
stiffened sheet 加强板
stiffening girder 加强梁

stiffness 刚性
stiffness driven oscillation 刚度驱动振动
stiffness factor 刚度系数
stiffness matrix 刚度矩阵
still 仍然；更；静止地
still air 静止空气
still water region 死水区
stimulator 激流器，激励装置
sting 尾撑；支杆
sting balance 尾撑天平
sting mounted model 尾撑模型
stirring 活跃的；搅拌；摇动
stirring motion 湍（涡）流；紊流（漩涡）运动
stitch bonding 自动点焊；跳焊；针脚式接合
stitch weld 缝焊
stochastic approach 随机方法
stochastic loading 随机载荷
stochastic nature 随机性
stochastic process 随机过程
stochastic stability boundary 随机稳定边界
stochastic vibration 随机振动
stochastic wind load 随机风载
stock 坯料；螺旋纹板
stockbridge damper 防震锤
Stokes flow 斯托克斯流
Stokes formula 斯托克斯公式
Stokes regime 斯托克斯流态
Stokes stream function 斯托克斯流函数
Stokes wave 斯托克斯波
stone quarry 采石场
stop 停机，停止，阻止；制动器
stop and go valve 起停阀
stop button 停止按钮
stop sign 停止信号
stop wind speed 停止风速

storage 贮藏；保管；贮藏库
storage battery 蓄电池
storage cell 蓄电池
storage condition 贮存条件
storage element 蓄电池
stored energy 储能
storm 暴(风)雨；(十级)狂风(24.5～28.4米/秒)
storm duration 风暴期；暴雨历时
storm front 风暴锋
storm gale 暴风(十一级风，风速28.5～32.6米/秒)
storm intensity 风暴强度
storm surge 风暴大浪；风暴潮；风暴汹涌
storm's wake 风暴尾迹
story （楼）层
straight 直的；连续的
straight bevel gear 直齿锥齿轮
straight through tube 直流式风洞（无回路的）
straight through wind tunnel 直流式风洞；开路式风洞
straightener （风洞）整流段；（风洞风扇）止旋片
strain 应变
strain aged embrittlement 应变时效脆性
strain clamp 耐张线夹；耐拉线夹
strain concentration 应变集中
strain energy 应变能
strain force 变形力
strain gauge 应变计；应变规，应变仪；拉力计
strain gauge transducer 应变传感器
strain hardenability 金属材料的应变硬化性；受力变硬性
strain hardening 形变硬化；冷加工硬化；应变硬化
strain hardening coefficient 应变硬化系数

strain hardening exponent 应变硬化指数
strain increment 应变增量
strain relief 应变消余；应变消除
strain relief tempering 消除应力回火
strain response 应变响应
strain tensor 应变张量
strake 条纹；箍条
stranded cable 股铰缆索
stranded conductor 绞线
strands of cotton yarn 面纱簇
strap 皮带；垫片
stratification 分层结构；层理；成层
stratification of wind 风层
stratified 分层的
stratified atmosphere turbulence 分层大气湍流
stratified flow 分层流
stratified fluid 分层流体
stratified inner layer 分层内层
stratified shear flow 分层剪切流
stratocumulus (SC) 层积云
stratocumulus castellanus (SC cas) 堡状层积云
stratocumulus cumulogenitus (SC cug) 积云性层积云
stratocumulus opacus (SC op) 遮光层积云
stratocumulus radiatus (SC ra) 辐状层积云
stratocumulus translucidus (SC tr) 透光层积云
stratopause 平流层顶
stratosphere 平流层
stratospheric 平流层
stratospheric cloud 平流层云
stratospheric coupling 平流层耦合作用
stratospheric fallout 平流层落尘
stratospheric ozone 平流层臭氧
stratospheric polar vortex 平流层极涡
stratospheric pollution 平流层污染
stray capacitance 寄生电容
stray current 杂散电流；涡流；漏泄电流
streak line 流动条纹线；脉线；染色线
stream 气流；溪流
stream angle 气流(偏)角；流线角
stream condition 气流状态
stream filament 流丝
stream form 流型
stream function 流函数
stream line 流(通量)线
stream over airfoil 翼型绕流
stream pattern 流线谱
stream surface 流面
stream swirl 涡流；气流涡流
stream temperature 气流温度
stream tube 流管
stream turbulence 气流湍流(度)
stream velocity 流速
streamline 流线；流线型的
streamline angle 流线偏角
streamline body 流线体
streamline coincident 流线重合
streamline contour 流线型外廓
streamline curvature effect 流线弯曲效应
streamline diagram 流线图
streamline fairing 流线型罩
streamline flow 流线型流
streamline flow resistance 流线型流动阻力
streamline form 流线型
streamline of motion 流线型运动
streamline pattern 流线型
streamline profile 流线型轮廓

streamline section 流线型截面
streamline shape 流线型
streamline squeezing effect 流线汇集效应
streamlined 流线型的
streamlined blade 流线型叶片
streamlined body 流线体
streamlined nosing 流线型头部整流罩
streamlining effect 流线效果
streamwise component 流向分量
streamwise pressure gradient 流向压力梯度
streamwise vorticity 流向涡量
strength 强度
strength safety coefficient 强度安全系数
stress accommodation 拉紧装置；张紧夹具
stress carring(stressed) covering 应力蒙皮
stress concentration 应力集中
stress-cycle diagram 应力循环图
stress deprivation 应力消失
stress field 应力场
street level wind 街道风
stress relieved 应力消失
stress relieving 应力消除
stress sheet 应力表
stress strain diagram 应力应变图
stress tensor 应力张量
stressed covering 受力蒙皮
stressed part 承力部件
stressed skin 应力蒙皮
stressed skin structure 承力蒙皮结构
strip chart recorder 条纸记录器；纸带记录器
strip theory 切片理论
strong 强烈的
strong breeze 强风（六级风）
strong current control 强电控制
strong gale 烈风（九级风，风速 20.8～24.4米/秒）
strong grid 强电网
strong wind 强风
strong wind boundary layer 强风边界层
strong wind fumigation plume 强风下垂型羽流；强风下沉型烟羽
Strouhal frequency 斯特哈鲁频率
Strouhal number 斯特哈鲁数
structural 结构上的；建筑的
structural (structure) part 结构件
structural attachment 结构附件
structural buckling 结构屈曲
structural build-up 结构累积
structural damping 结构阻尼
structural damping characteristic 结构阻尼特性
structural density 结构密度
structural dynamic similarity 结构动力相似
structural dynamics 结构动力学
structural element 结构部件
structural flexibility 结构柔度
structural frame 结构框架
structural integrity 结构完整程度
structural mode 结构模态；结构模式
structural modulus 结构模量
structural of turbulence 湍流结构
structural oscillation 结构振动
structural reliability 结构可靠性
structural response 结构响应
structural rigidity 结构刚度
structural stiffness 结构刚度
structural strengthening 结构加强
structural transfer function 结构传递函数
structural viscous damping 结构黏性阻尼
structure 结构物

structure analogue 结构模拟
structure chord 结构弦
structure damage 结构损坏
structure diagram 结构图
structure failure 结构损坏
structure flexibility 结构挠性；结构弹性
structure of aggregation 聚集态结构
structure of soil mass 土体结构
structure of turbulence 湍流结构
structure oscillation 结构振动
strut 支杆
strut support （风洞）支架；支柱架
STS (static transfer switch) 静态开关；静态转换开关
stud welding 螺栓焊接
stuffing box 填充体；填料盒
styling 造型；样式
subadiabatic atmosphere 亚绝热大气
subcritical flow regime 亚临界流动状态
subcritical hardening 低温硬化；亚临界温度淬火
subcritical Reynolds number 亚临界雷诺数
subcritical vortex shedding 亚临界旋涡脱落
subcritical wind speed 亚临界风速
subgrid-scale motion 次网格尺度运动
subharmonics 次谐波；亚谐波
sublimation cooling 升华冷却
submegawatt wind turbine 次兆瓦级的风电机组
submersible transformer 地下式变压器
submicron aerosol 亚微米气溶胶
subpolar low-pressure belt 副极地低压带
subsidence 沉降；下沉
subsidence inversion 下沉逆温
subsidiary equation 辅助方程
subsidiary spar 辅助（翼）梁
subsonic airfoil 亚音速机翼
subsonic wind tunnel 亚音速风洞
subspan 次跨
subspan galloping 次跨驰振
subspan oscillation 次跨振动
subspan wake-induced galloping 尾流诱导次跨驰振
subsynchronous resonance (SSR) 同步谐振
subsynchronous speed 次同步速率
subsynoptic scale 次综观尺度
subsynoptic scale weather system 次天气尺度系统
subsystem 子系统
suburb 市郊
suburban terrain 郊区地形
suburban wind exposure 郊区地貌风开敞度
suck down wind tunnel 下吸式风洞
suction 吸力；负压；吸气
suction anemometer 吸管式风速计
suction coefficient 负压系数；吸力系数
suction fan 吸风机
suction flow 吸入流
suction load 吸力荷载
suction peak 吸力峰；吸力峰值
suction side 吸力侧
suction type airfoil 吸入式机翼
suction vortex 吸入旋涡
summing circuit 总和线路；反馈系统中的比较环节
summit 山顶；顶点
sun 太阳
sun axle 中心轴
sun gear 太阳齿轮；中心齿轮
sun visor 遮阳板

sun-and-planet gear 行星齿轮
sun-and-planet gearing 行星齿轮传动装置
sunblind 百叶窗
sunlight 日光
sunlight wind machine 日照型风力机
sun's meridian altitude 太阳中天高度
sunshine duration 日照持续时间
sunshine hour 日照小时数
super 特级的；极好的
super pressure balloon 过压气球
superadiabatic atmosphere 超绝热大气
superadiabatic lapse rate 超绝热递减率
superadiabatic layer 超绝热层
superadiabatic state 超绝热状态
superconducting magnetic energy storage system 超导磁贮能装置；超导磁贮能系统
superconductive cable 超导电缆
superconductor 超导体
supercritical accident 超临界事故
supercritical flow regime 超临界流动状态
supercritical Reynolds number 超临界雷诺数
supercritical tower 硬塔架；刚性塔架
supercritical vortex shedding 超临界旋涡脱落
superelevation 超高
superelevation ramp 超高的坡道
superelevation slope 超高顺坡
supergeostrophic wind 超地转风
supergradient wind 超梯度风
superior air 高空下降空气
superposition 叠加
superposition method 叠加法
superposition of directive pattern 方向图叠加
superposition of flow patterns 流型叠加
superposition of vortex 旋涡叠加

superposition theorem 叠加定理
supersaturation 过饱和(度)
supersonic aerodynamic 超音速空气动力学
supersonic contoured nozzle 超音速型面喷嘴
superstructure 上层建筑
supersynchronous speed 超同步转速
supervision 检测；观察；监控
supervisory control and data acquisition (SCADA) 数据采集与监视控制系统
supplementary power 补充电力
support 支持；支撑
support interference 支架干扰
support interference correction 支架干扰修正
support structure (for wind turbines) （风电机组）支撑结构
support tower 支撑塔
support wire 支撑张线
supporting 支持的；辅助性的；次要的
supporting strut 支杆；支柱
suppressor grid 遏止栅极；抑制栅极；遏止栅
surface action 集肤作用；表面作用
surface air 地面空气；大气底层
surface blowhole 表面气孔
surface boundary layer 表面边界层；近地边界层
surface carburization 表面渗碳
surface colour 表面色
surface concentration 地面浓度；表面浓度
surface conditioner 底材表面处理剂
surface conduct 表面接触
surface configuration 地表形态
surface contour map 等高线地形图
surface coolant 表面冷却剂
surface cooler 表面冷却器
surface cooling 表面冷却

surface coordinate 面坐标；曲面坐标
surface creep 表面蠕动
surface crust 地表层；地表硬壳
surface curve 表面曲线
surface damage 表面损伤；机械损伤
surface dent 表面凹坑；表面凹痕
surface depletion model 表面耗减模式
surface deposit 表层堆积物；表层沉积
surface detention 地面阻滞
surface drag coefficient 表面阻力系数
surface erodibility 地面可蚀性；表面侵蚀度
surface finish 表面修整
surface flow visualization 表面流动显示
surface friction 表面摩擦力；地面摩擦力
surface friction velocity 地面摩擦速度；表面摩擦速度
surface geometry 地表几何形状；表面几何图形
surface geostrophic wind 地表地转风
surface gloss 表面光泽
surface gravity wave 表面重力波
surface hardening 表面硬化
surface heat flux 表面热通量
surface heat-transfer rate 表面热传导率
surface integrated pressure 表面积分压力
surface inversion 地面大气逆温
surface inversion layer 地面逆温层
surface irregularity 表面不平性
surface layer 地表层；近地层
surface of contact 接触面
surface oil film 表面油膜
surface preparation 表面预加工
surface pressure 表面压力
surface pressure coefficient 表面压力系数
surface pressure fluctuation 表面压力脉动
surface pressure pattern 表面压力分布图
surface quality 表面质量
surface rating 表面光洁度等级
surface resistance 表面阻力
surface ripple 地表波痕；表面波纹
surface roughness 地面粗糙度；表面粗糙度
surface roughness element 地面粗糙元；表面粗糙元
surface shear stress 表面剪应力
surface source 面源
surface streamline 表面流线
surface stress 表面应力
surface stress layer 表面应力层
surface sublayer 地表副层
surface temperature 表面温度
surface tension 表面张力
surface tension balance 表面张力计
surface tuft method 表面丝线法
surface turbulence 地面湍流
surface velocity 表面速度
surface wind 地面风；近地风
surface wind speed 表面风速
surge 浪涌；汹涌；巨涌
surge arrester 电涌放电器；过电压吸收器；避雷器
surge current 浪涌电流
surge diverter 避雷针
surge gap 脉冲放电器
surge line 喘振线；海边波涛线
surge protection 浪涌保护
surge suppressor 电涌抑制器；涌波抑制器
surrounding 周围的；附近的
surrounding building 周围建筑
surrounding wind 周围风
surveillance 监视；监测

survival wind speed 安全风速
suspended ash 悬浮灰分
suspended dust 悬浮尘埃
suspended particulate 悬浮粒子
suspended solid (SS) 悬浮固体
suspended structure 悬挂式结构物
suspender arm load 吊杆臂负荷
suspension 悬浮
suspension clamp 悬垂线夹
suspension load 悬浮荷重
suspension of snow particle 雪粒子悬浮
suspension ratio 悬浮比
sustained 持续的；持久的；持久不变的
sustained fault 持续故障
sustained load 持续载荷
sustained oscillations 持续振荡
sustained stress 持续应力
sustained wind 持续风
Sutton diffusion parameter 苏通氏扩散参数
Sutton formula 苏通氏（大气污染）公式
Suydam criterion 赛达姆判据
SVC (static var compensator) 静止无功补偿器
SVSM (static voltage stability margin) 静态电压稳定裕度
sway 摆动
sway acceleration 摆动加速度
sway angle 摆动角
sway response 摆动响应
sway vibration 摆振
sweat cooling 蒸发冷却
SWECS (small scale wind energy conversion system) 小型风能转换系统
sweep 扫描
sweep angle （桨叶）后掠角
sweep oscillator 扫描振荡器

sweep outline 流线型轮廓
swept area 扫略面积
swept area of rotor 风轮扫掠面积
swift 急流；涡流；快的；快速的
swinging 摆动；摇摆
swirl flow 旋流
swirling 旋动；打旋；旋涡
swirling ascending 旋动上升
swirling flow 旋流
swirling vortex 旋涡
swirling wind 旋风
switch 开关；转换；切换
switch board 配电板
switch box 开关柜
switch control 开关控制
switch desk 控制台
switch frequency 切换频率；开关频率
switch in 接通
switch off 切断
switch on 接通
switch over 转变；切换
switch panel 控制屏
switchboard 配电板
switched capacitor 开关电容
switches & buttons 开关及按钮
switchgear 开关设备；开关装置
switching 通断；切换；开关；转换；整流；配电
switching current 开关电流；合闸电流
switching flow rate 变换流动速率
switching frequency 开关频率
switching operation 切换运行
switching performance 开闭性能；开关性能
switching sequence 开关顺序
switching system 转换系统
switching time 切换时间；开关时间

switchyard 开关站；户外配电装置
swiveling 回转；旋转的
swivelling vane 回转叶片
symmetrical 均匀的；对称的
symmetrical airfoil 对称翼型
symmetrical body 对称体
symmetrical expression 对称式
symmetrical matrix 对称矩阵
symmetrical mode 对称模态
symmetrical section 对称状分段
synchromotor 同步电动机
synchronism 同步；同步性
synchronization effect 同步效应
synchronized oscillations 同步振动
synchronizing coefficient 同步系数
synchronous 同步的
synchronous condenser 同步调相机
synchronous condenser operation 同步调相运行
synchronous generator 同步电机
synchronous inverter 同步逆变器
synchronous machine 同步电机
synchronous reactance 同步电抗
synchronous rolling 同步横摇
synchronous speed 同步转速
synchroscope 同步指示器
synoptic 天气的
synoptic correlation 天气相关
synoptic map 天气图
synoptic model 天气模型
synoptic report 天气报告
synoptic scale 天气尺度
synoptic weather chart 天气图
synthetic 合成的
synthetic fiber 合成纤维

synthetic gas stream 综合气流
synthetic pollution data 综合污染指数
synthetic resin 合成树脂
system 系统
system analysis 系统分析
system damping 系统阻尼
system demonstration 系统论证
system design 系统设计
system deviation 系统偏差
system dynamics 系统动态特性
system efficiency 系统效率
system engineering 系统工程
system error 系统误差
system failure 系统故障
system identification 系统识别
system international of units (SI) 国际单位制
system main power 电源主动力系统
system mistake 系统错误
system of unit 单位制
system optimization 系统优化
system program 系统程序
system response 系统响应
system selection 系统选择
system simulation 系统模拟
system software 系统软件
system start-up 系统起动
system test time 系统测试时间
system with effectively earthed neutral 中性点有效接地系统
system with non-effectively earthed neutral 中性点非有效接地系统

现代英汉风力发电工程

A Modern English-Chinese Dictionary of Wind Power Engineering

T

T & D (transmission and distribution) 输变电及配电
T & D equipment (transmission & distribution equipment) 输配电设备
TA (tangent angle) 正切角
tachograph 转速记录仪；转速记录图
tachometer 转速表
tack 抢风调向
tack weld 临时点焊；间断焊；预焊
tack welding 平头焊接
tagged element 标记元素
tail 尾巴
tail wheel 尾轮（尾舵上的多叶片风轮）
tail wind (TW) 顺风
tail wind component 风的顺风分量
tailplane spar 水平尾翼翼梁
take moment about a 对a点取力矩
take off valve 输出阀
tall 高的
tall building 高层建筑
tall capsule building 高层盒式建筑
tall frame shear wall structure 高层框架剪力墙结构
tall frame structure 高耸框架结构
tall shear wall structure 高层剪力墙结构
tall slender structure 细高结构
tall suspension building 高层悬挂建筑
tall tube building 高层筒体建筑
tall tube-in-tube building 高层筒中筒建筑；高管套管式建筑
tangent angle (TA) 正切角
tangent slope 切线斜率
tangential 切线的；正切的
tangential deformation 切向变形
tangential flow 切向流
tangential force 切向力
tangential velocity 切线速度
tangential wind 切向风
tank oil cooler 罐油冷却器
tank test 水槽试验
tap change transformer 分接头切换变压器
tap changer （变压器）抽头转换器；抽头开关；调压开关
tap changer control 分接头控制
tap changing switch 分接头切换开关
tap position information 分接头位置信息
tape recorder 磁带记录器
tape sampler 纸带取样器
tape winding 布带缠绕
taper angle 尖削角；锥角

taper ratio 锥度比
tapered 锥形的
tapered aerofoil 不等弦翼面
tapered blade 楔形叶片
tapered sheet covering 锥形板蒙皮
tapered structure 锥形结构
tapping 分接；开孔
tapping hole 螺纹孔；螺丝孔
tare 皮重；(风洞)支架阻力皮重
tare weight 皮重
target 目标
target flow characteristics （模拟）目标气流特性
task 任务；工作
task performance 工作特性；工作效能
taut 拉紧的；紧张的
taut strip model 拉条模型；紧带模型
taut wire 紧拉钢线；张线
taut wire traverse 张绳
tautline position-reference system 紧绳监测定位标准系统
taut line system 张紧绳装置
tautness meter 拉力计
Taylor's statistical theory 泰勒统计理论
T-connector T形线夹
TC (time constant) 时间常数
T/C ratio (thickness chord ratio) 翼型厚弦比
TCR (thyristor controlled reactor) 晶闸管控制电抗器
tear and wear 磨损；损坏
tear and wear allowance 磨损留量；允许磨耗
technic 术语；专门技术；工艺
technic economic analysis 技术经济分析
technical 技术的，工艺的，学术的，专业的，技术术语
technical analysis 技术分析
technical application 技术应用
technical atmosphere 工业大气压
technical characteristic 技术特性
technical code 技术规范
technical condition 技术条件
technical control 工艺控制
technical control board 技术控制板
technical data 技术数据
technical design 技术设计
technical economical comparison 技术经济比较
technical economical index 技术经济指标
technical evaluation criteria 技术评价标准
technical failure 技术故障
technical feasibility 技术可行性
technical handbook 技术手册
technical information 技术信息
technical know-how 技术知识；技术诀窍
technical lifetime 技术寿命
technical load 工艺负荷；技术负荷
technical maintenance 技术保养
technical open-air climate 技术室外气候
technical parameter 技术参数
technical performance 技术性能
technical reference 技术参考文献
technical regulation 技术规程
technical report 技术报告
technical requirement 技术要求
technical specification 技术说明
technical standard 技术标准
technical term 术语；专门名词
technical training 技术培训
technicality 技术性；技术细节

technicalization 技术专门化；技术化
technique 技术
technological 技术的；工艺的
technological characteristics 工艺特性
technological conditions 工艺条件
technological equipment 工艺设备
technological parameter 工艺参数
technological process 工艺过程
technology 技术
technology assessment 技术评价；评估
technology transformation 技术转移
technostructure 技术专家控制体制；技术专家体制
tee T形；三通管；丁字接头
tee bend T形接头；三通
tee branch 三通管
tee connection T形接头；三通
tee fitting 三通管；T形管；三通接头
tee girder T形梁
tee iron T字钢
tee joint T形接头；三通
tee junction T形接头；三通
tee pipe T形管；丁形管
tee tube T形管
teeter 摇摆器；跷板式结构
teeter angle 摇摆角
teeter bearing 摇摆轴承
teeter brake 跷板式制动器
teeter bumper 摇摆防撞器
teeter hinge 跷板铰链
teeter restraint 摇摆限制器；摇摆阻尼器
teeter restraint model 跷跷板控制模型
teetering 跷跷板
teetering angle 跷板角；摇摆角
teetering bearing type 摇摆轴承类型
teetering hinge 摇摆铰链
teetering hub 摇摆轮毂；跷板式桨毂
teetering rotor 跷跷板风轮；摇摆风轮
telecommunication 远程通信
telecommunication cable 远程通信电缆
telecontrol 遥控；远距离控制
telemonitor 遥控
telemonitoring 远程监视
telephone kiosk 电话亭；电话间
telequipment 遥控装置
television mast 电视塔；天线杆
television tower 电视塔
telltale 信号装置；指示器
temperate 温和的
temperate climate 温带气候
temperate climate zone 温带气候带
temperate discontinuity 温度不连续性
temperate field 温度场
temperate gradient 温度梯度
temperate inversion 逆温
temperate lapse rate 温度递减率
temperate latitude 中纬度
temperate low belt 温带低压带
temperate rainy climate 温带多雨气候
temperate zone 温带
temperature 温度
temperature absolute 绝对温度
temperature adjustment 温度调节
temperature alarm 温度报警；过热报警
temperature build-up 温度升高
temperature buzzer 温度蜂鸣报警器
temperature characteristic 温度特性
temperature coefficient 温度系数
temperature coefficient of viscosity 黏度温度系数

temperature compensation 温度补偿
temperature contour 等温线
temperature contrast 温度不均匀分布；温度差；温度对比
temperature control 温度控制
temperature control coating 调温涂层
temperature control device 温度控制装置
temperature control equipment 温控设备
temperature control relay 温控继电器
temperature control sensor 温度控制传感器
temperature control switch 温度控制开关
temperature control system 温度控制系统
temperature control unit 温度调节装置
temperature control valve 温度控制阀
temperature controlled material 温控材料
temperature controlled panel 温度控制板
temperature controlled relay 温控继电器
temperature controlled sensor 温度控制传感器
temperature controlled system 温度控制系统
temperature controlled system test set 温度控制系统测试设备
temperature controller 温度控制器；温度调节器
temperature cycle 温度循环
temperature departure 气温距平；温度偏差
temperature deviation 温度偏差
temperature difference 温差；温度梯度
temperature distortion 热变形
temperature drop 温度下降
temperature energy 热能
temperature equalisation system 均温系统
temperature expansion 温度膨胀
temperature extremes 温度极限
temperature fluctuation 温度波动
temperature fluctuation range 温度波动范围
temperature gradient 温度梯度
temperature on operation 运行温度
temperature on standby 待机温度
temperature plume 热羽流
temperature profile 温度廓线
temperature resistance 耐温性
temperature resistant coating 耐热敷层
temperature resistant material 耐热材料
temperature rise 温升
temperature sensing bulb 感温包
temperature sensing element 热敏元件
temperature sensor 温度传感器
temperature stratification 温度层结
temporal average 时间平均值
temporal randomness 时间随机性
temporary failure 暂时故障
temporary repair 小修
tensile 拉力的
tensile deformation 拉伸变形
tensile elongation 拉伸长度；延伸率
tensile strain 拉伸应变
tensile strength 抗张强度
tensile stress 拉应力
tension 张力；拉力
tension axis 张力轴
tension bolt 拉紧螺栓
tension intensity 拉力强度
tension load 张力荷载
tension readjusting spring 张力重新调整弹簧
tension strain 拉伸应变
terawatt hour (tWh) 太(拉瓦)时；万亿瓦(特时)；兆兆瓦(特)时
term 术语
term by term differentiation 逐次微分

terminal 终端；端子；接线柱；引线；接头；末端的；终点的
terminal block 线弧；接头排；接线盒；接线板；线夹
terminal box 接线盒
terminal connector 设备线夹
terminal point 端子
terminal speed 末速度；终端速率；极限速度
terminal voltage 端电压
terrain 地形
terrain category 地形分类；地形类别
terrain downwash 地形性下洗
terrain environment 地形环境
terrain height 地形高度
terrain induced mesoscale system 地形诱导中尺度系统
terrain roughness 地形粗糙度
terrestrial 陆地生物
terrestrial radition 地面辐射；地球辐射；大地辐射
terrestrial surface 地球表面
terrestrial transfer model 陆地转移模式
terrestrial whirl wind 陆旋风
terrestrial wind 陆成风
test 试验
test apparatus 试验仪器
test atmosphere 试验环境
test bed 试验台
test bench 试验台
test chamber （风洞）试验段；试验间
test conditions 试验工况；试验条件
test data 试验数据
test desk 试验台；测试台
test equipment 试验设备
test error 试验误差
test for flow 气流试验
test ground 试验场
test installation 试验装置
test load 试验负荷
test loop 测试回路
test method 测试方法
test on bed 台架试验
test performance 试验性能
test period 试验期
test plant 试验车间
test point 测点
test procedure 试验程序
test record 试验记录
test Reynolds number 试验雷诺数
test rig 试验台
test run 试运行
test sample 试样
test section （风洞）试验段
test site 试验场地
test stand 试验台
test stream （风洞）试验（段气）流
tester 试验装置；分析仪器
testing 试验；测试；检验
testing accuracy 测试精度；试验精度
testing condition 试验工况；试验条件
testing principle 试验原理
testing similar condition 实验相似条件
testing tank 试验池
tethered 范围；系链
tethered balloon 系留气球
tethersonde 系留探测气球
tetroon 等容气球
TFT (transversal filament tape) 横向纤维带
TFT winding (transversal filament tape winding) 横向纤维带缠绕技术

TF_PMSG (transversal flux permanent magnet synchronous generator) 横向磁通永磁同步发电机
the atmospheric boundary layer thickness 大气边界层厚度
the dielectric 电介质
the family of airfoil 翼型族
the power of inertia 惯性力；惰性力
the secondary shaft 副轴
Theodorsen's circulation function 西奥多森环量函数
Theodorsen's theory 西奥多森理论
theorem 原理；定理
theoretical 理论的，理论上的
theoretical calculation 理论计算
theoretical capacity 理论容量
theoretical condition 理论条件；理论工况
theoretical efficiency 理论效率
theoretical equation 理论公式
theoretical error 理论误差
theoretical flow 理论流量
theoretical flow coefficient 理论流量系数
theoretical flow rate 理论流量
theoretical fluid 理论流体
theoretical horsepower 理论功率
theoretical hydrodynamics 流量流体动力学
theoretical power 理论功率
theoretical value 理论值
theoretical velocity 理论速度
theory 理论；学说
theory of aeroelasticity 空气弹性理论
theory of cross-flow 横向流动理论
theory of equivalent slope 等倾度理论
theory of probability 概率论
theory of similarity 相似理论

theory of turbulence 湍流理论
thermal 热的
thermal absorber 吸热器
thermal accomodation coefficient 热调节系数
thermal analysis 热分析
thermal anemometer 热风速仪
thermal anomaly 热的反常；热反常
thermal baffle 隔热板
thermal balance 热量平衡
thermal boundary layer 热边界层；温度边界层
thermal boundary layer thickness 热边界层厚度
thermal breakdown 热破坏
thermal brittleness 热脆性
thermal budget 热收支
thermal capacity 热容量
thermal circulation 热环流
thermal climate 热型气候
thermal comfort 热舒适性
thermal conductivity 导热率
thermal constant 热常数
thermal control 温度控制
thermal convection 热对流
thermal decomposition 热分解
thermal diffusion 热扩散
thermal diffusivity 热扩散率
thermal dilatation 热膨胀
thermal effect 热效应
thermal effluent 热废气；热废液
thermal element 热元件；热敏元件
thermal energy storage 热贮能
thermal environment 热环境
thermal equilibrium 热平衡
thermal gradient 热梯度

thermal inertia 热惯性；热质量
thermal insulation 隔热
thermal insulation structure 保温结构
thermal internal boundary layer (TIBL) 热力内边界层
thermal inversion 逆温
thermal inversion layer 逆温层
thermal island 热岛
thermal jump 热落差
thermal mass 热质量；热惯性
thermal mixing 热混合
thermal output 热量功率传热量
thermal over-current 热过电流
thermal over-current relay 热力式过载继电器
thermal performance 热力性能
thermal performance of building 建筑热性能
thermal plume 热羽流
thermal pollution 热污染
thermal protection structure 热防护结构
thermal radiation 热辐射
thermal relay 热继电器
thermal resistance 热阻
thermal similarity 热相似性
thermal source 热源
thermal stability 热稳定性
thermal steering 热成引导
thermal strain 热应变
thermal stratification 热层结
thermal structural coupling 热结构耦合
thermal structure 热结构
thermal tracer 热示踪物
thermal transmission 热传播
thermal turbulence 热湍流
thermal viscoelastic properties 热黏弹性特性
thermal wind 热成风

thermistor 热敏电阻
thermo anemometer 温差式风速计
thermo compression bonding 热压焊接热压接接合
thermo compression wire bonder 热压接引线接合机
thermo power station 火力发电站
thermo quenching 热浴淬火
thermocline 斜温层；变温层；温跃层
thermocouple 热电偶
thermodynamic diagram 热力学图
thermodynamic similarity 热力学相似
thermodynamic stability 热力学稳定性
thermodynamics 热力学
thermo electric power station 热电站
thermometer 温度计
thermometer support 温度计架
thermometric conductivity 导温系数
thermometric constant 热量常数；温度常数
thermometric element 测温元件
thermometric scale 温标
thermometrograph 温度记录器
thermopause 热成层顶
thermoplastic behaviour 热塑性
thermosphere 热层；电离层
thermostat 自动调温器；恒温器
thermostat controlled ventilation arrangement 恒温控制的通风设备
thick 厚的
thick airfoil 厚翼型
thickness 厚度
thickness chord ratio (T/C ratio) 翼型厚弦比
thickness function of airfoil 翼型厚度函数
thickness of airfoil 翼型厚度

thimble 心形环
thimble coupling 套筒联轴节；套管联结器
thin 薄的
thin airfoil 薄翼型
thin airfoil theory 薄翼型理论
thin wall construction 薄壁结构
third harmonic voltage 三次谐波电压
THM (top head mass) 机顶质量
thorough metal construction 全金属结构
threaded connection 螺纹连接
threaded flange 螺纹法兰
threaded pipe 螺纹管
three caliper disc brake 三卡钳盘式制动器
three cup anemometer 三杯风速表
three dimensional equation 三维方程
three dimensional flow 三维流动
three dimensional flutter 三维颤振
three dimensional pitot probe 三维皮托管
three dimensional vortex street 三维涡街
three dimensional wing 三维机翼
three island ship 三岛式船
three jaw chuck 三爪卡盘
three phase alternating current 三相交流电
three phase diode bridge 三相二极管桥
three phase full wave rectifier 三相全波整流器
three phase grid 三相电网
three phase insulated copper winding 三相绝缘铜线绕组
three phase stator winding 三相定子绕组
three phase symmetrically distributed stator winding wound 三相对称分布定子绕组
three phase variable frequency bi-directional back-to-back four-quadrant PEC 三相变频双向背靠背四象限 PEC

three point roller bearing 三点滚子轴承
three point single bearing 三点式单轴承
three space diagram 立体图
three stage slewing planetary gear unit 三级回转行星齿轮装置
threshold 临界值；阈值
threshold concentration 阈浓度
threshold dose 阈剂量
threshold field 阈场
threshold inductance 阈电感；临界电感
threshold limit value 阈限值
threshold of comfort 舒适阈
threshold of danger 危险阈
threshold of discomfort 不舒适阈
threshold of perception 感受阈
threshold of unpleasant 不悦感阈
threshold speed 阈速度
threshold value 阈值；界限值
throat （低速）风洞试验段；咽喉；颈前部
throttle 节流阀
throttle apparatus 节流装置
throttle button 节流阀钮
throttle condition 节流条件
throttle control 节流控制
throttle governing 节流调节
throttle regulation 节流调节
throttle valve 节流阀
through flow 穿堂风
through flow ventilation 贯穿通风
through quenching 淬透
through repair 大修
through ventilation 穿堂风
throwing 倒伏状（植物风力指示）
thrust 推力
thrust anemometer 推力风速计

thrust ball 止推滚珠；推力球
thrust block 止推（承）座；推力轴承
thrust block plate 推力轴承底板
thrust block seat 推力轴承座
thrust coefficient 推力系数；拉力系数
thrust collar 止推垫圈；止推环；止推轴承定位环
thrust force 轴向力；推力
thrust load 推力载荷
thrust loading coefficient 推力负荷系数
thrust washer 推力垫圈；止推垫圈
thumping sound 重击声
thunderstorm 雷暴
thunderstorm day 雷暴日
thunderstorm downburst 雷暴下击
thunderstorm high 雷暴高压
thunderstorm outflow 雷暴外流
thunderstorm rain 雷暴雨
thunderstorm rainfall 雷暴降雨量
thunderstorm wind 雷暴风
thyristor 晶闸管
thyristor controlled reactor (TCR) 晶闸管控制电抗器
thyristor switched capacitor (TSC) 晶闸管投切电容器
thyristor switched reactor (TSR) 晶闸管投切电抗器
thyristor soft-starter 晶闸管软启动器
TIBL (thermal internal boundary layer) 热力内边界层
tidal 潮汐的
tidal current 潮流
tidal wind 潮汐风
tie bar 连接杆
tight alignment 精确调整

tilt 倾斜；侧倾
tilt angle 倾斜角
tilt angle of rotor 风轮仰角
tilt up tower 倾斜塔
tilting manometer 倾斜压力计
tilting moment 倾覆力矩
timber shelter 木棚
time 时间
time averaged value 时均值
time constant (TC) 时间常数
time controller 时间控制器；自动定时器
time correlation 时间相关
time cut-out 定时短路器；定时断路器
time deformation 随时间而发生的变形
time delay 延时
time delay relay 延时继电器
time delay unit 延时器
time dependent flow 非定常流动；随时间变化流动
time dependent load 非定常荷载；随时间变化荷载
time domain 时域
time history 随时间变化过程；时程
time independent 定常的
time integrated concentration 时间积分尺度
time interval 时段；时间间隔
time invariant 时不变的
time invariant windspeed 稳定风速
time lag 时滞
time lag action 延时作用（动作）
time limit failure 时间限制失败
time mean concentration 时均浓度
time phase 时间相位
time relay 时间继电器
time response 时间响应

time scale 时间尺度
time step 时间步长
time-varying windspeed 随时间变化风速
timing controller 定时器；时序控制器
tinker's dam 焊缝
tip 翼尖；尖端
tip angle 顶锥角；顶圆锥角；顶端角；顶尖角
tip brake 叶尖制动器
tip brake system 叶尖制动系统
tip chord 翼梢弦
tip circle 齿顶圆
tip clearance 叶片间隙
tip eddy 桨叶尖旋涡；叶梢涡流
tip feathering 叶尖顺桨
tip flap 叶尖襟翼；叶尖小翼
tip flap control 叶尖襟翼控制
tip loss factor 叶尖损失因子
tip losses 叶尖损失
tip noise 叶尖噪声
tip of blade 叶尖
tip of file 锉尖
tip over 翻载
tip seal （叶片）顶封
tip section pitch 叶尖段桨距
tip speed 叶尖速度
tip speed ratio 叶尖速度比（高速性系数）
tip stall 叶尖失速
tip vane 叶尖小翼
tip vane augmented wind turbine 叶尖翼增力型风力机
tip vortex 叶尖旋涡
tip wire 尘塞线
tipping 俯仰；倾翻；倾翻的；倾卸的
tipping height 仰倾高度；倾卸高度
TLG (torque limiting gearbox) 转矩限制齿轮箱
TMD (tuned mass damper) 可调质量阻尼器；调谐质量阻尼器
tolerable 可以的；可容忍的
tolerable backlash 容许齿隙游移
tolerable concentration 许可浓度
tolerable criterion 许可准则
tolerable level 许可水平
tolerance 公差
tolerance deviation 容许偏差
tolerance fit 公差配合
tolerance limit 容许限度
tonality 音值
tool maker vises 精密平口钳
tooth 齿
tooth addendum 齿端高；齿顶高
tooth dedendum 齿根高
tooth depth 齿高
tooth flank 齿面
tooth fracture 轮齿断裂
tooth gauge 齿规
tooth gear 齿轮
tooth head 齿顶高
tooth space 齿槽；齿间；齿距
tooth thickness 齿厚
toothed 有齿；锯齿状的
toothed belt 有齿传动带；牙轮皮带
toothed synchronous belt drive 齿形同步皮带驱动器
top flange 顶部法兰；上翼缘；上凸缘
top head mass (THM) 机顶质量
top overhaul 大修
top panel 顶盖；（风洞）顶壁顶面板
top wind speed 最大风速
topoclimate 地形气候

topoclimatology 地形气候学
topographic 地形测量的
topographic effect 地形效应
topographic element 地形因素
topographic exposure factor 地形开敞度
topographic feature 地形特征
topographic interference 地形干扰
topographic modification 地形改造
topographic relief 地形起伏
topography 地形；地势
topography channel wind 狭道风
topology 拓扑结构
topsoil 表土
topspin 上旋
tornado 龙卷风；陆龙风
tornado affected area 龙卷风影响区
tornado alley 龙卷风通道
tornado axis 龙卷风轴
tornado belt 龙卷风带
tornado center 龙卷风中心
tornado core 龙卷风核
tornado cyclone 龙卷风气旋
tornado damage area 龙卷风破坏区
tornado dust collector 龙卷风式集尘器
tornado effect 龙卷风效应
tornado F-scale 龙卷风 F 等级
tornado funnel 龙卷风漏斗体
tornado hazard 龙卷风灾害
tornado parameter 龙卷风参数
tornado path 龙卷风路径
tornado resistant design 抗龙卷风设计
tornado simulation model 龙卷风模拟模型
tornado simulator 龙卷风模拟器
tornado storm 飓风
tornado track 龙卷风轨迹

tornado type wind energy system 旋风型风能系统
tornado vortex 龙卷涡
tornado wind load 龙卷风载
torque 力矩
torque allowance 转矩允许误差
torque arm 支耳；扭力杆；扭力臂；扭矩臂；转矩臂
torque balance 扭矩平衡
torque balance converter 转矩均衡换流器
torque balance device 扭矩平衡装置
torque balance element 扭矩天平元件
torque coefficient 力矩系数；转矩系数
torque density 转矩密度
torque fluctuation 转矩波动
torque force 转矩力
torque free body 自由旋转体
torque limiter 转矩限制器
torque limiting gearbox (TLG) 转矩限制齿轮箱
torque meter 扭矩计
torque reaction bar of gearbox 齿轮箱转矩从动杆
torque spanner 力矩扳手
torque speed characteristic 转矩—转速特性
torque spike 转矩脉冲；扭矩峰值
torque transducer 扭矩传感器
torque wrench 转矩扳手
torrid zone 热带
torsion 扭转
torsion balance 扭力天平
torsion bar 扭力杆
torsion beam 抗扭梁
torsion mode 扭转模态
torsion(al) strength 抗扭强度

torsional 扭转的
torsional buffeting 扭转抖振
torsional compliance 扭力柔度
torsional damping 扭转阻尼
torsional divergence 扭转发散
torsional flexibility 扭转挠性；扭转柔度
torsional flutter 扭转颤振
torsional frequency 扭转频率
torsional galloping 扭转驰振
torsional mode 扭转模态
torsional oscillation 扭转振动
torsional rigidity 抗扭刚度
torsional stall flutter 扭转失速颤振
torsional stiffness 扭转刚度；抗扭刚度
total 总数；合计
total absorption coefficient 总吸收系数
total backlash 总间隙
total body dose 全身剂量
total damping 总阻尼
total drag 总阻力
total enthalpy 总焓
total entropy 总熵
total environment 总（体）环境
total flow 总流量
total head 总压头
total installed capacity 总装机容量
total length 总长度
total mass airflow 空气质量总流量
total motion resistance 总运动阻力
total plume rise 羽流总抬升
total power 总功率
total power consumption 总功耗
total power loss 总功率损耗
total pressure 总压
total pressure pitot tube 总压皮托管

total pressure probe 总压探针
total vorticity （涡系）总涡量
touch voltage 接触电压
touchdown （羽流）着地
toughness 韧性
toughness test 韧性试验
tow phase flow 两相流
tower 塔架
tower base 塔基
tower base electronic controller 塔基电子控制器
tower block 塔形大厦
tower building 塔式建筑；摩天楼
tower climb ladder 塔架爬梯
tower crane 塔吊；塔式起重机
tower height 塔架高度
tower leg 塔架腿
tower like structure 塔形结构
tower nodding 塔架摆动
tower section 塔节
tower shadow 塔架阴影
tower shadowing effect 塔影效应
tower type 塔架类型
tower type building 塔式建筑
towing hook 牵引钩
towing plate 牵引板
town fog 城市雾
town planning 城市规划
town refuse 城市垃圾
toxic fume 毒烟
toxic gas 毒气
toxic pollution 有毒污染物；毒性污染
toxic smog 毒雾
trace amount 痕量
tracer 示踪物

tracer agent 示踪剂
tracer dye 示踪染色水
tracer element 示踪元素
tracer experiment 示踪实验
tracer gas 示踪气体
tracer material 示踪材料
tracer technique 示踪技术
tracing streamline 示踪流线
track 轨迹;轨道;轮距
track cant angle 轨道倾角
track gauge 轨距
tracking filter 跟踪滤波器
trackway 轨道;踏出来的路;行迹
trade off study 权衡分析;折中研究
trade wind 信风
trade wind belt 信风带
traditional 传统的
traditional windmill 传统风车
trail 后缘;拖曳物;尾;痕迹
trailing 拖尾
trailing bogie 后转向架
trailing edge 后缘
trailing edge angle 后缘角
trailing edge cap (叶片)后缘帽
trailing edge flap 后缘襟翼
trailing edge spar 后梁;尾梁
trailing edge vortex 后缘涡
trailing vortex 后缘涡流;尾涡;拖曳涡
trailing vortex sheet 尾涡面
train of gears 齿轮系
trajectory 流轨;轨道;轨线
transcritical 跨临界
transcritical flow regime 过临界流动状态;跨临界流动状态
transcritical Reynolds number 过临界雷诺数;跨临界雷诺数
transducer 传感器
transducer response 传感器响应
transfer 转让;转移
transfer function 传递函数
transfer of axis 轴系转换
transfer ratio 传递比
transformation 转化;转变
transformation of coordinate 坐标系变换
transformer 变压器
transformer fitted with OLTC 有载调压变压器
transformer inductance 变压器电感
transient aerodynamic 瞬变空气动力学
transient behaviour 暂态行为
transient characteristic 过渡特性;瞬态特性;暂态特性
transient condition 过渡工况;瞬时条件;瞬变工况
transient current 瞬时电流;瞬态电流;暂态电流
transient disturbance 瞬时扰动
transient event 暂态事件;暂现事件
transient fault 瞬态故障
transient flow 瞬间流动;瞬变流动;暂态流
transient interruption 瞬断
transient load 瞬态荷载
transient loading 瞬时荷载
transient performance 瞬态性能
transient response 瞬态响应
transient rotor 瞬态电流
transient stability 瞬时稳定性
transient vibration 瞬态振动
transient vortex 瞬变涡;瞬时涡;非稳定旋涡
transient wind speed 瞬时风速

transistor 晶体管；晶体三极管
transistor ageing 晶体管老化
transistor base 晶体管基极
transistor switch 晶体管开关
transition 过渡；转变
transition flow regime 过渡流态
transition height （风速廓线）转换高度；过渡高
transition of height difference 过渡高差
transition point 转捩点
transition Reynolds number 转捩雷诺数
transition section （风洞）过渡段
transition strip 转捩带
transition wire 转捩绊线
transition zone 转捩区，过渡区
translating gear 变换齿轮
translation 平移
translation of axes 轴系平移
translational 平移的；直移的
translational speed of tornado vortex 龙卷涡移动速度
translational speed of vortex street 涡街移动速度
translational vibration 平移振动
translational wind machine 平移型风力机
transmission 传动装置；传递；传送
transmission & distribution equipment (T & D equipment) 输配电设备
transmission accuracy 传动精度
transmission and distribution (T & D) 输变电及配电
transmission case 变速箱
transmission coefficient 传递系数
transmission error 传动误差
transmission gauge 传动齿轮

transmission gear 传动齿轮
transmission line 输电线
transmission line inductance 传输线电感
transmission loss 传输损耗；传导损失；穿透损失
transmission ratio 传动比
transmission system operator (TSO) 输电系统运营商
transmission tower 输电杆塔；输电塔；铁塔
transpiration 蒸腾
transpiration cooling 蒸发冷却
transpiration wind tunnel 蒸腾风洞
transport 输运；迁移
transport characteristics of wind 风力输运特性
transport container 运输集装箱
transport equation 输运方程
transport equipment 运输设备
transport mode 输运模式
transport of momentum 动量输运
transport of pollutants 污染物运输
transport wind speed 输运风速
transportation 运输
transportation condition 运输条件
transversal 横向的
transversal axis 横轴
transversal filament tape (TFT) 横向纤维带
transversal filament tape winding (TFT winding) 横向纤维带缠绕技术
transversal flux permanent magnet synchronous generator (TF-PMSG) 横向磁通永磁同步发电机
transversal galloping 横向驰振
transversal loading 横向荷载
transversal oscillation 横向振动

transversal plunging 横向浮沉
transversal stability 横向稳定性
transverse 横向的；横断的
transverse bending 横向弯曲
transverse correlation 横向相关
transverse current 横流
transverse fiber belt winding 横向纤维带缠绕
transverse flow 横流；横向流动
transverse galloping 横向驰振
transverse loading 横向荷载
transverse metacenter 横稳心
transverse oscillation 横向振动
transverse plunging 横向浮沉
transverse section 横截面
transverse shear 横向剪切
transverse stability 横向稳定性
trapezoidal wing tip 梯形翼尖
trapped vortex 脱体涡
trapped wave 陷波
trapping 捕捉；收集
trapping snow 挡雪
transient blade-flapping behaviour 叶片挥舞运动的过渡状态
travelling 移动的
travelling highs 移动性高气压
travelling lows 移动性低气压
travelling wave 行波
traverse gear 横向转动齿轮；横动装置；回转装置
traversing bar 横梁
traversing carriage 移动支架
traversing probe 横移探测管
trellis 格架
TRIAC (tri-electrode AC switch) 三端双向可控硅开关

trial and error method 尝试误差法；试误法
trial and error theory 尝试误差说
triangular symbol 三角符号
tridimensional flow 三维流动
tri-electrode AC swith (TRIAC) 三端双向可控硅开关
trigger 起动装置
trigger action 触发作用
triggered gaps 电花隙避雷器
triggering device 触发装置
triggering disturbance 触发扰动
trigonometric transformations 三角变换
trim 配平；纵倾
trim angle 纵倾角；俯仰角
trim tab 配平片；纵倾调整片
trimmed state 配平状态
trimming inductance 微调电感
trimming moment 纵倾力矩；配平力矩
trip 绊；绊倒
trip coil 跳闸线圈
trip free 自由脱扣
trip free mechanism 自由脱扣机构
trip free relay 自由脱扣继电器；自动跳闸继电器
triple 三个的
triple blade windmill 三叶片风车
triple cambered aerofoil 三曲翼面
tripped breaker 断开的断路器
tripping 切断装置；止动闸；挡器；卡爪
tripping device 释放装置；跳脱装置
tripping fence 跳闸栅栏
tripping pulse 切断脉冲；触发脉冲；跳闸脉冲
tripping rib 跳闸肋
tripping strip 脱扣带
tripping wire 绊线；绊网

tropic 热带；回归线

tropic of cancer 北回归线

tropic of capricorn 南回归线

tropical 热带的

tropical air (mass) 热带气团

tropical climate 热带气候

tropical climatic zone 热带气候带

tropical cyclone 热带气旋

tropical revolving storm 热带风暴；热带旋转风暴

tropical storm 热带风暴

tropopause 对流层顶

tropopause funnel 漏斗状对流层顶

tropopause inversion 对流层顶逆温

troposphere 对流层

troposphere stratosphere exchange 对流层平流层交换

tropospheric aerosol 对流层气溶胶

tropospheric dust suspension 对流层悬浮尘

tropospheric fallout 对流层沉淀物；对流层落尘

trouble 故障；事故

trouble free operation 无故障运行

trouble light 故障信号灯

trouble shoot 排除故障

trouble shooting 故障检修

trouble zone 故障区

troubleshooting data 修理指南；检修指南

trough 低压槽；水槽

true angle of attack 真实迎角

truncated chimney 截顶烟囱

truncated normal distribution 截断常态分布

truncated spectral model 截谱模式

truncated spectrum 截谱

truncation error 截断误差

trundle 滚动；移动

trunnion 凸耳；耳轴

truss 桁架

truss action 桁架作用

trussed beam 桁架式梁

TSC (thyristor switched capacitor) 晶闸管投切电容器

TSR (thyristor switched reactor) 晶闸管投切电抗器

TSO (transmission system operator) 输电系统运营商

tube 风洞管；电子管；隧道

tube building 筒体建筑

tube in tube building 筒中筒建筑；套管式建筑

tubular 管状的

tubular bridge 管桁；管桁桥

tubular concrete section 混凝土圆筒节

tubular cooler 管状散热器；管式冷却器

tubular frame 管式桁架

tubular segment 筒节

tubular tower 圆筒式塔架

tuft method 丝线法

tufted flow study 丝线法气流研究

tumble home 舷缘内倾

tumbling motion 滚动

tuned damper 可调阻尼器；调谐阻尼器

tuned mass damper (TMD) 可调质量阻尼器；调谐质量阻尼器

tuner 调谐器

tuning inductance 调谐电感

tuning mass 调整质量

tunnel 隧道

tunnel balance test 风洞天平试验

tunnel blockage 风洞阻塞

tunnel blockage correction 风洞阻塞修正

tunnel boundary 风洞边界；风洞洞壁
tunnel ceiling 风洞顶壁
tunnel constraint 风洞壁约束
tunnel induced velocity 风洞诱导速度
tunnel inlet 风洞进气口
tunnel intake 风洞进气段
tunnel interference correction 洞壁干扰修正
tunnel reference pressure 风洞参考压力
tunnel return flow 风洞回流
tunnel shape factor 风洞（试验段）形状因子
tunnel speed control 风洞风速控制
tunnel speed hole 风洞洞壁测速孔
tunnel technique 风洞试验技术
tunnel test 风洞试验
tunnel turbulence 风洞湍流（度）
tunnel ventilation 隧道通风
tunnel wall interference 洞壁干扰
turbidity factor 混浊因子
turbine 涡轮
turbine type wind machine 涡轮式风力机
turbopause 湍流层顶
turbosphere 湍流层
turbulence 紊流；湍流
turbulence admittance factor 湍流导纳因子
turbulence closure model 湍流封闭模型
turbulence component 湍流分量
turbulence decay 湍流衰减
turbulence-development fetch 湍流发展风程
turbulence distortion 湍流畸变
turbulence effect 湍流效应
turbulence energy 湍流能量
turbulence energy content 湍流能量含量
turbulence energy spectrum 湍流能谱
turbulence excitation 湍流激励
turbulence factor 湍流因子

turbulence field 湍流场
turbulence generation mechanism 湍流发生机理
turbulence generator 湍流发生器
turbulence grid 湍流格网
turbulence induced vibration 湍流致振
turbulence intensity 湍流强度；扰动强度
turbulence inversion 湍流逆温
turbulence length scale 湍流长度尺度
turbulence level 湍流度
turbulence nonuniformity 湍流不均匀性
turbulence number 湍流度；湍流数
turbulence report 湍流报告
turbulence resistance 湍流阻力
turbulence response 湍流响应
turbulence Reynolds number 湍流雷诺数
turbulence scale 湍流尺度
turbulence scale parameter 湍流尺度参数
turbulence scatter 紊流散布
turbulence screen 湍流网；（风洞）阻尼网
turbulence spectrum 湍流谱
turbulence sphere 湍流球
turbulence structure 湍流结构
turbulence transition 湍流转捩
turbulence viscosity 湍流黏性
turbulent 混乱的
turbulent boundary flow 紊流边界流动
turbulent boundary layer 湍流边界层
turbulent buffeting 湍流抖振
turbulent bursting 湍流猝发
turbulent condition 紊流状态
turbulent convection 湍流对流
turbulent deposition velocity 湍流沉降速度
turbulent diffusion 湍流扩散
turbulent dispersion 湍流弥散

turbulent dissipation 湍流耗散
turbulent drag 湍流阻力
turbulent eddy 湍涡
turbulent Ekman layer 埃克曼湍流层
turbulent energy 湍流能量
turbulent entrainment 湍流卷挟
turbulent exchange 湍流交换
turbulent flow 湍流
turbulent flow regime 湍流状态
turbulent flow theory 湍流理论
turbulent flow velocity 湍流速度
turbulent fluctuation 湍流脉动
turbulent friction 湍流摩擦
turbulent gust 湍流阵风
turbulent heat exchange 湍流热交换
turbulent intensity 湍流强度
turbulent inversion 湍流逆温
turbulent Lewis number 湍流刘易斯数
turbulent mass exchange 湍流质量交换
turbulent microstructure 湍流微结构
turbulent mixing 湍流混合
turbulent motion 扰动；湍流运动
turbulent pattern 湍流结构
turbulent plume 湍流羽流
turbulent Prandtl number 湍流普朗特数
turbulent property 湍流性
turbulent reattachment 湍流再附着
turbulent region 湍流区
turbulent Reynolds number 湍流雷诺数
turbulent ring 湍流环
turbulent Schmidt number 湍流施密特数
turbulent separation 湍流分离
turbulent shear 湍流剪切；湍流切变
turbulent shear layer 湍流剪切层
turbulent stream 湍流

turbulent transfer 湍流传递
turbulent transport 湍流输运
turbulent transport theory 湍流输运理论
turbulent viscosity 湍流黏性
turbulent vortex 湍流涡；湍流旋涡
turbulent vortex street 湍流涡街
turbulent wake 湍流尾流
turbulent wind 湍流风
turbulent wind field 湍流风场
turbulent wind fluctuation 湍流风脉动
turbulent zone 湍流区
turbulivity 湍流度；湍流系数
turn about the axis 绕轴旋转
turn buckle 花篮螺栓；花篮螺丝；索具；紧线扣
turnaround efficient 周转效率
turnaround time 周转时间
turning vane （风洞）导流片；导流板
turn off characteristic 关断特性
turn off transient 关断瞬态
turn on transient 接通瞬态
turns ratio 变比；匝比
turns ratio of the rotor-to-stator winding 定子与转子绕组的匝数比
turntable （风洞）转盘
turntable bearing 回转轴承；滚动轴承
tuyere notch 风嘴孔；风口
TW (tail wind) 顺风
twin box girder 双箱梁
twin I girder 双工字梁
twin rectangular box girder 双矩形箱梁
twin tower 双柱塔
twist 扭动；旋转；转动
twist aerofoil 扭曲翼剖面
twist angle 扭转角
twist angle of blade 扭角

twist axis 扭转轴
twist blade 扭转叶片
twist drill 螺旋钻头
twist moment 扭转力矩
twist of blade 叶片扭角
twist vector 扭矢
twisted blade 扭曲叶片
twisting force 扭力
twisting moment 扭矩
twisting stiffness 抗扭刚度
twisting strength 抗扭强度
two bladed wind turbine 双叶风电机组
two caliper disc brake 二卡钳盘式制动器
two color laser Doppler anemometer 双色激光多普勒风速计
two dimensional airfoil 二维翼型
two dimensional boundary layer 二维边界层
two dimensional flow 二维流动
two dimensional jet 二维射流
two dimensional plume 二维羽流
two dimensional puff 二维喷团
two dimensional space 二维空间
two dimensional water tunnel 二元水洞
two dimensional wind tunnel 二元风洞
two dimensional wing 二维机翼
two sided spectral density function 双边谱密度函数
two vertical planes system 双竖向平面系
two way configuration 二线制
type 类型
type of blade 叶片类型
type of connection 连接方式
type of flow 流动方式；流动状态
type of load 负荷类型
typhoon 台风
typical value 标准值；典型值；代表值

现代英汉风力发电工程

A Modern English-Chinese Dictionary of Wind Power Engineering

U

U-bolt U形螺钉
UCTE (Union for Coordination for Transmission Electricity) 欧洲输电联盟
ultimate 最终的；极限的；根本的
ultimate bearing capacity 极限承载量
ultimate capacity 最大功率
ultimate compressive strength 极限抗压强度
ultimate design 极限设计
ultimate design resisting moment 极限设计抗力矩
ultimate elongation 极限伸长
ultimate factor of safety 极限安全系数
ultimate height of plume 羽流最终高度
ultimate limit state 最大极限状态；极限限制状态
ultimate load 极限荷载
ultimate load design 极限载荷设计
ultimate set 相对伸长
ultimate shear stress (U.S.S.) 极限剪应力
ultimate strength 极限强度
ultimate value 极限值
ultra capacitor 超级电容器
ultra high frequency 特高频
ultra vires 超越权限
ultrasonic 超声波
ultrasonic anemometer 超声波风速计
ultrasonic bonding 超声波焊接
ultrasonic examination 超声波探伤
ultrasonic flaw detector 超声波探伤器
ultrasonic investigation 超声波探测；超声波探查
ultraviolet lamp 紫外线灯
umbrella 烟囱顶罩；保护伞；庇护；雨伞；伞形结构
unacceptable 不能接受的
unacceptable criterion 不可接受性判据
unattackable 耐腐蚀的
unbalance 不平衡
unbalanced factor 不平衡因素
unbalanced flow 不平衡流动
unbalanced load 不平衡负载
unbiased 公正的；无偏见的
unbiased estimators 无偏估计量；无偏估计值
uncertain 含糊的
uncertain variability of wind 风的不确定性
uncertainty 误差；不确定性；不可靠
uncertainty in measurement 测量误差
unclad structure 无围护结构
unconfined 自由的；松散的
unconfined vortex wind machine 非约束涡型

风力机

unconstrained air 非约束空气
unconstrained body 非约束体
uncontrolled 不受控制的
uncontrolled device 不可控器件
uncorrected 未修正的
uncoupling 解耦
uncut grass 未剪草地
undamped 不减弱的
undamped oscillations 无阻尼振荡
undamped resonant frequency 无阻尼共振频率；无阻尼谐振频率
under allowance 尺寸下偏差
under construction （正在）修建中
under frequency 低频率
under speed 欠速
under voltage 欠电压；低电压
undercurrent 底流；潜流
underestimate 低估
underground pumped hydrostorage 地下抽水储能
underlying surface 下垫面
underpan 底壳；底盘
underprediction 过低预计；过低预报
underpressure 负压；抽空；压力不足
underside 下面；阴暗面
underspin 下旋
undisturbed 未被扰的
undisturbed airflow 未受扰气流
undisturbed boundary layer 未扰动边界层
undisturbed flow 未扰动气流
undisturbed wind 未扰动风
undisturbed wind speed 未受扰风速
uneven terrain 不平坦地形
unfurling 扬帆

unidirectional 单向性的
unidirectional current 单向电流；直流电
unidirectional flow 单向流动
unidirectional wind 单向风
uniform 统一的；一致的
uniform beam 等截面梁
uniform blade adjustment 整体叶片调整
uniform continuity 均匀连续性
uniform distribution 均匀分布
uniform elongation 均匀伸长
uniform flow 均匀流
uniform flow wind tunnel 均匀流风洞
uniform load 均布载荷
uniform section 均匀截面
uniform steady wind 均匀定常风
uniform vorticity 均匀涡量
uniform wind 均匀风
uniformity coefficient 均匀系数
uniformity modulus 均匀模数
uninterrupted power supply (UPS) 不间断供电电源
union coupling 联轴节；管连接；联管节
Union for Coordination for Transmission Electricity (UCTE) 欧洲输电联盟
unit area 单位面积
unit commitment 机组组合
unit control 单元控制
unity power factor 单位功率因数
universal 普遍的
universal chuck 万能卡盘
universal coupling 万向联轴器
universal gas constant 通用气体常数
universal meter 万用表
universal milling machines 万能铣床
universal motor 交直流两用电动机

unlimited 无限量的
unlimited stream 无限流；无约束流
unload 卸载
unloading 卸载
unloading device 卸载装置
unloading equipment 卸载设备
unloading valve 释荷阀；卸载阀
unmanned 无人值守
unobstructed 没有障碍的；畅通无阻的
unobstructed airflow 自由气流
unpleasant smell 不悦气味
unpleasant wind condition 不悦风况
unpowered ascent 无动力上升
unpowered bogie 无动力转向架
unpowered control 无助力操纵
unpowered flight 无动力飞行
unpowered rotor 自转旋翼的转速
unsaturated zone 未饱和区
unscheduled maintenance 非计划性维修
unseparated flow 未分离流动
unserviceability 不适用性；运转不安全性；使用不可靠性
unshrouded rotor 无罩风轮
unstability 不稳定性
unstability of speed 转速不稳定度
unstable 不稳定的
unstable atmosphere 不稳定大气
unstable characteristic 不稳定振动特性
unstable condition 非稳定状态
unstable equilibrium 不稳定平衡
unstable galloping 不稳定驰振
unstable oversteer 不稳定转向
unstable regime 不稳定状态
unstable stratification 不稳定层结
unstationary flow 不稳定流

unstationary state 非固定状态
unsteadiness 非定常性；不稳定
unsteady airfoil theory 非定常翼型理论
unsteady Bernoulli's equation 非定常伯努利方程
unsteady boundary layer 非定常边界层
unsteady flow 非定常流
unsteady state 非稳态
unsteady state condition 不稳定工况
untwist 解缆
untwist the cables 电缆解绕
up current 上升气流
up draft 上升气流
up valley wind 进谷风
upflow 上升流
upkeep 维持；保养；维修
uplift 举起；抬起
uplift coefficient 上升系数；上提系数
uplift wind load 上吸风载
upper air 高层大气；高层空气；高空
upper air circulation 高空环流
upper atmosphere 高空大气
upper hub 上轮毂
upper inversion 高空逆温
upper level cyclone 高空气旋
upper level wind data 高空风数据
upper surface brake 上翼面减速板
upper wind 高空风
upright resistance 抗拔力
UPS (uninterrupted power supply) 不间断供电电源
upsetting arm 倾覆力臂
upsetting level 倾覆水平
upsetting moment 倾覆力矩
upslope wind 上坡风

upstream 上游
upstream influence 上游影响
upstream pressure 上游压力；进口压力；阀前压力
upstream pressure limit 上游压力限制
upstream windspeed 上游风速
uptake 上升烟道；上升井；上风口
upward air current 向上气流
upward current 上升气流
upward current of air 上升气流
upwarping 向上挠曲
upwash 上洗；气流上洗
upwind (UW) 迎风；顶风；上升气流；上风向
upwind effect 上吹效应
upwind fetch 上风吹程
upwind rotor 上风式风轮
upwind type of WECS 上风式风能转换系统
upwind wind turbine 迎风式风电机组
urban air pollution 城市空气污染
urban air pollution source 城市空气污染源
urban area 市区
urban atmosphere 城市大气
urban boundary layer 城市边界层
urban canyon 城市街谷；都市峡谷
urban climate 城市气候
urban complex 城市建筑群
urban diffusion parameter 城市扩散参数
urban domestic heating 城市居民区增温
urban effect 城市效应
urban environment 城市环境
urban freeway 城市高速道路；都会区高速公路
urban heat island 城市热岛
urban heat plume 城市热羽流
urban heating rate 城市加热率
urban industry heating 城市工业增温

urban mixing layer 城市混合层
urban planning 城市规划
urban renewal 城市更新
urban terrain 城市地形
urban traffic 城市交通
urban wind 市区风
urbanization 城市化
U.S. Environment Protection Agency (EPA or US EPA) 美国环境保护局
useful 有用的
useful energy 有效能
useful height 有效高度
useful life 使用寿命；使用期限；有效期
useful life period 使用寿命期限
useful load 有效负荷；实用负载
useful output 有效输出；有效功率
useful output power 有效输出功率
useful power 有效功率
useful range 有效范围
useful time 有效寿命
useful work 有效功
U.S.S (ultimate shear stress) 极限剪应力
utility 实用的；通用的
utility grid 公用电网
utility power 公用电力；公用电源
utilization 利用；应用
utilization efficiency 利用效率
utilization of energy 能量利用
utilization ratio 利用率
UW (upwind) 迎风；顶风；上升气流；上风向

现代英汉风力发电工程

A Modern English-Chinese Dictionary of Wind Power Engineering

V

vacuum 真空；真空的
vacuum arrester 真空避雷器
vacuum augmentor 真空增强器
vacuum bag 真空袋
vacuum evaporation 真空蒸发
vacuum evaporator 真空蒸发器
vacuum fan 抽风机；真空风扇
vacuum gauge 真空表；真空计
vacuum gauge pressure 真空表压力
vacuum lighting protector 真空避雷器
vacuum tube 真空管；电子管
vacuum tube accelerometer 电子管加速计
vacuum tube adapter 电子管适配器
vacuum tube microammeter 电子管微安计
vacuum tube millivoltammeter 电子管毫伏表
vacuum tube relay 电子管继电器
validity 有效性
valley 谷；山谷
valley breeze 谷风
valley floor 谷底
valley flow 山谷气流
valley fog 谷雾
valley slope 山谷坡；河谷坡降
valley slope wind 谷坡风
valley wind 谷风

valley wind circulation 谷风环流
valuation 评价
value 值
valve 阀；闸门
valve amplifier 电子管放大器
valve arrester 电子管避雷器；阀式避雷器
vane 风向标；（风洞）导流片；尾舵
vane anemometer 风杯风速计；叶轮风速计
vane control 叶片调节
vane tip 叶片尖端
vane twist 叶片的扭曲度
vane type anemometer 翼式风速计
vapor 蒸汽；烟雾
vapor cloud 蒸汽云
vapor cycle condenser 蒸汽循环冷凝器
vapor dispersion 蒸汽弥散
vapor jump 蒸汽跃迁
vapor loss 蒸汽损失
vapor nozzle 蒸汽喷嘴
vapor phase 汽相
vapor plume 蒸汽羽流
vapor pressure 蒸汽压
vapor rise 蒸汽抬升
vaporization cooling 蒸发冷却
vapour concentration 蒸汽浓度；水汽浓度；水

蒸气密度

var (volt-ampere reactive) 乏；无功伏安

variability 多变性；变率

variable acceleration 可变加速度

variable blade 可变安装角的叶片

variable camber 可变弯度

variable camber flap 变弯度襟翼

variable chord blade 变截面叶片

variable coning 可变锥角

variable cross section wind tunnel 可变截面风洞

variable density wind tunnel 变密度风洞

variable error 可变误差

variable frequency generator 变频发电机

variable generator speed fixed-pitch (VGS-FP) 变速定桨距

variable generator speed fixed-pitch wind turbine 变速定桨距风电机组；VGS-FP风电机组

variable inductance 可变电感

variable load 可变负荷

variable of state 状态变量

variable pitch blade 变距叶片；变距桨叶

variable pitch rotor 可变桨距风轮

variable quantity 可变量

variable reactive compensator 可变无功补偿器

variable resistive load control 可变电阻性负载控制

variable resistor 变阻器；可变电阻

variable rigidity 可变刚度

variable rotor resistance 转子变阻器

variable speed 可变速；无级变速

variable speed constant frequency generator 变速恒频发电机

variable speed control 变速调节

variable speed device 变速装置

variable speed doubly fed induction generator 变速双馈异步发电机

variable speed electrical generator 变速发电机

variable speed gear 变速齿轮

variable speed motor 变速电动机

variable speed rotor 变速风轮

variable speed wind turbine 变速风电机组

variable speed wound rotor induction generator 变速绕线式转子异步发电机

variable wind 多变风

variance 方差

variant scalar 变量标量

variation 变化；变动；变量

variation in load 负荷变化

varnish 清漆

varying lift 变升力

VAWT (vertical axis wind turbine) 垂直轴风轮机

V-belt 三角皮带

vector 矢量

vector addition 矢量加法

vector analysis 矢量分析

vector area 矢面积

vector balancing 矢量平衡

vector control 矢量控制

vector diagram 矢量图

vector equation 矢量方程

vector phase control 矢量相位控制

vector quantity 矢量

vector triangle 矢量三角形

vector wind field 向量风场

vectorial lift 矢量升力

vectorial resultant 矢量合成
veering 调风；对风；顺时针转向
veering tendency 自动转弯趋势
veering wind 顺时针旋风
velocity 速度
velocity amplitude 速度幅值
velocity contour 速度分布图；等流速线
velocity defect 速度亏损
velocity defect law （边界层）速度亏损律
velocity deficit decay 速度亏损衰减
velocity deficit profile 速度亏损廓线
velocity duration curve 速度持续曲线
velocity energy 动能；速度能
velocity field 速度场
velocity fluctuation 速度脉动
velocity gradient 速度梯度
velocity head 速度头
velocity modulation 调速
velocity of flow 流速
velocity potential 速度势
velocity pressure 速压
velocity profile 速度廓线
velocity reduction 减速
velocity resolution 速度分解
velocity shock 速度冲击
velocity space 速度空间
velocity spectrum 速度谱
velocity stratification 速度层结
velocity time diagram 速度时间图
velocity triangle 速度三角形
velocity vector (VV) 速度矢量
velocity vector component 速度矢量分量
vent 通气孔；排气口
vent gas plume 排出气羽流
vent hole 通气孔
vent patch 排气孔补片
vent stack 通风井道；排泄烟道
vented structure 排风结构
ventilation 通风
ventilation capacity 通风量
ventilation coefficient 通风系数
ventilation duct 通风管
ventilation hole 通风孔
ventilation louver 气窗；通风百叶窗
ventilation resistance 通风阻力
venting rate 通风率
Venturi 文丘里
Venturi effect 文丘里效应
Venturi meter 文丘里流量计
Venturi tube 文丘里管；文丘里流速计
Venturi type wind machine 文丘里型风力机
verbal assessment 口述评价
vertex chamber 涡流室
vertical 垂直的；直立的
vertical axis 垂直轴
vertical axis Darrieus turbine 达里厄垂直轴风力机
vertical axis rotor 垂直轴风轮
vertical axis rotor WECS 垂直轴风能转换系统
vertical axis wind machine 垂直轴风力机
vertical axis wind turbine (VAWT) 垂直轴风轮机
vertical bending mode 垂直弯曲模态
vertical bracing 竖向支撑；垂直支撑
vertical circulation cell 垂直环流圈
vertical diffusion 垂直扩散
vertical direction 垂直方向
vertical double-column type machining centers 立式双柱加工中心
vertical fin 垂直尾翼；垂尾

vertical gradient 垂直梯度
vertical machining centers 立式加工中心
vertical mixing 垂直混合
vertical panel 竖直面板
vertical plume 垂直羽流
vertical plume growth 羽流垂直增长
vertical rise 垂直上升
vertical speed (VSP) 垂直速度
vertical speed brake 垂直速度制动器
vertical stability 垂直稳定性
vertical symmetry (VS) 垂直对称
vertical wind shear 垂直风切变
vertical wind tunnel 立式风洞
vertiginous current 旋流
VGS-FP (variable generator speed fixed-pitch) 变速定桨距
VHN 维氏硬度
vibration 振动
vibration absorbed base 振动吸收基地
vibration absorbent 消振
vibration absorber 减振器；吸振器
vibration absorption 吸振
vibration amplitude 振幅
vibration behaviour 振动特性
vibration damper 振动阻尼器；吸振器
vibration frequency 振动频率
vibration indicator 测振仪
vibration isolation 隔振
vibration isolation joint 防振接头；隔振接头
vibration isolation rubber 防振橡皮；隔振橡胶
vibration isolator 减震器
vibration meter 振动计
vibration mode 振型
vibration proof 防振
vibration sensors 振动传感器
vibration source 振动源
vibration spring isolator 弹簧减震器
vibration stopper 减震器
vibration suppressor 振动衰减器
vibration switch 振动开关
vibration tests 振动试验
vibration velocity 振动速度
vibrator 振动器；振荡器
vibroshock 减振器；缓冲器
vice jaw 虎钳口
Vickers 维氏硬度计
Vickers diamond hardness 维氏硬度
Vickers hardness test 维氏硬度试验
view and size 外形和尺寸
vigorous vertical mixing 强垂直混合
violent storm 十一级风；暴风；暴风骤雨；急风暴雨
violent tornado 强龙卷
violent vortices of air 强空气旋涡系
virtual 虚拟的
virtual airfoil 有效翼型
virtual angle of attack 虚迎角
virtual camber 虚弯度
virtual camber effect 虚弯度效应
virtual center of balance （塔式机械）天平虚中心
virtual deformation 潜变形；假变形；虚（拟）变形
virtual diffusion coefficient 虚拟扩散系数
virtual displacement 虚位移
virtual incidence 虚迎角
virtual kinetic energy 虚动能
virtual mass 虚质量
virtual source 虚源

virtual temperature 虚温
virtual work 虚功
viscid 黏质的
viscid theory 黏性流体理论
viscidity 黏性
viscoelastic 黏弹性的
viscoelastic damper 黏弹性阻尼器
visoelastic model 黏弹性模型
viscoelastic property 黏弹性能
viscoelasticity 黏弹性
viscometer 黏度计
viscose glue 胶水
viscosimeter 黏度计
viscosity 黏度
viscosity coefficient 黏性系数
viscosity constant 黏度常数
viscosity effect 黏性效应
viscosity fluid 黏性流体
viscosity friction 黏性摩擦
viscosity index 黏性指数
viscosity resistance 黏滞阻力
viscous 黏性的
viscous core 黏性核
viscous damping 黏滞阻尼
viscous damping structure 黏性阻尼结构
viscous dissipation 黏性耗散
viscous drag 黏滞阻力
viscous flow 黏性流动
viscous fluid 黏性流体
viscous force 黏性力
viscous fracture 黏性破坏
viscous friction 黏性摩擦；黏滞摩擦
viscous friction coefficient 黏滞摩擦系数
viscous friction force 黏性摩擦力
viscous layer 黏性层

viscous pressure resistance 黏性压力阻力
viscous resistance 黏性阻力
viscous shear 黏性剪应力
viscous stress 黏性应力
viscous sub-layer 黏性底层
viscous surface layer 黏性表面层
viscous vortex 黏性旋涡
viscous wake 黏性尾流
vise 虎钳
vise jaw 虎钳钳口
visibility 能见度
visible plume 可见羽流
visible region 可见区
visible tracer 可见示踪物
visual 视觉的
visual acuity 视觉灵敏度
visual angle 视角
visual assessment 目测评价
visual estimation 目测
visual impact 视觉影响；视觉冲击；视觉震撼；视觉效果
visual inspection 外观检查
visual measurement 目测
visual meteorological conditions 目测气象条件；能见气象条件
visual monitor 目视监视器
visual observation 目测；目视观察
visual persistence 视觉暂留
visual pollution 视觉污染
visual range 视程
visual signal 目视信号；视觉信号
visualization 目测，可视化
V-joint V形连接
void 空隙；孔隙
void content 孔隙度；空隙度；孔隙量

void fraction 空隙率；空隙分数；空隙组分
void ratio 孔隙比；空隙比；孔隙率，空隙率
void space 孔隙空间；空隙空间
volcanic debris 火山碎屑
volcanic dust 火山尘
volcanic eruption 火山喷发
volcanic plume 火山羽流
volcanic smoke 火山烟
voltage 电压
voltage across the terminals 端电压
voltage alarm 电压报警
voltage change factor 电压变化系数
voltage collapse 电压崩溃
voltage control 电压控制
voltage control system 电压控制系统
voltage dip 电压骤降；电压突跌；电压下降
voltage drop 电压降落；电压下降；电压降
voltage fluctuation 电压波动
voltage grade 电压等级
voltage imbalance 电压不平衡
voltage level 电压等级
voltage loop 电压波腹
voltage node 电压波节
voltage peak 电压峰值
voltage phasor 电压相量
voltage phasor diagram 电压相量图
voltage polarity 电压极性
voltage profile 电压分布
voltage source 电压源
voltage source inverter (VSI) 电压源逆变器
voltage stability margin 电压稳定裕度
voltage stabilizer 稳压器
voltage swell 电压骤升；电压突升
voltage to earth 对地电压
voltage transient 电压瞬变

voltage sag 电压暂降；电压骤降；电压凹陷
voltameter 电量表
volt-ampere 伏安
volt-ampere characteristic 伏安特性
volt-ampere curve 伏安特性曲线
volt-ampere output characteristic 伏输出特性
volt-ampere reactive (var) 乏；无功伏安
voltmeter 电压表
volume 量；流量
volume differentiation 体积微分
volume flow 体积流量；体积流率
volume in volume 体积百分数
volume rate of flow 体积流量
volume source 体源
volume weight 容重
volumetric 体积的；容积的
volumetric efficiency 容量效率
volumetric expansion 体积膨胀
volumetric factor 容量因数
volumetric flask 容量瓶
volumetric flow 体积流量
volumetric flow meter 容积式流量计；容积流量计
volumetric flow rate 体积流量；体积流率
volumetric flux 容积通量
volute 螺旋形；涡形；涡螺；涡螺壳
vortex 旋涡；涡流（面）；涡旋（体）
vortex anemometer 旋涡风速计
vortex augmentor 旋涡增强装置
vortex axis 旋涡轴
vortex band 涡流带
vortex breakdown 旋涡破碎
vortex bursting 旋涡猝发
vortex cavity 旋涡空穴
vortex circulation 旋涡环量

vortex cluster 旋涡群
vortex concentrator device 旋涡集聚式风能装置
vortex cone 涡流锥
vortex core 涡核
vortex decay 涡旋衰减
vortex density 涡流密度
vortex drag 涡阻
vortex eddy 涡流；旋涡
vortex equation 涡流方程
vortex excitation 涡激励
vortex excited oscillation 涡激振荡
vortex excited response 涡激响应
vortex filament 涡旋线；涡（旋）丝
vortex flow 涡流
vortex flux 涡通量
vortex formation region 旋涡形成区
vortex free motion 无旋运动
vortex frequency 涡旋频率
vortex generator 旋涡发生器
vortex induced acoustic vibration 涡致声振
vortex induced displacement 涡致位移
vortex induced noise 涡致噪声
vortex induced oscillation 涡致振荡
vortex induced response 涡致响应
vortex induced vibration 涡致振动
vortex intensity 旋涡强度；涡流强度
vortex interference 旋涡干扰
vortex invariant 涡旋不变量
vortex lattice 涡流栅
vortex layer 旋涡层
vortex lift 涡升力
vortex line 涡线
vortex lattice method 涡格法
vortex location 涡流位置
vortex lock-in 锁涡
vortex lock-on 锁涡
vortex pair 涡对
vortex path 涡迹
vortex pattern 涡谱；涡流分布
vortex plane 涡面
vortex plane rolling-up 涡面卷起
vortex response 涡响应
vortex rhythm 旋涡节拍
vortex ring 涡环
vortex row 涡列
vortex sensor 涡流检测器
vortex separation 旋涡分离
vortex shedding 旋涡脱落
vortex shedding anemometer 旋涡脱落风速计
vortex shedding frequency 旋涡脱落频率
vortex shedding period 旋涡脱落周期
vortex shedding rhythm 旋涡脱落节拍
vortex sheet rolling-up 涡面卷起
vortex street 涡街；涡道
vortex strength 旋涡强度
vortex strip 涡带
vortex structure 旋涡结构
vortex trail 涡旋尾迹
vortex train 涡列
vortex trajectory 旋涡轨迹
vortex trunk 涡线；旋涡螺丝
vortex tube 涡管；涡流管
vortex wake 尾涡流
vortical singularity 漩涡奇点
vortical surface 旋回面
vortices 涡系
vorticity 涡量；涡度；涡旋
vorticity effect 涡流影响

vorticity equation 涡流方程
vorticity in isotropic turbulence 各向同性紊流的涡量
vorticity source 涡源
vorticity trajectory 旋度轨迹

VS (vertical symmetry) 垂直对称
VSI (voltage source inverter) 电压源逆变器
VSP (vertical speed) 垂直方向速度
VV (velocity vector) 速度矢量

现代英汉风力发电工程

A Modern English-Chinese Dictionary of Wind Power Engineering

W

wake 尾流
wake blockage 尾流阻塞
wake boundary 尾流边界
wake bubble 尾流气泡
wake buffeting 尾流抖振
wake capture 尾流捕获
wake cavity region 尾流空穴区
wake circulation 尾流环量
wake closure 尾流封闭区
wake decay 尾流衰减
wake drag 尾流阻力
wake effect losses 尾流效应损失
wake energy 尾流能量
wake entrainment 尾流卷挟
wake excitation 尾流激励
wake excited crosswind response 尾流横风响应
wake expansion 尾流扩展
wake flutter 尾流颤振
wake front 尾流前沿
wake galloping 尾流驰振
wake growth 尾流增长
wake induced vibration 尾流致振
wake losses 尾流损失
wake momentum thickness 尾流动量厚度
wake oscillator model 尾流振子模型
wake recirculation region 尾流回流区
wake region 尾流区
wake relaxation 尾流松弛
wake resistance 尾流阻力
wake shadow 尾流阴影
wake spread 尾流扩展
wake stream 尾流
wake strength 尾流强度
wake Strouhal number 尾流斯特劳哈尔数
wake suction 尾流吸力
wake survey 尾流测量
wake turbulence 尾流湍流（度）
wake vortex 尾流旋涡
wake vortex resonance 尾流旋涡共振
wake vortex system 尾流涡系
wake vorticity 尾流涡量
wake width 尾迹宽度
wake-oscillator model 尾流振子模型
wall 墙；墙壁；墙体；壁
wall boundary layer （风洞）洞壁边界层
wall cladding 外墙围护结构
wall constraint （风洞）洞壁约束
wall correction 洞壁修正
wall friction 壁面摩擦（力）

wall interference 洞壁干扰
wall jet 贴壁射流
wall opening ratio 墙面开洞比
wall paper 壁纸
wall pressure 壁压
wall pressure hole 洞壁测压孔
wall radiant heating 墙面辐射采暖
wall radiant tube 墙壁辐射管
wall radiator 壁挂式散热器
wall receptacle 墙插座
wall register 墙面风口
wall roughness 壁面粗糙度
wall sampler 墙壁取样器
wall shear stress 壁面剪应力
wall shearing stress 壁剪应力
wall stress 壁应力
wall turbulence 壁面湍流
wall up 把……封住；把……关住
warm 温暖的
warm air 热空气
warm air mass 暖气团
warm anticyclone 暖性反气旋
warm braw 暖风
warm bubble 热泡
warm climate 温暖气候
warm cloud 暖云
warm current 暖流
warm cyclone 暖性气旋
warm front 暖锋
warm-cold air tunnel 冷热空气风洞
warner 报警器
warning 警告
warning apparatus 报警装置
warning indicator 警告指示器；报警器
warning lamp 警告灯；指示灯

warning light 报警灯
warning signal 报警信号
warning system 警告系统；报警系统
warping 拖曳；扭曲；翘曲
warping drum 卷缆筒
warping moment 翘曲力矩
warping stress 翘曲应力
wash 洗；冲洗
wash away 冲走；清洗；清除
wash down 下洗；冲洗
wash off 清除掉；清洗掉
wash out 破产；淘汰
wash port 排水口
washer 垫圈
washout coefficient 冲洗系数；冲蚀系数
washout deposition rate 冲洗沉积率；冲蚀沉积率
washout factor 冲洗因子
washout parameter 冲洗参数
washout rate 冲洗率；清除速率
washout ratio 冲洗比；冲刷率
wastage allowance 材料许可损耗率
waste gas 废气
waste gas emission standard 废气排放标准
wasted power 耗散功率
water 水
water channel 水槽；水洞
water chilled 冷冻水
water coolant 冷却水
water evaporation cooling 水蒸发冷却
water front 岸线
water hardening 水淬
water head accumulator 液压蓄能器
water line 吃水线
water pump 水泵
water pumping windmill 抽水风车

water quenching	水淬火
water repellent envelope	防水外壳
water screw	推水螺旋桨
water tank	水槽；水箱
water tower	水塔
water tube condenser	水管冷凝器
water tunnel	水洞
water's saturated vapor pressure	水的饱和蒸汽压
watt	瓦特
watt component	有功分量
watt hour	瓦时
watt hour efficiency	瓦时效率（能量效率）
wattage dissipation	功率耗散
watt-hour meter	电度表
wattless component	无功分量
wattless power	无功功率
wattmeter	电（力）表；功率表
wave	波；波动
wave age	波龄
wave current	波动电流；波流
wave drag	波阻
wave field	波场
wave flow	波形流
wave force	波力
wave height	波高
wave number	波数
wave pressure	波压
wave profile	波剖面
wave propagation	波传播
wave steepness	波的陡度；波陡；波浪陡度
waveguide	波导；波导管
wavelet	小浪；微波
WD (wind direction)	风向
weak current control	弱电控制
weak grid	弱电网
weak interaction regime	弱相互作用状态
weak wind boundary layer	弱风边界层
wear	磨损；耐久性
wear ability	耐磨性
wear allowance	磨损余量
wear and tear	磨损；自然磨损；正常损耗；磨损和毁坏
wear free	无磨损
wear hardness	耐磨性
wear rate	磨损率
wear well	耐用
wearing capacity	耐磨性
wearing depth	磨损深度
wearing part	易损件
wearing test	磨损试验
weather	天气；气候作用
weather analysis	天气分析
weather anchor	抗风锚
weather anomaly	天气异常
weather base	气象站
weather board	防风雨板
weather chart	气象图
weather cock	风标；风向标
weather condition	气象条件
weather constituent	气象要素
weather data	气候数据；气象参数
weather element	气象要素
weather flag	气象预报信号
weather hazard	天气灾害
weather map scale	天气图尺度
weather pattern	气象图
weather phenomena	天气现象
weather proof	耐风雨
weather resistance	耐风雨侵蚀能力；耐气候性；

耐风化性
weather sensitive load 天气敏感荷载
weather service station 气象服务站
weather sphere 天气层
weather vane 风向标
weather warning (WW) 气象警报
weather wind tunnel 气象风洞
weathering 侵蚀；风化；雨蚀
weathering quality 耐风蚀性；耐气候性
weathering steel 耐候钢
weathervane 风标
weathervaning 调向；对风
weathervaning stability 调向稳定性
web 梁腹；腹板
web girder 腹梁
web type spar 腹板(翼)梁
Weber number 韦伯数
WEC (wind energy conversion) 风能转换
WECS (wind energy conversion system) 风能转换系统
wedge boarding 楔镶板
wedge style 楔形造型
weekly average concentration 周平均浓度
weep hole 泄水孔；排水孔
Weibull distribution 威布尔分布
Weibull form and scale parameter 威布尔形状和尺度参数
Weibull function 威布尔函数
Weibull mode 威布尔模态
Weibull parameter 威布尔参数
Weibull scale parameter 威布尔尺度参数
Weibull shape parameter 威布尔形状参数
Weibull slope 威布尔斜率
weight 重量
weight empty 空重

weight factor 加权因子
weighted accumulator 重力式蓄能器
weighted average 加权平均
weighted deviation 加权偏差
weighted factor 加权因子
weighted function 加权函数
weighted sound pressure level 加权声压级
weld 焊接
weld all around 围焊
weld bond 熔合线
weld crosswise 交叉焊接
weld flux 焊药；焊剂
weld joint 焊缝；焊接接头
weld junction 熔合线
weld machined flush 削平补强的焊缝
weld metal 焊缝金属
weld neck flange 带颈对焊法兰；焊接颈状法兰
weld puddle 熔池
weld reinforcement 焊缝补强
welded elbow 焊接弯头；焊接机械肘
welded fissure 焊接裂缝；焊合的裂缝；焊接裂纹
welded flange 焊接法兰
welded joint 焊接接头
welded main frame 焊接主框架
welded pipe 焊接管
welded steel 焊接钢
welding pool 焊接熔池
welding rod 焊条
welding slag 焊渣
well organized vortex 规则旋涡
well separated flow 完全分离流动
wet adiabatic change 湿绝热变化
wet adiabatic curve 湿绝热曲线

wet adiabatic lapse rate 湿绝热递减率
wet bulb temperature 湿球温度
wet cooling tower 湿式冷却塔
wet deposition 湿沉降
wet lay-up 湿法成型；湿法敷涂层
wet mechanical draft cooling tower 湿式机械通风冷却塔
wet natural draft cooling tower 湿式自然通风冷却塔
wet plume 湿羽流
wet snow zone 湿雪地带
wetted area 浸湿面积；受潮面积
WFEO(World Federation of Engineering Organizations) 世界工程组织联合会
whale back dune 鲸背型沙丘
whirl 旋转；涡流
whirl core 旋动核；旋风核心
whirl flutter 旋动颤振
whirl tube 风洞；风道
whirl velocity 涡流速度
whirling 旋转的；涡流的
whirling current 旋涡流
whirling speed 旋转速度
whirling vortices 旋转涡流系
whirlpool 旋涡
whirlwind 旋风
white body 白体
white noise excitation 白噪声激励
white noise field 白噪声速度场；白噪声场
white noise spectrum 白噪声谱
whiteout 乳白天空
whole gale 狂风（十级风，风速 24.5~28.4 米/秒）
wide band excitation 宽带激励
wide range 宽波段的；宽量程的；广范围；宽范围
widen 加宽
wideth 宽度
wind 风
wind abeam 侧风；横风
wind abrasion 风磨蚀
wind aloft 高空风
wind aloft report 高空风报告
wind angle 风迎角；风角
wind approach angle 来流风迎角；风接近角
wind area 迎风面积
wind arrow 风矢；风向及风力指示符号；风向指针
wind assessment 风场评价
wind axes 风轴
wind axes system 风轴系；气流坐标系
wind axis wind machine 风轴风力机
wind azimuth 风方位角
wind barometer table 风速—气压换算表
wind belt 风带
wind blown debris 风吹碎片
wind blown dust 风吹沙
wind blown sand 风积沙
wind blown snow 风吹雪
wind blown soil 风积土
wind borne dust 风载尘
wind borne material 风载物质
wind borne pollution 风载污染物
wind borne sediment 风成沉积物
wind borne snow 风载雪
wind bracing 风撑；抗风支撑
wind break 风障
wind break barrier 防风墙
wind break fence 防风篱
wind break forest 防风林

wind break system 防风林系统
wind calm 静风
wind cap 防风帽
wind carrier sand 风积沙
wind catching surface 获风表面
wind caused catastrophic 风灾
wind channel 风洞；风道
wind characteristic 风特性
wind chart 风图
wind chill 寒风
wind circulation 风环流；气流循环
wind class 风级
wind climate 风气候
wind cock 风向标
wind comfort level 风舒适度
wind concentrator 集风器
wind cone 风向袋
wind conversion efficiency 风能转换效率
wind corrosion 风蚀
wind cowl （烟囱）风帽
wind current 风流；气流
wind damage 风害
wind data 风数据；风资料
wind data logger 风数据记录器
wind deflection 风向偏转
wind deflector 导风板；导风装置
wind deposit 风沉降
wind deposited soil 风积土
wind description 风描述
wind desiccation 风干作用
wind diffuser 扩风器
wind direction (WD) 风向
wind direction fluctuation 风向波动
wind direction frequency 风向频数
wind direction meter 风向测定器

wind direction recorder 风向自记器
wind direction sensor 风向感受器
wind discontinuity 风不连续性
wind dispersion 风弥散作用
wind divide 风向界线
wind drag 风阻
wind drag load 风阻荷载
wind drift 飞砂；漂流；风偏流
wind drift sand 流沙；风沙
wind driven circulation 风引起的环流
wind driven brine pump 风力盐井泵
wind driven generator 风力发电机
wind driven irrigation 风力提灌
wind driven pump 风力驱动泵
wind driven sawmill 风力锯木厂
wind driven snow 风吹雪
wind driven wave 风浪
wind duration 风期；风时
wind dynamic load 动态风载
wind eddy 风涡
wind effect 风效应
wind effect on agriculture 风对农业影响
wind effect on air pollution 风对空气污染影响
wind effect on building 风对建筑物的影响
wind effect on pedestrian 风对行人的影响
wind effect on structure 风对结构物影响
wind effected phenomena 风力效应现象
wind energy 风能
wind energy collector 风能收集装置
wind energy content 风能蕴藏能
wind energy conversion (WEC) 风能转换
wind energy conversion system (WECS) 风能转换系统
wind energy converter 风能转换装置

wind energy density	风能密度
wind energy engineering	风能工程
wind energy extraction limit	风能获取极限值
wind energy farm	风电场
wind energy flux	风能通量
wind energy pattern factor	风能能型因子
wind energy potentiality	风能潜力
wind energy region	风能区划
wind energy resource	风能资源
wind energy rose	风能玫瑰图
wind energy spectrum	风能谱
wind energy system	风能转换系统
wind engineer	风工程师
wind engineering	风工程
wind environment	风环境
wind erosion	风蚀
wind erosion basin	风蚀盆地
wind excitation	风激励
wind excited acceleration	风激加速度
wind excited oscillation	风激振荡
wind exposure area	受风面积
wind fairing	风嘴[公路科技]
wind farm	风电场
wind fence	风篱；风栅栏
wind fetch	风程
wind field	风场
wind flag	风旗
wind flow	风流；气流
wind flume	风槽
wind flutter noise	风颤振噪声
wind fluttering factor	风振系数
wind following	风向跟踪；对风
wind force	风力
wind force coefficient	风荷载系数
wind force scale	风力级；风级
wind frame	抗风构架
wind frequency curve	风频数曲线
wind furnace	自然通风炉；自然通风式炉；通风炉
wind gap	风口；风隙
wind gauge	风速表；风力计；风压计
wind generated noise	风致噪声
wind generated sea wave	风致海浪
wind generated wave	风生波
wind generation	风力发电
wind generator	风力发电机
wind generator set	风力发电机组
wind gorge	风峡
wind gradient	风速梯度
wind gradient exponent	风速梯度指数
wind gust	阵风
wind hole	风穴
wind indicator	风力指示物
wind induced acceleration	风致加速度
wind induced circulation	风引起的环流
wind induced discomfort	风致不舒适
wind induced heat loss	风致热损失
wind induced internal pressure	风致内压
wind induced load	风致荷载
wind induced oscillation	风致振荡
wind induced pressure	风压
wind induced vibration	风致振动
wind industry	风电行业
wind information	风资料
wind intensity	风强度
wind laid deposits	风沉积；风积物
wind laid soil	风积土
wind layer	风层
wind load	风荷载

wind load spectrum 风载谱
wind loading chain 风载分析链
wind loading code 风载规范
wind loading design criterion 风载设计准则
wind loading standard 风载标准
wind loss 风害损失
wind lull 风速暂减；风暂息
wind machine 风力机
wind map 风图
wind measurement mast 测风塔
wind meter 风速计
wind microzonation 风力微区划
wind mill 风力机
wind mixing 风混合
wind moment 风力矩；风载力矩
wind natural ventilation 自然通风
wind noise 风噪声；气流噪声
wind nose 风嘴
wind of Beaufort force 2 （蒲福风级）二级风
wind off test 无风试验
wind on test 有风试验
wind path 风迹
wind porch 风廊
wind power 风电；风功率
wind power climatology 风能气候学
wind power density 风功率密度
wind power duration curve 风能持续时间曲线
wind power generation 风力发电
wind power installation 风能装置
wind power penetration 风电穿透
wind power penetration limit 风电穿透功率极限
wind power plant (WPP) 风力发电厂
wind power potential 风能潜力
wind power profile 风能廓线
wind power project 风电项目
wind power station 风电场
wind power system 风能发电系统
wind power technology 风力发电技术
wind powered aeration system 风力曝气系统
wind powered aerator 风力曝气机
wind powered generation 风力发电
wind powered generator 风力发电机
wind powered heat pump 风力热泵
wind powered installation 风电装置；风电安装
wind powered irrigation system 风力提灌系统
wind powered machinery 风力机械
wind pressure 风压
wind pressure coefficient 风力系数
wind pressure factor 风压系数
wind pressure out service 无效风压
wind pressure measuring hole 风压测量孔
wind profile 风廓线
wind profile wind shear law 风廓线风切变律
wind proof capacity 抗风力
wind proof construction 防风结构
wind proof design 耐风设计
wind proof performance 耐风性能
wind propulsion system 风力助航系统
wind protection plantation 防风林
wind protection screen 防风板
wind pump 风力泵
wind reference pressure 基本风压
wind regime 风况
wind resistance 抗风；抗风性；风阻力
wind resistance loss 风阻损耗
wind resistant column 抗风柱

wind resistant design 抗风设计
wind resistant feature 抗风特性
wind resource 风力资源
wind resource assessment 风力资源评价
wind resource map 风资源地图
wind rib 抗风肋
wind riddle 风选筛
wind right 风权
wind ripple 风成波痕
wind roadway 风道
wind rose 风向频率图；风向风速图；风向玫瑰图；风玫瑰；风图
wind rotation 风向转变
wind rotor 风轮
wind run 风程
wind rush noise 风急速流动噪声
wind scale 风力级
wind sector 风区
wind sensitive building 风敏感建筑
wind sensitive structure 风敏感结构
wind sensor 风感受器
wind sensor assembly 风况传感组件
wind shadow 风影；背风区
wind shaft 风矢杆
wind shear 风切变
wind shear exponent 风切变指数
wind shear law 风切变律
wind shield wiper 挡风玻璃刮水器
wind shift 风向转变
wind shift line 风变线
wind site 风场
wind slash 风害迹地
wind sleeve 风向袋
wind sock 风向袋
wind spectrum 风谱

wind speed 风速
wind speed alarm 风速警报器
wind speed and direction (WS&D) 风速与风向
wind speed and direction setter 风速风向给定器
wind speed counter 风速计数器
wind speed distribution 风速分布
wind speed duration curve 风速持续时间曲线
wind speed frequency 风速频率
wind speed frequency curve 风速频率曲线
wind speed profile 风速廓线
wind speed Reynolds number 风速雷诺数
wind speed scale 风速等级；风速刻度盘
wind speed spectrum 风速谱
wind spilling 风能溢出
wind spilling type of governor 风能溢出式限速装置；风能溢出式调速器
wind spun vortex 风动涡旋
wind statistics 风统计学
wind strength 风力
wind stress 风应力
wind stripping 防风带
wind strom 风暴
wind structure 风结构
wind suction 风吸力
wind survey 测风
wind swept area 受风区；风扫略面积
wind system 风系
wind thrust 风推力
wind trajectory 风迹
wind transducer 风向传感器
wind truss 抗风桁架
wind tunnel 风洞

wind tunnel air supply 风洞空气供给
wind tunnel axes system 风洞轴系
wind tunnel bottom 风洞底壁
wind tunnel ceiling 风洞顶壁
wind tunnel centerline 风洞轴线；风洞中心线
wind tunnel choking 风洞壅塞
wind tunnel circuit 风洞回流道；风洞回路
wind tunnel contraction 风洞收缩段；风洞收缩比
wind tunnel control 风洞控制
wind tunnel data (WTD) 风洞数据
wind tunnel determination 风洞实验测定
wind tunnel diffuser 风洞扩散段
wind tunnel diffuser end 风洞扩散段端
wind tunnel drive 风洞驱动
wind tunnel efficiency 风洞效率
wind tunnel energy ratio 风洞能量比
wind tunnel fan 风洞风扇
wind tunnel fetch 风洞吹程
wind tunnel floor 风洞底壁
wind tunnel geometry 风洞几何形状
wind tunnel heat regenerator 风洞回热器
wind tunnel intake 风洞进气口
wind tunnel laboratory 风洞实验室
wind tunnel measuring techniques 风洞（实验）测量方法
wind tunnel model 风洞模型
wind tunnel model tests 风洞模型试验
wind tunnel modeling technique 风洞模拟技术
wind tunnel noise 风洞噪声
wind tunnel nozzle 风洞喷管
wind tunnel observation 风洞观测
wind tunnel performance 风洞性能
wind tunnel plant 风洞设备
wind tunnel radius 风洞半径
wind tunnel Reynolds number 风洞雷诺数
wind tunnel roof 风洞顶壁
wind tunnel shape 风洞形状
wind tunnel simulation 风洞模拟
wind tunnel solid boundary 风洞实壁边界
wind tunnel speed 风洞风速
wind tunnel study 风洞研究
wind tunnel suspension system 风洞模型悬挂系统
wind tunnel techniques 风洞实验技术
wind tunnel test 风洞试验
wind tunnel test section 风洞试验段
wind tunnel time 风洞吹风时间；风洞占用时间
wind tunnel turbulence 风洞湍流
wind tunnel turbulence factor 风洞紊流系数
wind tunnel turning vane 风洞导流片
wind tunnel turntable 风洞转盘
wind tunnel wall 风洞洞壁
wind tunnel wall effect 风洞洞壁效应
wind tunnel wall interference 风洞洞壁干扰
wind tunnel window 风洞试验段观察窗
wind tunnel working section 风洞试验段
wind turbine 风力机；风电机组
wind turbine cost effectiveness 风电机组的经济效益
wind turbine generating 风力发电
wind turbine generator (WTG) 风力发电机
wind turbine generator system (WTGS) 风力发电机组
wind turbine international standard 风电机组国际标准
wind turbine penetration 风力机渗透度
wind turbine simulator 风力机模拟器

wind turbine terminals 风力机端口	windmill 风车
wind turbulence 风湍流（度）	windmill anemometer 风车式风速表
wind unpredictability 风不可预计性	windmill curve 风车线
wind up 上紧发条；使结束	windmill generator 风力发电机
wind uplift 风浮力	windmill pump 风车（力）泵
wind valley 风谷	windmill sail 风车轮叶
wind vane 风向标；对风尾舵	window counter tube 窗形计数管
wind variability 风变性	window groove fastener 窗户挡风毡条座
wind vector 风矢量	window states word 窗口状态字
wind veering 风向转变	windstream direction 气流方向
wind velocity 风速	windstream power density 风功率密度
wind velocity gradient 风速梯度	windstream turbulence 风流湍流（度）
wind velocity profile 风速廓线	windward 上风向的；迎风向的
wind wake 风尾流	windward face 向风面
wind water channel 风水槽	windward side 迎风面
wind water interface 风水交界面	windward slope 迎风坡
wind water-lifting set 风力提水机组	windward wall 迎风壁
wind wave 风浪	windwheel 风轮
wind wave tank 风浪水槽，风浪模拟水池	windy district 多风区
wind wheel anemometer 风轮风速仪	windy site 多风地带
wind yaw angle 风偏航角	wing 翼
wind zone 风带	wing bending 机翼弯曲
windage 风阻；风力影响；风偏修正；风偏	wing butt 翼根
windage losses 风阻损失	wing cellule 翼组
windage resistance 风阻	wing chord 翼弦
windbreak 防风林；防风物	wing contour 机翼外形
windiness 多风性；招风；多风，有风	wing covering 机翼蒙皮
winding 绕组	wing drop 机翼下坠
winding displacement 绕组位移；排线	wing rib 翼肋
winding distribution 绕组分布	wing rock 机翼摇摆
winding factor 绕组系数	wing root 翼根
winding loss 绕组铜损耗	wing root interference effect 翼根干扰影响
winding pitch 绕组（线圈）节距	wing section 翼剖面
winding type internal rotor 绕线式内部转子	wing shaped core 翼状型心
winding up 清理	wing spar 翼梁

wing tip effect 翼尖影响
wing twisting 翼扭转
winglet 小翼
winter oil 耐冻油
winter proofing 防冻
wiper 刮水器
wiper arm 刮水器臂
wiper blade 刮水器刮片
wire 导线；电线
wire coil 线盘；线卷
wire connector 接线器
wire fuse 保险丝
wire mesh 金属丝网目；金属丝网
wire netting structure 铁丝网结构
wire rope 钢丝绳
wire screen 金属丝网
wire strain gauge 金属丝应变计
wire terminal 电线接头
wire tripping device 丝脱扣装置
wireless SCADA system 无线数据采集与监控系统
WMO (World Meteorological Organisation) 世界气象组织
wobbler action 偏心作用
wooden envelope 木质蒙皮
wool tuft technique 贴线法
work hardening 加工硬化
work holding jaw 工件夹爪；钳口
work input 输入功
work ratio 工作效率
working condition 工况；工作条件
working depth 有效齿高
working earthing 工作接地
working expense 运行费用
working flank 工作齿面

working fluid 工作流体
working fluid characteristics 工作流体特性；工质特性
working life 工作寿命
working medium 工作介质；工质
working parameter 工作参数；运行参数
working pressure 工作压力；试验段压力
working section 试验段
working specification 操作规程
working voltage alternative current (WVAC) 交流工作电压
working voltage direct current (WVDC) 直流工作电压
working wind speed 工作风速
World Federation of Engineering Organizations (WFEO) 世界工程组织联合会
World Meteorological Organisation (WMO) 世界气象组织
World Wind Energy Association (WWEA) 世界风能协会
worm 蜗杆
worm wheel 蜗轮
worst case design 最不利情况设计
wound internal rotor 绕线式内部转子
wound rotor 绕线式转子
wound rotor induction generator (WRIG) 绕线式转子异步发电机
wound rotor synchronous generator (WRSG) 绕线式转子同步发电机
wound stator 绕线式定子
WPP (wind power plant) 风力发电厂
wrapped cable 缠绕缆索
wrench 扳手；扳钳
WRIG (wound rotor induction generator) 绕线式转子异步发电机

wrong indication 错误显示
WRSG (wound rotor synchronous generator) 绕线式转子同步发电机
WS & D (wind speed and direction) 风速与风向
WTD (wind tunnel data) 风洞数据
WTG (wind turbine generator) 风力发电机
WTGS (wind turbine generator system) 风力发电机组

WVAC (working voltage alternative current) 交流工作电压
WVDC (working voltage direct current) 直流工作电压
WW (weather warning) 气象警报
WWEA (World Wind Energy Association) 世界风能协会
wye-connected 星形连接

现代英汉风力发电工程

A Modern English-Chinese Dictionary of Wind Power Engineering

x-axis X轴；X坐标轴
x-coordinate X坐标；横坐标
x-ray detection X射线探伤法
x-ray flaw detector X射线探伤器
x-ray inspection X射线检查

现代英汉风力发电工程

A Modern English-Chinese Dictionary of Wind Power Engineering

Y

yaw 偏航
yaw acceleration 偏航加速度；侧滑加速度；横摆加速度
yaw action rotor 主动调向风轮
yaw and pitch 偏航与俯仰
yaw angle 偏航角；侧滑角；横偏角
yaw angle of rotor shaft 风轮偏角
yaw angle sensor 偏航角传感器
yaw base 偏航基座；偏航盘
yaw bearing 偏航轴承
yaw brake 偏航制动器
yaw control 偏航控制
yaw control mechanism 偏航控制机构
yaw coupling 偏航耦合
yaw coupling parameter 偏航耦合参数
yaw drive 偏航驱动
yaw drive dead band 调向死区
yaw drive pinion ratio 偏航驱动小齿轮传动比
yaw driven upwind wind turbine 偏航驱动迎风式风电机组
yaw error 偏航误差
yaw error amplifier 偏航误差放大器
yaw fixed rotor 定向风轮
yaw gear 偏航齿轮
yaw inertia 对风惯性

yaw mechanism 偏航机构
yaw misalignment 偏航角误差；偏航失调
yaw motor 偏航马达
yaw passive rotor 被动调向风轮；被动对风风轮
yaw probe 方向探头
yaw rate 偏航速率
yaw ring 偏航齿圈
yaw sensor 偏航传感器
yaw system 偏航系统
yaw turntable 偏航转盘
yaw vane 调向尾舵；对风尾舵；偏航翼
yawed 偏航的
yawer 偏航控制器
yawing 偏航
yawing angle of rotor shaft 风轮偏航角
yawing device 调向装置；对风装置
yawing driven 偏航驱动
yawing mechanism 偏航机构
yawing moment 偏航力矩
yawing moment of inertia 横摆惯性矩；偏航惯性矩
yawing movement 偏航运动
yawing orientation 偏航调向；偏航对风
yawing speed 偏航速度

yawmeter 偏航计；偏航指示器

y-axis Y 轴

y-azimuth 基准方向角

y-coordinate 纵坐标；Y坐标

y-Darrieus rotor Y型达里厄式风轮

y-Δ connected step-up transformer Y-Δ连接升压变压器

year to year temperature difference 年际温差

yearly average wind speed 年平均风速

yearly load curve 年负荷曲线

yearly maintenance 年度保养；年度维修

yearly maximum load 年最大负荷

yearly mean efficiency 年平均效率

yearly wind speed 年平均风速

yield limit 屈服极限

yield point 屈服点，降伏点

yield strength 屈服强度

yield stress 屈服应力

yield stress ratio 屈服应力比

yoke plate 联板；轭板

yoke ring 轭环

yoke type support 叉形支杆；轭式支架

Young's modulus 杨氏模量

现代英汉风力发电工程

A Modern English-Chinese Dictionary of Wind Power Engineering

Z

z-axis Z轴
zenith angle 天顶角
zero allowance 无公差；零容差
zero balance 零位调整
zero balance reading 天平读零
zero bias 零偏压
zero bit 零位
zero displacement height 零风面位移高度
zero frequency 零频率
zero frequency component 直流分量；零频分量
zero frequency current 零频电流
zero frequency gain 零频率增益
zero lift (ZL) 零升力
zero lift angle of attack 零升力攻角
zero lift chord (ZLC) 零升力弦
zero lift drag 零升阻力
zero lift drag rise 零升力时的阻力增量
zero lift line 零升力线
zero lift moment 零升力矩
zero line (ZL) 零线；基准线
zero plane displacement 零风面位移；零面位移
zero power 零功率
zero reading correction 零读数修正
zero sequence current 零序电流
zero setting 调零；置零
zero shift technique 零点位移法
zero signal 零信号
zero voltage 零电压
zero wind drag coefficient 无风阻力系数
zinc plated 镀锌的
ZL (zero lift) 零升力
ZL (zero line) 零线；基准线
ZLC (zero lift chord) 零升力弦
zone 区域；带
zone load 区域负荷
zone of comfort 舒适区
zone of flow establishment 流动形成区
zone of negative pressure 负压区
zone of positive pressure 正压区
zone of reserve flow 逆流区
zone of stagnation 驻区；滞流区

附　　录

附录一　希腊字母表

表1-1　希腊字母表

希腊字母		英文读音	希腊字母		英文读音
大写	小写		大写	小写	
A	α	alpha ['ælfə]	N	ν	nu [nju:]
B	β	beta ['bi:tə]	Ξ	ξ	xi [ksai]
Γ	γ	gamma ['gæmə]	O	o	omicron [əu'mkairən]
Δ	δ	delta ['deltə]	Π	π	pi [pai]
E	ε	epsilon ['epsilən]	P	ρ	rho [rəu]
Z	ζ	zeta ['zi:tə]	Σ	σ	sigma ['sigmə]
H	η	eta ['i:tə]	T	τ	tau [tau]
Θ	θ	theta ['θi:tə]	Υ	υ	upsilon ['ju:psilɔn]
I	ι	iota ['aiəutə]	Φ	φ	phi [fai]
K	κ	kappa ['kæpə]	X	χ	chi [kai]
Λ	λ	lambda ['læmdə]	Ψ	ψ	psi [psi:]
M	μ	mu [mju:]	Ω	ω	omega ['əumigə]

附录二　化学元素表

表2-1　化学元素表

英文	元素名	读音	符号	原子序数	英文	元素名	读音	符号	原子序数
Actinium	锕	阿	Ac	89	Carbon	碳	炭	C	6
Aluminum	铝	吕	Al	13	Cerium	铈	市	Ce	58
Americium	镅	眉	Am	95	Cesium (caseium)	铯	色	Cs	55
Antimony	锑	梯	Sb	51	Chlorine	氯	绿	Cl	17
Argon	氩	亚	Ar(A)	18	Chromium	铬	各	Cr	24
Arsenic	砷	申	As	33	Cobalt	钴	古	Co	27
Astatine	砹	艾	At	85	Copper	铜	同	Cu	29
Barium	钡	贝	Ba	56	Curium	锔	局	Cm	96
Berkelium	锫	陪	Bk	97	Deuterium	氘	刀	D	
Beryllium (glucinium)	铍	皮	Be(Gl)	4	Dysprosium	镝	滴	Dy	66
Bismuth	铋	必	Bi	83	Einsteinium (athenium)	锿	哀	Es(An)	99
Boron	硼	朋	B	5	Erbium	铒	耳	Er	68
Bromine	溴	秀	Br	35	Europium	铕	有	Eu	63
Cadmium	镉	隔	Cd	48	Fermium (centuryum)	镄	费	Fm(Ct)	100
Calcium	钙	丐	Ca	20	Fluorine	氟	弗	F	9
Califormium	锎	开	Cf	98	Francium	钫	方	Fr	87

续表

英 文	元素名	读音	符号	原子序数	英 文	元素名	读音	符号	原子序数
Gadolinium	钆	轧	Gd	64	Potassium	钾	甲	K	19
Gallium	镓	家	Ga	31	Praseodymium	镨	普	Pr	59
Germanium	锗	者	Ge	32	Promethium (illinium)	钷	颇	Pm(Il)	61
Gold	金	今	Au	79	Protaotinium (protoactinium)	镤	仆	Pa	91
Hafnium (celtium)	铪	哈	Hf (Ct)	72	Protium	氕	撇	H	
Helium	氦	亥	He	2	Radium	镭	雷	Ra	88
Holmium	钬	火	Ho	67	Radon (niton)	氡	冬	Rn(Nt)	86
Hydrogen	氢	轻	H	1	rhenium	铼	来	Re	75
Indium	铟	因	In	49	Rhodium	铑	老	Rh	45
Iodine	碘	典	J(I)	53	Rubidium	铷	如	Rb	37
Iridium	铱	衣	Ir	77	Ruthenium	钌	了	Ru	44
Iron	铁	铁	Fe	26	Samarium	钐	杉	Sm(Sa)	62
Krypton	氪	克	Kr	36	Scandium	钪	亢	Sc	21
Lanthanum	镧	栏	La	57	Selenium	硒	西	Re	34
Lawrencium	铹	劳	Lr	103	Silicon	硅	归	Si	14
Lead	铅	千	Pb	82	Silver	银	银	Ag	47
Lithium	锂	里	Li	3	Sodium	钠	钠	Na	11
Lutecium (cassiopeium)	镥	鲁	Lu (Cp)	71	Strontium	锶	思	Sr	38
Magnesium	镁	美	Mg	12	Sulfur (sulphur)	硫	流	S	16
Manganese	锰	猛	Mn	25	Tantalum	钽	坦	Ta	73
Mendelevium	钔	门	Md	101	Technetium (masurium)	锝	得	Tc(ma)	43
Mercury	汞	拱	Hg	80	Tellurium	碲	帝	Te	52
Molybdenum	钼	目	Mo	42	Terbium	铽	忒	Tb	65
Neodymium	钕	女	Nd	60	Thallium	铊	它	Tl	81
Neon	氖	乃	Ne	10	Thorium	钍	土	Th	90
Neptunium	镎	拿	Np	93	Thulium	铥	丢	Tu(tm)	69
Nickel	镍	臬	Ni	28	Tin	锡	析	Sn	50
Niobium (columbium)	铌	尼	Nb(Cb)	41	Titanium	钛	太	Ti	22
Nitrogen	氮	淡	N	7	Tritium	氚	川	T	
Nobelium	锘	诺	No	102	Uranium	铀	由	U	92
Osmium	锇	鹅	Os	76	Vanadium	钒	凡	V	23
Oxygen	氧	养	O	8	Wolfram (tungsten)	钨	乌	W	74
Palladium	钯	把	Pd	46	Xenon	氙	仙	Xe(x)	54
Phosphorus	磷	邻	P	15	Ytterbium	镱	意	Yb	70
Platinum	铂	博	Pt	78	Yttrium	钇	乙	Y(Yt)	39
Plutonium	钚	不	Pu	94	Zinc	锌	辛	Zn	30
Polonium	钋	泼	Po	84	Zirconium	锆	告	Zr	40

附录三 世界部分国家和地区名称及其符号表

表 3-1 世界部分国家和地区名称及其符号表

符 号	英文名称	中文名称	符 号	英文名称	中文名称
AL	Albania	阿尔巴尼亚	ES	Spain	西班牙
AR	Argentina	阿根廷	FI	Finland	芬兰
AT	Austria	奥地利	FR	France	法国
Au	Australia	澳大利亚	UK	United Kingdom	英国
AZ	Azerbaijan	阿塞拜疆	GR	Greece	希腊
BE	Belgium	比利时	HK	Hong Kong	香港
BG	Bulgaria	保加利亚	HU	Hungary	匈牙利
BR	Brazil	巴西	ID	Indonesia	印度尼西亚
CA	Canada	加拿大	IE	Ireland	爱尔兰
CH	Switzerland	瑞士	IL	Israel	以色列
CI	Ivory coast	象牙海岸	IN	India	印度
CL	Chile	智利	IQ	Iraq	伊拉克
CM	Cameroon	喀麦隆	IR	Iran	伊朗
CN	China	中国	IT	Italy	意大利
CO	Colombia	哥伦比亚	JQ	Jordan	约旦
CU	Cuba	古巴	JP	Japan	日本
CZ	Czech Republic	捷克共和国	KE	Kenya	肯尼亚
GE	Germany	德国	KR	Korea	韩国
DK	Denmark	丹麦	KW	Kuwait	科威特
DZ	Algeria	阿尔及利亚	LB	Lebanon	黎巴嫩
EG	Egypt	埃及	LU	Luxemburg	卢森堡
MA	Morocco	摩洛哥	SE	Sweden	瑞典
MG	Madagascar	马达加斯加	SG	Singapore	新加坡
MX	Mexico	墨西哥	SI	Slovenia	斯洛文尼亚
MY	Malaysia	马来西亚	SK	Slovakia	斯洛伐克
NL	The Netherlands	荷兰	SN	Senegal	塞内加尔
NO	Norway	挪威	TG	Togo	多哥
NZ	New Zealand	新西兰	TH	Thailand	泰国
PE	Peru	秘鲁	TN	Tunisia	突尼斯
PK	Pakistan	巴基斯坦	TR	Turkey	土耳其
PL	Poland	波兰	TW	Taiwan	中国台湾
PT	Portugal	葡萄牙	UA	Ukraine	乌克兰
RO	Rumania	罗马尼亚	US	USA	美国
RU	Russian Federation	俄罗斯	VE	Venezuela	委内瑞拉
SD	Sudan	苏丹	VN	Vietnam	越南
ZA	South Africa	南非			

附录四 蒲福（Beaufort）风力等级表

表4-1 蒲福（Beaufort）风力等级表

英文名称	中文名称	风 级	风 速 (m/s)
Calm Calm	静风	0级风	0.0～0.2
Light air	软风	1级风	0.3～1.5
Light breeze	轻风	2级风	1.6～3.3
Gentle breeze	微风	3级风	3.4～5.4
Moderate breeze	和风	4级风	5.5～7.9
Fresh breeze	清劲风	5级风	8.0～10.7
Strong breeze	强风	6级风	10.8～13.8
Near gale	疾风	7级风	13.9～17.1
Gale	大风	8级风	17.2～20.7
Strong gale	烈风	9级风	20.8～24.4
Storm	狂风	10级风	24.5～28.4
Violent storm	暴风	11级风	28.5～32.6
Hurricane	飓风	12级风	＞32.6

注：风速均为平地10m高处风速 (m/s)。

附录五 风工程常用无量纲数

表5-1 风工程常用无量纲数

名称 英文	名称 中文	符号	定 义	说 明
Eckert number	埃克特数	Ec	$Ec = V^2 / Cp\Delta T$	能量耗散速率/能量对流传输速率
Euler number	欧拉数	Eu	$Eu = P / \rho V^2$	压力/惯性力
Froude number	弗劳德数	Fr	$Fr = V / \sqrt{Lg}$	惯性力/重力
Prandtl number	普朗特数	Pr	$Pr = \mu Cp / k$	动量传输速率/热量传输速率
Reynolds number	雷诺数	Re	$Re = \rho V L / \mu = V L / \nu$	惯性力/黏性力
Richardson number	理查森数	Ri	$Ri = (\Delta T / T)/(Lg / V^2)$	热浮力/惯性力
Rossby number	罗斯贝数	Ro	$Ro = V / L\Omega$	惯性力/科氏力
Strouhal number	斯托拉哈数	St	$St = nL / V$	刻画旋涡脱落频率的参数
Schmidt number	施密特数	Sc	$Sc = \nu / L$	动量扩散/质量扩散
Scruton number	斯克拉顿数	Scr	$Scr = 2M\delta / \gamma L^2$	度量结构气动力不稳定性的参数
Cauony number	柯西数	Ca	$Ca = \rho V^2 / E$	惯性力/弹性力
Theodorson number	西奥多森数	T	$T = n_T L / V$	刻画结构扭转振动的参数

注：Cp—定压比热；E—弹性模量；g—重力加速度；L—特征长度；M—单位长度质量；n—结构振动或旋涡脱落频率；n_T—扭转基频；P—压力；T—温度；V—流速；γ—气温干绝热递减率；δ—结构阻尼比；k—传热系数；μ—动力黏性系数；ν—运动黏性系数；ρ—流体密度；Ω—地球自转角速度。

附录六 单位换算表

1. 力的单位换算表

表6-1 力的单位换算表（一）

已知单位	所求单位					
	sn	tf	tonf	US tonf	kgf	gf
1斯坦 sn	1	1.01972×10	1.00361×10	1.12405×10	1.01972×10^2	1.01972×10^6
1吨力 tf	9.80665	1	9.84207×10	1.10231	1×10^3	1×10^6
1英吨力 tonf	9.96402	1.01605	1	1.12	1.01605×10^3	1.01605×10^6
1美吨力 US tonf	8.89644	9.07188×10	8.92857×10	1	9.07188×10^2	9.07188×10^5
1千克力① kgf	9.80665×10^{-3}	1×10^{-3}	9.84207×10^{-4}	1.10231×10^{-3}	1	1×10^3
1克力 gf	9.80665×10^{-6}	1×10^{-6}	9.84207×10^{-7}	1.10231×10^{-6}	1×10^{-3}	1
1牛 N	1×10^{-3}	1.01972×10^{-4}	1.00361×10^{-4}	1.12405×10^{-4}	1.01972×10	1.01972×10^2
1达因 dyn	1×10^{-8}	1.01972×10^{-9}	1.00361×10^{-9}	1.12405×10^{-9}	1.01972×10	1.01972×10^{-3}
1磅力 lbf	4.44822×10^{-3}	4.53592×10^{-4}	0.466429×10^{-4}	5×10^{-4}	4.53592×10^{-2}	4.53592×10^2
1磅达 pdl	1.38255×10^{-5}	1.40981×10^{-5}	1.38754×10^{-5}	1.55405×10^{-5}	1.40981×10^{-2}	1.40981×10
1开皮 kip	4.44822	4.53592×10	4.46429×10	5	4.53592×10^2	4.53592×10^5

① 1千克力即1千克质量的重量（取标准重力加速度g=9.80665米/秒²）。许多文献中也混用千克（kg）代表千克力（kgf），用时要区分清楚。

表6-2 力的单位换算表（二）

已知单位	所求单位				
	N	dyn	lbf	pdl	kip
1斯坦 sn	1×10^3	1×10^8	2.24809×10^2	7.23301×10^3	2.44809×10^{-1}
1吨力 tf	9.80665×10^3	9.80665×10^8	2.20462×10^3	7.09316×10^4	2.20462
1英吨力 tonf	9.96402×10^3		2.24×10^3	7.20699×10^4	2.240
1美吨力 US tonf	8.89644×10^3	8.89644×10^8	2×10^3	6.43481×10^4	2
1千克力 kgf	9.80665	9.80665×10^5	2.20462	7.09316×10	2.20462×10^{-3}
1克力 gf	9.80665×10^{-3}	9.80665×10^2	2.20462×10^{-3}	7.09316×10^{-2}	2.20462×10^{-6}
1牛① N	1	1×10^5	2.24809×10^{-4}	7.23301	2.24809×10^{-4}
1达因 dyn	1×10^{-5}	1	2.24809×10^{-6}	7.23301×10^{-5}	2.24809×10^{-9}
1磅力 lbf	4.44822	4.44822×10^5	1	3.21740×10	1×10^{-3}
1磅达 pdl	1.38255×10^{-1}	1.38255×10^4	3.10810×10^{-2}	1	3.10809×10^{-5}
1开皮② (kip)	4.44822×10^3	4.44822×10^8	1×10^3	3.21741×10^4	1

① 1牛=1千克·米/秒² (1N=1kg·m/s²)。
② 此单位常在美国使用。

2. 动量单位换算表

表 6-3　动量单位换算表

已知单位	所求单位			
	kg·m/s	g·cm/s	lb·ft/s	lb·in/s
1 千克·米 / 秒　kg·m/s	1	1×10^5	7.23301	8.67962×10
1 克·厘米 / 秒　g·cm/s	1×10^{-5}	1	7.23301×10^{-5}	8.67962×10^{-4}
1 磅·英尺 / 秒　lb·ft/s	1.38255×10^{-1}	1.38255×10^4	1	1.2×10
1 磅·英寸 / 秒　lb·in/s	1.115212×10^{-2}	1.15212×10^3	8.33333×10^{-2}	1

3. 压力单位换算表

表 6-4　压力单位换算表（一）

已知单位	所求单位				
	kgf/m^2	kgf/cm^2	atm	$dyne/cm^2$	$Pa(N/m^2)$
1 千克力 / 米^2　kgf/m^2	1	1×10^{-4}	9.67841×10^{-5}	9.80665×10	9.80665
1 千克力 / 厘米^{2}①　kgf/cm^2	1×10^4	1	9.67841×10^{-1}	9.80665×10^5	9.80665×10^4
1 标准大气压　atm	1.03323×10^4	1.03323	1	1.01325×10^6	1.013250×10^5
1 达因 / 厘米^2　$dyne/cm^2$	1.01972×10^{-2}	1.01972×10^{-6}	9.86923×10^{-7}	1	1×10^{-1}
1 帕斯卡　$Pa(N/m^2)$	1.01972×10^{-1}	1.01972×10^{-5}	9.86923×10^{-6}	1×10	1
1 斯坦 / 米^2　sn/m^2	1.01972×10^2	1.01972×10^{-2}	9.86923×10^{-3}	1×10^4	1×10^3
1 牛 / 毫米^2　N/mm^2	1.01972×10^5	1.01972×10	9.86923	1×10^7	1×10^6
1 百巴　hbar	1.01972×10^6	1.01972×10^2	9.86923×10	1×10^8	1×10^7
1 磅达 / 英尺^2　pdl/ft^2	1.51750×10^{-1}	1.51750×10^{-5}	1.46870×10^{-5}	1.48816×10	1.48816
1 磅力 / 英寸^2　lbf/in^2	7.03070×10^2	7.03070×10^{-2}	6.80462×10^{-2}	6.89476×10^4	6.89476×10^3

① 在德国等一些欧陆国家，常用符号 kp 代替 kgf，故 1 千克力 / 厘米^2=1kgf/cm^2=1kp/cm^2，千克力 / 厘米^2 也被定义为"工程大气压"，用符号"at"表示。

表 6-5　压力单位换算表（二）

已知单位	所求单位				
	sn/m^2	N/mm^2	hbar	pdl/ft^2	lbf/in^2
1 千克力 / 米^2　kgf/m^2	9.80665×10^{-3}	9.80665×10^{-6}	9.80665×10^{-7}	6.58976	1.42233×10^{-3}
1 千克力 / 厘米^2　kgf/cm^2	9.80665×10	9.80665×10^{-2}	9.80665×10^{-3}	6.58976×10^4	1.42333×10
1 标准大气压　atm.	1.01325×10^2	1.01325×10^{-1}	1.01325×10^{-2}	6.80874×10^4	1.46959×10
1 达因 / 厘米^2　$dyne/cm^2$	1×10^{-4}	1×10^{-7}	1×10^{-8}	6.71969×10^{-2}	1.45038×10^{-5}
1 帕斯卡　$Pa(N/m^2)$	1×10^{-3}	1×10^{-6}	1×10^{-7}	6.71969×10^{-1}	1.45038×10^{-4}
1 斯坦 / 米^2　sn/m^2	1	1×10^{-3}	1×10^{-4}	6.71969×10^2	1.45038×10^{-1}
1 牛 / 毫米^2　N/mm^2	1×10^3	1	1×10^{-1}	6.71969×10^5	1.45038×10^2
1 百巴　hbar	1×10^4	1×10	1	6.71969×10^6	1.45038×10^3
1 磅达 / 英尺^2　pdl/ft^2	1.48816×10^{-3}	1.48816×10^{-6}	1.48816×10^{-7}	1	2.15840×10^{-4}
1 磅力 / 英寸^2　lbf/in^2	6.89476	6.89476×10^{-3}	6.89476×10^{-4}	4.63306×10^3	1

注：真空度用压力单位表示，一般有两种表示方法。
1. 用毫米水银柱表示，例：1mmHg 真空度表示"-1mmHg 的表压"。
2. 用百分数表示，例：百分之一的真空度表示"所用基准大气压的负百分之一的表压"。

表 6-6　压力单位换算表（三）

已知单位	所求单位				
	kgl/cm²	dyne/cm²	Pa	bar	mbar
1 千克力 / 厘米 ²[1] kgf/cm²	1	9.80655×10^5	9.80665×10^4	9.80665×10^{-1}	9.80665×10^2
1 达因 / 厘米 ²[2] dyne/cm²	1.01972×10^{-6}	1	1×10^{-1}	1×10^{-6}	1×10^{-3}
1 帕斯卡 Pa (N/m²)	1.01972×10^{-5}	1×10	1	1×10^{-5}	1×10^{-2}
1 巴　bar	1.01972	1×10^6	1×10^5	1	1×10^3
1 毫巴[2] mbar	1.01972×10^{-3}	1×10^3	1×10^2	1×10^{-3}	1
1 标准大气压[3] atm	1.03323	1.01325×10^6	1.01325×10^5	1.01325	1.01325×10^3
1 托[4] torr	1.35951×10^{-3}	1.33322×10^3	1.33322×10^2	1.33322×10^{-3}	1.33322
1 英吨力 / 英寸 ² tonf/in²	1.57488×10^2	1.54443×10^8	1.54443×10^7	1.54443×10^2	1.54443×10^5
1 英吨力 / 英尺 ² tonf/ft²	1.09366	1.07252×10^6	1.07252×10^5	1.07252	1.07252×10^3
1 美吨力 / 英尺 ² US tonf/ft²	9.76484×10^{-1}	9.57604×10^5	9.57604×10^4	9.57604×10^{-1}	9.57604×10^2

① 在德国等一些欧陆国家，常用符号 kp 代替 kgf，故 1 千克力 / 厘米 ²=1kgf/cm²=1kp/cm²。
② 毫巴可缩写为 mb，常用在气象学中。
③ 1 工程大气压用 1at. 表示，等于 1 千克力 / 厘米 ²，即 1at.=1kgf/cm²=1kp/cm²=98066.5Pa。
④ 1 托 =1 毫米水银柱高。

表 6-7　压力单位换算表（四）

已知单位	所求单位				
	atm	torr	tonf/in²	tonf/ft²	US tonf/ft²
1 千克力 / 厘米 ² kgf/cm²	9.67841×10^{-1}	7.35559×10^2	6.34971×10^{-3}	9.14358×10^{-1}	1.02408
1 达因 / 厘米 ² dyne/cm²	9.86923×10^{-7}	7.50062×10^{-4}	6.47490×10^{-9}	9.32385×10^{-7}	1.04427×10^{-6}
1 帕斯卡 Pa (N/m²)	9.86923×10^{-6}	7.50062×10^{-3}	6.47490×10^{-8}	9.32385×10^{-6}	1.04427×10^{-5}
1 巴　bar	9.86923×10^{-1}	7.50062×10^2	6.47490×10^{-3}	9.32385×10^{-1}	1.04427
1 毫巴　mbar	9.86923×10^{-4}	7.50062×10^{-1}	6.47490×10^{-6}	9.32385×10^{-4}	1.04427×10^{-3}
1 标准大气压　atm	1	7.60×10^2	6.56072×10^{-3}	9.44742×10^{-1}	1.05811
1 托　torr	1.31579×10^{-3}	1	8.63249×10^{-6}	1.24308×10^{-3}	1.39225×10^{-3}
1 英吨力 / 英寸 ² tonf/in²	1.52423×10^2	1.15842×10^5	1	1.44×10^2	1.6128×10^2
1 英吨力 / 英尺 ² tonf/ft²	1.05849	8.04452×10^2	6.94444×10^{-3}	1	1.12
1 美吨力 / 英尺 ² US tonf/ft²	9.45083×10^{-1}	7.18263×10^2	6.20040×10^{-3}	8.92857×10^{-1}	1

注：真空度用压力单位表示，一般有两种表示方法：
　1. 用毫米水银柱表示，例：1mmHg 真空度表示 "-1mmHg 的表压"。
　2. 用百分数表示，例：百分之一的真空度表示 "所用基准大气压的负百分之一的表压"。

表 6-8　压力单位换算表（五）

已知单位	所求单位				
	kgf/cm²	kgf/m²	dyne/cm²	atm	Pa
1 千克力 / 厘米 ²[1] kgf/cm²	1	1×10^4	9.80665×10^5	9.67841×10^{-1}	9.80665×10^4
1 千克力 / 米 ² kgf/m²	1×10^{-4}	1	9.80665×10	9.67841×10^{-5}	9.80665
1 达因 / 厘米 ² dyne/cm²	1.01972×10^{-6}	1.01972×10^{-2}	1	9.86923×10^{-7}	1×10^{-1}

续表

已知单位	所求单位				
	kgf/cm²	kgf/m²	dyne/cm²	atm	Pa (N/m²)
1标准大气压[②] atm	1.03323	1.03323×10⁴	1.01325×10⁶	1	1.013250×10⁵
1帕斯卡 Pa (N/m²)	1.01972×10⁻⁵	1.01972×10⁻²	1×10	9.86923×10⁻⁶	1
1毫巴[③] mbar	1.01972×10⁻³	1.01972×10	1×10³	9.86923×10⁻⁴	1×10²
1磅力/英尺² lbf/ft²	4.88243×10⁻⁴	4.88243	4.78803×10²	4.72542×10⁻⁴	4.78803×10
1英寸水柱高 in. H₂O	2.54×10⁻³	2.54×10	2.49089×10³	2.45832×10⁻³	2.49089×10²
1毫米汞柱高 mmHg	1.35951×10⁻³	1.35951×10	1.33332×10³	1.31579×10⁻³	1.33322×10²
1英寸汞柱高 in. Hg	3.45316×10⁻²	3.45316×10²	3.38639×10⁴	3.34211×10⁻²	3.38639×10³

[①] 在德国等一些欧陆国家，常用符号 kp 代替 kgf，故 1 千克力/厘米²=1kgf/cm²=1kp/cm²。
[②] 1 工程大气压用 1at. 表示，等于 1 千克力/厘米²，即 1at.=1kgf/cm²=1kp/cm²=98066.5Pa。
[③] 毫巴可缩写为 mb，常用在气象学中。

表 6-9 压力单位换算表（六）

已知单位	所求单位				
	mbar	lbf/ft²	in. H₂O	mmHg	in. Hg
1千克力/厘米² kgf/cm²	9.80665×10²	2.04816×10³	3.93701×10²	7.35559×10²	2.89590×10
1千克力/米² kgf/m²	9.80665×10⁻²	2.04816×10⁻¹	3.93701×10⁻¹	7.355559×10⁻¹	2.89590×10⁻³
1达因/厘米² dyne/cm²	1×10⁻³	2.08854×10⁻³	4.01463×10⁻⁴	7.50062×10⁻⁴	2.95300×10⁻⁵
1标准大气压 atm	1.013250×10³	2.11621×10³	4.06782×10²	7.60×10²	2.99213×10
1帕斯卡	1×10⁻²	2.08854×10⁻²	4.01463×10⁻³	7.50062×10⁻³	2.95300×10⁻⁴
1毫巴 mbar	1	2.08854	4.01463×10⁻¹	7.50062×10⁻¹	2.95300×10⁻²
1磅力/英尺² lbf/ft²	4.78803×10⁻¹	1	1.92222×10⁻¹	3.59131×10⁻¹	1.41390×10⁻²
1英寸水柱高 in. H₂O	2.49089	5.20233	1	1.86832	7.35559×10⁻²
1毫米汞柱高 mmHg	1.33322	2.78450	5.35240×10⁻¹	1	3.93701×10⁻²
1英寸汞柱高 in. Hg	3.38639×10	7.07262×10	1.35951×10	2.54×10	1

4. 速度单位换算表

表 6-10 速度单位换算表（一）

已知单位	所求单位				
	m/s	cm/s	m/min	m/h	km/h
1米/秒 m/s	1	1×10²	6×10	3.6×10³	3.6
1厘米/秒 cm/s	1×10⁻²	1	6×10⁻¹	3.6×10	3.6×10⁻²
1米/分 m/min	1.66667×10⁻²	1.66667	1	6.0×10	6×10⁻²
1米/时 m/h	2.77778×10⁻⁴	2.77778×10⁻²	1.66667×10⁻²	1	1×10⁻³
1千米/时 km/h	2.77778×10⁻¹	2.77778×10	1.66667×10	1×10³	1
1英尺/秒 ft/s	3.048×10⁻¹	3.048×10	1.82880×10	1.09728×10³	1.09728
1英里/时 mile/h	4.4704×10⁻¹	4.4704×10	2.68224×10	1.609344×10³	1.609344
1国际海里/时 kn	5.14444×10⁻¹	5.14444×10	3.08666×10	1.852×10³	1.852
1英海里/时 UK knot[①]	5.14773×10⁻¹	5.14773×10	3.08664×10	1.85318×10³	1.85318
1市里/时	1.38889×10⁻¹	1.38889×10	8.33333	5×10²	5×10⁻¹

[①] 此单位已不再通用，每小时航行 1 英海里的速度又叫 1 节。

表 6-11 速度单位换算表（二）

已知单位	所求单位				
	ft/s	mile/h	kn	UK knot	市里/时
1米/秒 m/s	3.28084	2.23694	1.94384	1.94260	7.2
1厘米/秒 cm/s	3.28084×10^{-2}	2.3694×10^{-2}	1.94384×10^{-2}	1.94260×10^{-2}	0.072
1米/分 m/min	5.46807×10^{-2}	3.72823×10^{-2}	3.23973×10^{-2}	3.23767×10^{-2}	0.12
1米/时 m/h	9.11344×10^{-4}	6.21371×10^{-4}	5.39957×10^{-4}	5.39612×10^{-3}	2×10^{-3}
1千米/时 km/h	9.11344×10^{-1}	6.21371×10^{-1}	5.39957×10^{-1}	0.539612	2
1英尺/秒 ft/s	1	6.81818×10^{-1}	5.92484×10^{-1}	0.592105	2.19456
1英里/时 mile/h	1.46667	1	8.68976×10^{-1}	0.868421	3.218688
1国际海里/时 kn	1.68781	1.15078	1	0.999361	3.704
1英海里/时 UK knot	1.68889	1.15152	1.00064	1	3.706368
1市里/时	4.55672×10^{-1}	3.10686×10^{-1}	2.69978×10^{-1}	0.269806	1

5. 功、能及热量单位换算表

表 6-12 功、能及热量单位换算表（一）

已知单位	所求单位				
	J	erg	kW·h	kgf·m	m³·atm
1焦① J	1	1×10^{7}	2.77778×10^{-7}	1.01972×10^{-1}	9.86923×10^{6}
1尔格② erg	1×10^{-7}	1	2.77778×10^{-14}	1.01972×10^{-8}	9.86923×10^{-13}
1千瓦·时 kW·h	3.6×10^{6}	3.6×10^{13}	1	3.67098×10^{5}	3.55292×10
1千克力·米 kgf·m	9.80665	9.80665×10^{7}	2.72407×10^{-6}	1	9.67841×10^{-5}
1米³·大气压③ m³·atm	1.01325×10^{5}	1.01325×10^{12}	2.81458×10^{-2}	1.03323×10^{4}	1
1英尺·磅达 ft·pdl	4.21401×10^{-2}	4.21401×10^{5}	1.17056×10^{-8}	4.29710×10^{-3}	4.15891×10^{-7}
1英尺·磅力 ft·lbf	1.35582	1.35582×10^{7}	3.76616×10^{-7}	1.38255×10^{-1}	1.33809×10^{-5}
1马力·时 Hp·h	2.64780×10^{6}	2.64780×10^{13}	7.35500×10^{-1}	2.7×10^{5}	2.61317×10
1英马力·时 Hp·h	2.68452×10^{6}	2.68452×10^{13}	7.45700×10^{-1}	2.73745×10^{5}	2.64941×10
1电子伏特 eV	1.60219×10^{-19}	1.60219×10^{-12}	4.45053×10^{-26}	1.63378×10^{-20}	1.58124×10^{-24}

① 1焦=1牛·米。
② 1尔格=1达因·厘米。
③ 标准大气压与工程大气压的关系：1米³·工程大气压=0.967841米³·标准大气压。

表 6-13 功、能及热量单位换算表（二）

已知单位	所求单位				
	ft·pdl	ft·lbf	Hp·h	Hp·h	eV
1焦 J	2.37304×10	7.37562×10^{-1}	3.77672×10^{-7}	3.72506×10^{-7}	6.24146×10^{18}
1尔格 erg	2.37304×10^{-6}	7.37562×10^{-8}	3.77672×10^{-14}	3.72506×10^{-14}	6.24146×10^{11}
1千瓦·时 kW·h	8.54293×10^{7}	2.65522×10^{6}	1.35962	1.34102	2.24693×10^{25}
1千克力·米 kgf·m	2.32715×10^{2}	7.23301	3.70370×10^{-6}	3.65304×10^{-6}	6.12078×10^{19}
1米³·大气压 m³·atm	2.40448×10^{6}	7.47335×10^{4}	3.82676×10^{-2}	3.77442×10^{-2}	6.32416×10^{23}

续表

已知单位	所求单位				
	ft·pdl	ft·lbf	Hp·h	Hp·h	eV
1英尺·磅达 ft·pdl	1	3.10810×10⁻²	1.59151×10⁻⁸	1.56974×10⁻⁸	2.63016×10¹⁷
1英尺·磅力 ft·lbf	3.21740×10	1	5.12055×10⁻⁷	5.05051×10⁻⁷	8.46230×10¹⁸
1马力·时	6.28334×10⁷	1.95292×10⁶	1	9.86321×10⁻¹	1.65261×10²⁵
1英马力·时 Hp·h	6.37045×10⁷	1.98×10⁶	1.01387	1	1.67554×10²⁵
1电子伏特 eV	3.80206×10⁻¹⁸	1.18171×10⁻¹⁹	6.05102×10⁻²⁶	5.96825×10⁻²⁶	1

表 6-14 功、能及热量单位换算表（三）

已知单位	所求单位				
	J	erg	kW·h	kgf·m	l·atm
1焦 J	1	1×10⁷	2.77778×10⁻⁷	1.01972×10⁻¹	9.86923×10⁻³
1尔格 erg	1×10⁻⁷	1	2.77778×10⁻¹⁴	1.01972×10⁻⁸	9.86923×10⁻¹⁰
1千瓦·时 kW·h	3.6×10⁶	3.6×10¹³	1	3.67098×10⁵	3.55292×10⁴
1千克力·米 kgf·m	9.80665	9.80665×10⁷	2.72407×10⁻⁶	1	9.67841×10⁻²
1升·大气压 l·atm	1.01325×10²	1.01325×10⁹	2.81458×10⁻⁵	1.03323×10	1
1英尺·磅力 ft·lbf	1.35582	1.35582×10⁷	3.76616×10⁻⁷	1.38255×10⁻¹	1.33809×10⁻²
1马力·时	2.64780×10⁶	2.64780×10¹³	7.35500×10⁻¹	2.7×10⁵	2.61317×10⁴
1卡 cal	4.1868	4.1868×10⁷	1.163×10⁻⁶	4.26936×10⁻¹	4.13205×10⁻²
1英热单位 Btu	1.05506×10³	1.05506×10¹⁰	2.93071×10⁻⁴	1.07587×10²	1.04126×10
1百度热单位 CHU	1.89911×10³	1.89911×10¹⁰	5.27531×10⁻⁴	1.93656×10²	1.87428×10

注：动能与功的关系式：1千克·米²/秒²=0.101972千克力·米。

表 6-15 功、能及热量单位换算表（四）

已知单位	所求单位				
	ft·lbf	马力·时	cal	Btu	CHU
1焦 J	7.37562×10⁻¹	3.77672×10⁻⁷	2.38846×10⁻¹	9.47817×10⁻⁴	5.26565×10⁻⁴
1尔格 erg	7.37562×10⁻⁸	3.77672×10⁻¹⁴	2.38846×10⁻⁸	9.47817×10⁻¹¹	5.26565×10⁻¹¹
1千瓦·时 kW·h	2.65522×10⁶	1.35962	8.59845×10⁵	3.41214×10³	1.89563×10³
1千克力·米 kgf·m	7.23301	3.70370×10⁻⁶	2.34228	9.29491×10⁻³	5.16384×10⁻³
1升·大气压 l·atm	7.47335×10	3.82676×10⁻⁵	2.42011×10	9.60376×10⁻²	5.33542×10⁻²
1英尺·磅力 ft·lbf	1	5.12055×10⁻⁷	3.23832×10⁻¹	1.28507×10⁻³	7.13927×10⁻⁴
1马力·时	1.95292×10⁶	1	6.32416×10⁵	2.50963×10³	1.39424×10³
1卡 cal	3.08803	1.58124×10⁻⁶	1	3.96832×10⁻³	2.20462×10⁻³
1英热单位 Btu	7.78169×10²	3.98467×10⁻⁴	2.51996×10²	1	5.55556×10⁻¹
1百度热单位 CHU	1.40071×10³	7.17241×10⁻⁴	4.53595×10²	1.8	1

6. 功率单位换算表

表6-16　功率单位换算表（一）

已知单位	所求单位					
	W	kW	erg/s	kgf·m/s	米制马力	ft·lbf/s
1瓦① W	1	1×10^{-3}	1×10^{7}	1.01972×10^{-1}	1.35962×10^{-3}	7.37562×10^{-1}
1千瓦② kW	1×10^{3}	1	1×10^{10}	1.01972×10^{2}	1.35962	7.37562×10^{2}
1尔格/秒 erg/s	1×10^{-7}	1×10^{-10}	1	1.01972×10^{-6}	1.35962×10^{-10}	7.37562×10^{-8}
1千克力·米/秒 kgf·m/s	9.80665	9.80665×10^{-3}	9.80665×10^{7}	1	1.33333×10^{-2}	7.23301
1米制马力	7.35499×10^{2}	7.35499×10^{-1}	7.35499×10^{9}	7.5×10	1	5.42476×10^{2}
1英尺·磅力/秒 ft·lbf/s	1.35582	1.35582×10^{-3}	1.35582×10^{7}	1.38255×10^{-1}	1.84340×10^{-3}	1
1英制马力 Hp	7.45700×10^{2}	7.45700×10^{-1}	7.45700×10^{9}	7.60402×10	1.01387	5.50×10^{2}
1卡/秒 cal/s	4.1868	4.1868×10^{-3}	4.1868×10^{7}	4.26935×10^{-1}	5.69246×10^{-3}	3.08803
1千卡/时 kcal/h	1.163	1.163×10^{-3}	1.163×10^{7}	1.18593×10^{-1}	1.58124×10^{-3}	8.57785×10^{-1}
1英热单位/时 Btu/h	2.93071×10^{-1}	2.93071×10^{-4}	2.93071×10^{6}	2.98849×10^{-2}	3.98466×10^{-4}	2.16158×10^{-1}
1百度热单位/时 CHU/h	5.27530×10^{-1}	5.27530×10^{-4}	5.27530×10^{6}	5.37933×10^{-2}	7.17240×10^{-4}	3.89086×10^{-1}

① 1瓦=1焦耳/秒=1安培·伏特=1米^{2}·千克·秒$^{-3}$。
② 1千瓦(kW)=6.11832×10^{3} kgf·m/min。

表6-17　功率单位换算表（二）

已知单位	所求单位				
	Hp	cal/s	kcal/h	Btu/h	CHU/h
1瓦 W	1.34102×10^{-3}	2.38846×10^{-1}	8.59845×10^{-1}	3.41214	1.89563
1千瓦 kW	1.34102	2.38846×10^{2}	8.59845×10^{2}	3.41214×10^{3}	1.89563×10^{3}
1尔格/秒 erg/s	1.34102×10^{-10}	2.38846×10^{-8}	8.59845×10^{-8}	3.41214×10^{-7}	1.89563×10^{-7}
1千克力·米/秒 kgf·m/s	1.31509×10^{-2}	2.34228	8.43220	3.34617×10	1.85897×10
1米制马力 Hp	9.86320×10^{-1}	1.75671×10^{2}	6.32415×10^{2}	2.50963×10^{3}	1.39423×10^{3}
1英尺·磅力/秒 ft·lbf/s	1.81818×10^{-3}	3.23832×10^{-1}	1.16579	4.62624	2.57013
1英制马力 Hp	1	1.78107×10^{2}	6.41186×10^{2}	2.54443×10^{3}	1.41357×10^{3}
1卡/秒 cal/s	5.61459×10^{-3}	1	3.6	1.42860×10	7.93662
1千卡/时 kcal/h	1.55961×10^{-3}	2.77778×10^{-1}	1	3.96832	2.20461
1英热单位/时 Btu/h	3.93015×10^{-4}	6.99988×10^{-2}	2.51996×10^{-1}	1	5.55556×10^{-1}
1百度热单位/时 CHU/h	7.07428×10^{-4}	1.25998×10^{-1}	4.53594×10^{-1}	1.8	1

7. 各种温度单位换算表

表6-18 各种温度单位换算表

已知单位	所求单位				
	T/K	$t/℃$	$t_F/°F$	$t_R/°R$	$t_R/°R$
热力学温度 T,K	1	$T/K-273.15$	$(9/5)T/K-459.67$	$(9/5)T/K$	$(4/5)T/K-218.52$
摄氏温度 $t,℃$	$t/℃+273.15$	1	$(9/5)t/℃+32$	$(9/5)t/℃+491.67$	$(4/5)t/℃$
华氏温度 t_F [1]，$°F$	$(5/9)(t_F/°F-32)$	$(5/9)(t_F/°F+459.67)$	1	$t_F/°F+459.67$	$(4/9)(t_F/°F-32)$
兰氏温度 t_F [2]，$°R$	$(5/9)t_R/°R-273.15$	$t_R/°R-459.67$	1		$(4/9)t_R/°R-218.52$
列氏温度 t_R，$°R$	$(5/9)t_R/°R-273.15$	$(5/9)t_R/°R$	$(9/4)t_R/°R+32$	$(9/4)t_R/°R+491.67$	1

① $1°F = \frac{5}{9}(℃) = \frac{5}{9}(K)$。

② 兰氏与列氏温度往往混用符号"R"，故使用时应加以注意。